Plant Genes, Genomes and Genetics

Plant Genes, Genomes and Genetics

Erich Grotewold

The Ohio State University, USA

Joseph Chappell

University of Kentucky, USA

Elizabeth A. Kellogg

Donald Danforth Plant Science Center, St. Louis, Missouri, USA

WILEY Blackwell

Library of Congress Cataloging-in-Publication Data

Grotewold, Erich, author.
  Plant genes, genomes, and genetics / Erich Grotewold, Joseph Chappell, Elizabeth Kellogg.
    pages cm
  Includes bibliographical references and index.
  ISBN 978-1-119-99888-4 (cloth) – ISBN 978-1-119-99887-7 (pbk.) 1. Plant molecular genetics. 2. Plant gene expression.
3. Genomics. I. Chappell, Joseph, author. II. Kellogg, Elizabeth Anne, author. III. Title.
  [DNLM: 1. Plants – genetics. 2. Genomics. 3. Plant Physiological Phenomena. 4. RNA, Plant – genetics. QK 981]
  QK981.4.G76 2015
  572.8′2 – dc23

                                                            2014028955

A catalogue record for this book is available from the British Library.

Wiley also publishes its books in a variety of electronic formats. Some content that appears in print may not be available in electronic books.

Cover illustration by Debbie Maizels

Typeset in 10/12pt Minion by Laserwords Private Limited, Chennai, India.
Printed and bound in Singapore by Markono Print Media Pte Ltd.

1 2015

Contents

Acknowledgements

In writing this book, we have enormously benefited from the advice and valuable comments from many colleagues, to whom we are truly indebted for their multiple suggestions and contributions. These colleagues include the following: Biao Ding (The Ohio State University, USA); Sherry Flint-Garcia (USDA-ARS, Columbia, Missouri, USA); Irene Gentzel (The Ohio State University, USA); Venkat Gopalan (The Ohio State University, USA); Art Hunt (University of Kentucky, USA); Rebecca Lamb (The Ohio State University, USA); Pal Maliga (Rutgers University, USA); Michael McMullen (USDA-ARS, Columbia, Missouri, USA); Craig Pikaard (Indiana University, USA); Mark Rausher (Duke University, USA); Keith Slotkin (The Ohio State University, USA); Jan Smalle (University of Kentucky, USA); David Somers (The Ohio State University, USA); Dan Voytas (University of Minnesota, USA); and Ling Yuan (University of Kentucky, USA). We also thank the anonymous reviewers who contributed with their experience in the classroom to improve the overall utility of this book for students and professors.

We also want to especially thank the current and past members of our research groups for their many important contributions to the growth of our knowledge of genes, genomes and genetics. Without their support, this book would not have been possible. Last but not least, we want to thank the funding agencies, particularly the National Science Foundation, the US Department of Agriculture, and the National Institutes of Health, for their continuous support of the research conducted in our laboratories.

Introduction

The word "plant" has many meanings

One goal of this book is to highlight the aspects of molecular biology that are unique to plants, and that represent mechanisms that cannot be understood simply by studying animals, yeast or bacteria. We therefore need to spend some time discussing what we mean by the word "plant", which, perhaps surprisingly, does not have a simple or universally accepted definition.

When most people think of a plant, they generally immediately come up with an image of a tomato plant, or a petunia, or corn. A scientist might think of *Arabidopsis thaliana*, the tiny weed that has been domesticated by molecular biologists. All these are examples of **flowering plants (angiosperms)**, which are the dominant forms of land plants on Earth today. The flowering plants represent a large group that originated in the early Cretaceous (~140 million years ago, although the exact date is subject to much current debate); the group has subsequently diversified to produce most trees, shrubs, and herbs. The flowering plants include more than 300 000 species; only a few thousand are cultivated, and surprisingly, only a few of these – fewer than twenty – produce the vast majority of the food for all of humanity.

The term "plant" is often used to mean "**land plant**", a much larger group that includes the flowering plants, but also the gymnosperms, ferns, lycophytes, mosses, hornworts and liverworts. This large group is monophyletic, a term that refers to all being descendants of a common ancestor, and is often called the **Embryophytes** because all members produce embryos retained on the parent plant. A phylogeny of the Embryophyta is presented in Figure 1, which is assembled on the basis of the main characteristics that define the major groups of plants. Clades (or groups) within the land plants include the **seed plants** (flowering plants plus gymnosperms, distinguished by how they bear their seeds) and other **vascular plants** [ferns (pteridophytes) and lycophytes], in which the

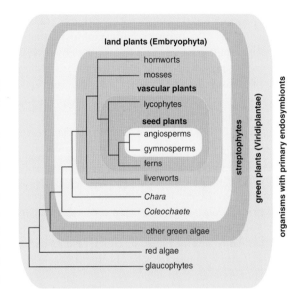

Figure 1 Phylogeny of organisms that originated with the primary endosymbiosis, in which a eukaryote acquired a symbiotic cyanobacterium. Superimposed on the phylogeny is a Venn diagram of major groups. While the green plants (Viridiplantae) and streptophytes are sometimes called "plants", in this book we will use the term "plant" to refer to the land plants, the group shaded in green. Subgroups within the land plants are also indicated.

diploid sporophyte forms on the independent gametophyte, and dispersal occurs via spores. In contrast, the non-vascular plants (hornworts and liverworts) are distinguished not only by the absence of phloem and xylem vessels, but by having a dominant gametophytic (haploid) stage of life and only a short lived sporophytic (diploid) stage.

Another possible definition of "plant" is the group known as the **Streptophytes**, which includes the land plants plus their immediate relatives, *Chara* and *Coleochaete* (both formerly considered green algae). The Streptophytes all share a peculiar method of cell division, the phragmoplast, and a unique structure of proteins to make cellulose (the sugar polymer that is the primary component of plant cell walls), the cellulose rosettes.

A third definition of "plant" corresponds to organisms that have chloroplasts and make chlorophyll a and b. These are known as the **Viridiplantae** (Latin for **green plants**). This group includes the Streptophytes (i.e., land plants plus *Coleochaete* and *Chara*) plus all the green algae. The latter group includes the well-studied single cell organism *Chlamydomonas*.

Finally, a fourth (and uncommon) definition of "plant" includes all organisms with chloroplasts that are the result of a **primary endosymbiosis**, that is organisms that acquired their chloroplasts by directly aquiring a cyanobacterium (see Chapter 5). Members of this group are Viridiplantae, the **red algae** (Rhodophyta) and the glaucophytes. Some data suggest that the primary endosymbiosis occurred only once, in the common ancestor of the Viridiplantae, red algae and glaucophytes. Evidence is accumulating to suggest that indeed Viridiplantae, red algae and glaucophytes are all part of a monophyletic group, which is sometimes called the Archaeplastida. However, each primary endosymbiotic event could be independent, with the capture of a cyanobacterium occurring independently several times. In either case, origin of plastids from cyanobacteria has been extremely rare in the history of life.

The plastid bearing organisms diverged from other eukaryote lineages, including animals plus fungi, at least 1 billion years ago (Knoll, 2003). Given this enormously long period of evolution, it is remarkable that there are any similarities at all in the cellular apparatus between animals (i.e., you), fungi (i.e., yeast), and any plants. There are many similarities of course, but we suggest here that they need to be demonstrated, not assumed. In other words, the fact that the transcriptional machinery is similar between animals and yeast, does not necessarily mean that it will also be similar in plants. In addition, the term similarity does not mean identity. Processes in common could have arisen because of convergent forces, and really the metric for similarity has become conservation in the DNA encoding these functions.

In the past, the term "plant" was sometimes applied to all photosynthetic organisms. However, such a broad use of the term is now rejected. Many organisms that are able to undergo photosynthesis have gained that ability by acquiring a red alga along with its plastid. In other words, the plastid is a symbiont in the red alga and the red alga is the symbiont in another (previously non-photosynthetic) organism. Such symbioses are known as **secondary endosymbioses** to distinguish them from the primary endosymbioses of the Archaeplastida. In organisms with a secondary endosymbiont, the structure of the membranes around the symbiont shows that it was once a separate organism that was picked up by its host. Organisms with secondary endosymbioses include the Stramenopiles, the group that includes the brown algae (e.g., *Fucus*, a common seaweed) and golden brown algae (which occur mostly in freshwater), the dinoflagellates, and the kinetoplastids (e.g., *Euglena*, trypanosomes, and the apicomplexans, which include the organisms that cause malaria). Each of these groups is as different from plants as animals are, and as different from animals as plants are. In these organisms in particular one might expect to find novel genes, proteins and cellular mechanisms. If the term "plant" were applied to all photosynthetic organisms, the ones with the secondary endosymbioses are so diverse and so totally unrelated (other than all being eukaryotes) that the term would be effectively meaningless.

In summary, the term plant is used to apply to many sets of organisms, the smallest of which is the land plants and the largest is all photosynthetic organisms. Most commonly, however, "plant" refers either to the entire green plant lineage (Viridiplantae), or to the land plants. In common parlance its use is even more restricted to refer informally to flowering plants. In this textbook we will use the term to refer to land plants. Most of the data we present come from flowering plants, so in most cases, the reader can assume that we are extrapolating, generally without evidence, from the flowering plants to the gymnosperms, ferns, lycophytes, mosses, liverworts and hornworts. If we have data from species outside the land plants, we will cite that explicitly.

The basic structure of a plant is deceptively simple

The processes described in this book can in theory occur in any cell in the plant. However, some familiarity with basic plant morphology is assumed. Plant growth occurs from dedicated sets of stem cells, known as **meristems**. These are active throughout the life of the plant, so that development is continuous and modular. This is quite different from the situation in animals, in which the entire organism develops in a coordinated fashion and then ceases development entirely at maturity. If a human were to grow like a plant, the fingers, toes and the top of the head might keep growing throughout the life of the human.

Meristems are organized during embryonic development. In the seed plants these initially consist of two clusters of cells, the **shoot apical meristem** and the **root apical meristem**, at opposite ends of the plant. These are the basis of the **bipolar embryo**, which is only found in the seed plants. Meristems in non-seed bearing vascular plants (ferns, lycophytes) consist of only a few cells, and the root apical meristem in particular develops late and on one side of the embryonic axis.

A flowering plant has an obvious above ground component, the **shoot**, and a below-ground component, generally the **root** (Figure 2). The apical meristem of the shoot produces leaves on its flanks. In the axil of each leaf, another meristem forms, the **axillary meristem**; this meristem is often dormant for a while but may grow out to form a branch. The root apical meristem forms the **primary root**. **Lateral roots** are not formed from the apical meristem, but rather are formed from meristems that arise de novo just outside the vascular tissue. In most eudicots, the primary root persists and forms a prominent below ground structure (think of a carrot or a dandelion root), whereas in most monocots, the primary root only lives for a few months and is replaced by roots forming from the very base of the shoot, near ground level (think of onion or grass roots). The **vascular tissue** connects all parts of the plant, transporting water, nutrients and some hormones up from the roots into the leaves and meristems of the shoot. At the same time carbohydrates and other hormones are transported both up and down from the leaves.

The basic tissues of the plant are obvious in cross sections of a leaf and a root (Figure 2). Unlike animals, which have an elaborate set of tissue types, plants have only three basic sorts of tissue – the **epidermis**, which covers all parts of the plant, the **vascular tissue**, and **ground tissue**, which includes everything else. The epidermis of a leaf is generally made up of flat translucent cells and is covered with a waxy layer, **cuticle** which prevents drying. Within the epidermis are specialized holes known as **stomata** (literally "mouths") that permit entry of CO_2 for photosynthesis and escape of O_2, the by-product of photosynthesis. The stomata also permit the escape of water vapor. As water escapes, it creates a gradient of water pressure that pulls water up through the vascular system from the roots and hydrates all the cells in the plant. If water is limited, however, the stomata close and prevent drying of the tissues. Stomatal opening and closing is caused by changes in the turgor pressure of the **guard cells**, which sit on either side of the opening. In addition, the leaf epidermis can also have hairs (also known as trichomes) and glands. Depending on the plant species, they can be unicellular (as in Arabidopsis) or multicellular (e.g., the glands that accumulate peppermintoil in *Mentha piperita*).

The epidermis of a root includes some cells that will develop long projections known as **root hairs**. Root hairs are thin-walled, and are central to the uptake of water and nutrients from the soil. In addition, they are the site of interaction with soil bacteria, such as *Rhizobia*, which interact and form symbioses with some species of plants. Cells that will form root hairs alternate with non-root-hair cells. The pattern of root-hair and non-root-hair cells varies between species but is quite stereotyped in the model plant *Arabidopsis* where the controls of patterning have been studied extensively. As the root pushes through the soil and grows in diameter, the epidermis is sloughed off and replaced by cells from inner layers of the root. Because of this process, root hairs are only present right behind the root apical meristem and are lost in older roots.

Vascular tissue is arranged in bundles of conducting cells that extend throughout the plant. The water conducting tissue is **xylem**, which consists mostly of cells that are dead at maturity; in the vascular bundle of a leaf, the xylem is generally on the top (adaxial) side. Because the water is pulled up the plant following a pressure gradient from the roots to the shoots, the xylem cell walls must be strong enough to withstand the tension on the water column and hence are generally lignified at maturity. The carbohydrate transport tissue is **phloem**; phloem cells are alive at maturity and in a leaf are generally found on the bottom (abaxial) side of the vascular bundle. Unlike water, which is pulled up the plant under tension, the phloem sap is pushed around the plant under pressure. Because many molecules are dissolved in phloem sap, it generally is hyperosmotic to the surrounding tissues and takes up water.

The cells inside the epidermis and outside the vascular tissue are part of the ground tissue. Depending on the organ and the stage of development, these cells may vary considerably throughout the plant. An example is shown in Figure 2 in the cross section of the leaf, where the ground tissue is known as **mesophyll**.

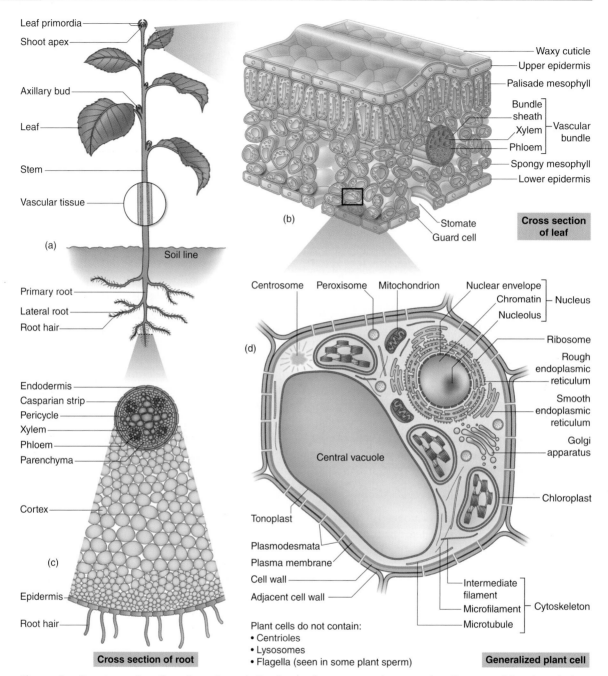

Figure 2   Structure of a flowering plant, indicating major organs, tissues and cell types. (a) Adapted from Taiz and Zeigler (1991). Reproduced with permission. (b) http://tpsbiology11student.wikispaces.com/Plants+-+Anatomy,+Growth,+and+Function. (c) Adapted from Taiz and Zeigler (1991). Reproduced with permission. (d) http://turfgrass.cas.psu.edu/education/turgeon/Modules/10_AnnualBluegrass/Annual_Bluegrass_Module/1.71%20root%20cross%20section.html. Reproduced with permission of A. J. Turgeon.

(The word mesophyll is simply Greek for "middle of the leaf"; meso = middle and phyll = leaf.) In many angiosperm leaves, the upper mesophyll cells form long, closely packed rectangles; because of their appearance in cross section they are known as the **palisade layer**. The lower mesophyll cells, in contrast, are less tightly packed and more isodiametric and are known as the **spongy mesophyll**. The cells of the spongy mesophyll cease cell division before those of the palisade, and are pulled apart as the leaf expands, creating air spaces between them.

In the root, the vascular tissue forms a solid cylinder in the center. It is surrounded by a ring of cells, the **endodermis**, which regulates the flow of water into and out of the vasculature. In most roots, water can enter through the cytoplasm of cells such as root hairs, or can enter the gaps between the cells. It then flows through or around the cells in the **cortex**. The pathway through cells is known as the **symplastic** pathway, whereas the pathway around cells is known as the **apoplastic** pathway. When the water reaches the endodermis, however, the water (and any ions or other substances) must go through the cells, that is, it is forced into a symplastic pathway. The endodermal cells are held in a tight ring by a layer of suberin, the **Casparian strip**, which prevents water or anything else from going around the cells; in other words, the Casparian strip blocks the apoplastic pathway. By controlling transporters and the osmotic force inside the cells, the endodermis thus controls which substances move in and out.

Individual plant cells have many of the same structures as other eukaryotic cells (Figure 2). Like all living cells (including Bacteria and Archaea), the plant cell is surrounded by a **plasma membrane**, uses **DNA** as its genetic material, and synthesizes proteins with **ribosomes**. Like all other Eukarya, the plant cell has a **nucleus** with a **nuclear membrane** that is contiguous with the **endoplasmic reticulum**, and has **mitochondria**, **peroxisomes**, and **Golgi apparatus**. The endoplasmic reticulum may be smooth or rough depending on whether ribosomes are attached to it. The cytoskeleton is made up of **microfilaments** (formed by the protein actin), **intermediate filaments**, and **microtubules** (formed by the protein tubulin).

Other structures are not shared with animals, although they may occur in other eukaryotes. Unlike most animals (e.g., any mammal), the plant cell is enclosed in a wall made up of **cellulose**. The wall often is penetrated by specialized tunnels known as **plasmodesmata**; the plasma membrane is continuous through these tunnels and the diameter of the plasmodesmata is tightly regulated by proteins that reside in the membrane. The plant cell contains **chloroplasts**, symbiotic bacteria that are the site of photosynthesis. Also many plant cells have a prominent **vacuole** surrounded by an independent membrane known as a **tonoplast**. In mesophyll cells the vacuole often fills up so much of the center of the cell that the cytoplasm and organelles are pressed to the edges against the plasma membrane. Compared with such cells, the vacuole drawn in Figure 2 is abnormally small.

This very brief introduction to plant structure and plant evolutionary history should provide a foundation for the rest of this book. In the chapters that follow, we will provide a view of biology focusing on aspects that characterize the many species of land plants. You have probably already read biology textbooks that focus on humans as a representative mammal, but recall that there are only about 5000 species of mammals, and among them humans have strikingly low genetic diversity. In contrast there may be almost 70–100 times as many species of land plants, although exact numbers are unknown. Plants dominate our environment, provide food, clothing and shelter, and create the air we breathe. Without land plants, there would be no land animals and certainly no humans. Plants thus support all of life on land; here we present their genes, genomes, and genetics.

Reference

Knoll, A.H. (2003) *Life on a Young Planet: The First Three Billion Years*, Princeton University Press.

About the Companion Website

Plant Genes, Genomes and Genetics is accompanied by a companion website:

www.wiley.com/go/grotewold/plantgenes

The website includes:
- Powerpoints of all figures from the book for downloading
- PDFs of all tables from the book for downloading
- Answers to the problems

Part I

Plant Genomes and Genes

Chapter 1
Plant genetic material

1.1 DNA is the genetic material of all living organisms, including plants

Like all living organisms, plants use deoxyribonucleic acid (DNA) as their genetic material. DNA is a polymer that consists of alternating sugars and phosphates with nitrogenous bases attached to the sugar moiety. More specifically, the nucleotide building block of DNA is a deoxyribose sugar with a phosphate group attached to carbon 5 (C-5) and a nitrogenous base to carbon 1 (C-1). Phosphodiester bonds connect the C-5 phosphate group of one nucleotide to the carbon 3 (C-3) of another, creating the alternating sugar–phosphate backbone of the DNA molecule. This means that one end of the chain is terminated by a C-5 phosphate, and is known as the 5′ end, whereas the other end is terminated by a C-3 hydroxyl, and is known as the 3′ end (Figure 1.1a). The idea that DNA molecules have a polarity is one that will be revisited over and over throughout this book.

Only four nitrogenous bases are used in a DNA molecule. Two of these, **cytosine** (C) and **thymine** (T), have a single aromatic ring consisting of four carbons and two nitrogen groups, and are classified as **pyrimidines**. The other two, **adenine** (A) and **guanine** (G), each have a double ring consisting of a pyrimidine ring fused to a 5-membered, heterocyclic ring, and are classified as **purines**. The bases form a linear molecule, a strand of DNA that interacts with the nitrogenous bases on the other strand.

Two DNA polymers, or strands, together form the iconic double helix, a structure like a twisted ladder that has come to symbolize life and its historical continuity (Figure 1.1b). Even viruses, many of which have genomes of single-stranded nucleic acids, must eventually pass through a double-stranded stage to reproduce. The strands are held together by hydrogen bonds (H-bonds) between the nitrogenous bases, with two bonds between A and T, and three between G and C (Figure 1.1c). Since more H-bonds between bases hold them together more tightly, it is significantly easier to denature a DNA molecule with many A-T base pairs than one with many C-G base pairs. The pairing rules for DNA are largely inflexible: A forms H-bonds with T and G with C. The strands are arranged in antiparallel fashion, so that the 5′ end base of one strand pairs with the 3′ end base of the other, and vice versa.

The structure of DNA is not unique to plants, but rather is shared among all three domains of life (Eukarya, Bacteria, and Archaea), as well as by viruses. The patterns of covalent bonds and H-bonds can thus be studied in any organism, and indeed much of what we know about DNA structure was originally worked out in bacteria, which are unicellular and prokaryotic (lacking a nucleus).

The four bases (A, C, G and T) are not present in equal amounts and can vary between genomes, parts of genomes, and species. For example, the A+T content of the chloroplast genome, an organellar genome discussed later in Chapter 5, is variable, but generally greater than 50% of the total. In contrast, nuclear genes of many grasses are enriched in G+C, a bias that is particularly noticeable in maize.

Plant Genes, Genomes and Genetics, First Edition. Erich Grotewold, Joseph Chappell and Elizabeth A. Kellogg.
© 2015 John Wiley & Sons, Ltd. Published 2015 by John Wiley & Sons, Ltd.
Companion Website: www.wiley.com/go/grotewold/plantgenes.

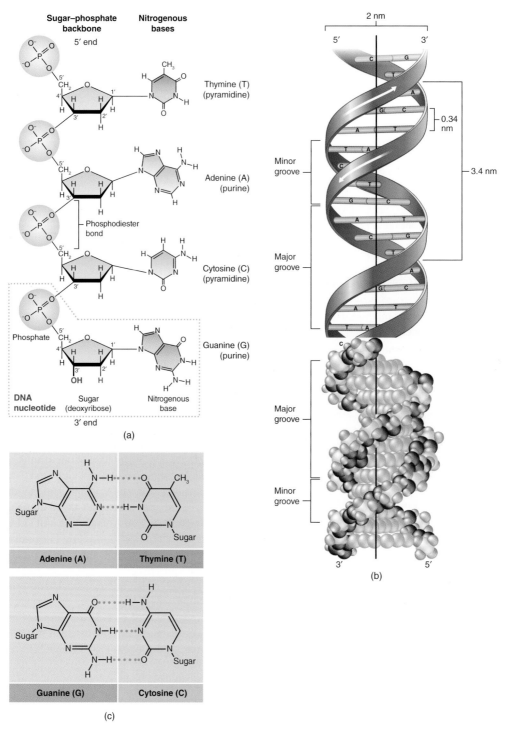

Figure 1.1 Structure of the DNA molecule. (a) Alternating phosphate and ribose groups make up the backbone of the DNA strand. The C-3 hydroxyl at the 3′ end is required for attaching new nucleotides to the chain, so the DNA strand always is extended from the 3′ end. The nitrogenous bases are attached to C-1 of the ribose. (b) The double helix, formed from two antiparallel strands of DNA. (c) The four nitrogenous bases and the hydrogen bonds between them. (a) and (c) Reece *et al.* (2011). (b) Adapted from Raven *et al.* (2011)

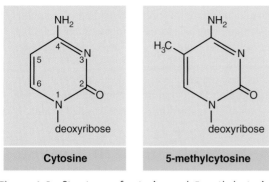

Figure 1.2   Structures of cytosine and 5-methylcytosine. http://en.wikipedia.org/wiki/File:Cytosine_chemical_ structure.svg. By Engineer gena (Own work) [Public domain], via Wikimedia Commons

In plants, the nucleotide bases may be modified by attachment of methyl (–CH_3) groups to particular sites. A common position for DNA methylation is on C-5 of C (Figure 1.2), although adenine methylation is also possible, particularly in bacteria, Archaea, and unicellular eukaryotes. This common modification of the DNA is known to affect transcription, and will be discussed in more detail in Chapter 12. While methylation is also common in mammals, it is relatively rare in yeast and in the fruit fly (*Drosophila melanogaster*). Other insects, however, have extensive DNA methylation, as is the case of honeybees. Many aspects of biotechnology exploit the basic structure of DNA, as described in the box "Working with DNA."

Working with DNA: biotechnology takes advantage of the properties of the DNA molecule

The polymerase chain reaction (PCR), a method used extensively in biotechnology for generating large numbers of similar or identical DNA fragments, relies on repeatedly increasing the temperature to separate the DNA strands and then decreasing it to allow primers to bind. It is thus important to know the temperature at which the strands of particular DNA molecules separate; this is known as the melting temperature, or T_m, and corresponds to the temperature at which half the DNA molecules are single-stranded (melted) and half are double-stranded. The T_m is controlled by several factors, but a major one is the fraction of G-C base pairs. Because G-C pairs are held together by three H-bonds (rather than two as in A-T pairs), breaking them requires more energy input, such as higher temperatures. A rough equation for the T_m of a short (<20 base pairs, bp) strand of DNA is:

$$T = 2(A + T) + 4(G + C)$$

where T is the temperature in degrees Centigrade, A+T is the total number of A-T base pairs and G+C is the total number of G-C base pairs.

Assuming a random nucleotide distribution, this rough equation makes two assumptions. The first is that one strand of the DNA is bound to a membrane, as it would be for a Southern blot, and that the blot is being probed with a short oligonucleotide (a single-stranded DNA molecule, generally <100 nucleotides long; the Greek prefix *oligo-* means "few"). With one strand immobilized, the DNA melts at a lower temperature (about 8°C less) than it would in solution. The second assumption is that the concentration of salt (e.g., NaCl) is 0.9 M, and that there is no chemical in the solution that would interfere with the formation of H-bonds between the bases (such as formamide, $HCONH_2$). The melting temperature of DNA increases with the $\log_{10}$ of the concentration of salt. This means that the higher the salt concentration, the more stable the DNA heteroduplex. It also decreases linearly with the concentration of formamide or other similar small molecules that interfere with DNA H-bond formation. Thus, a more complete equation is:

$$T_m = 81.5 + 16.6 \log M + 41(XG + XC) - 500/L - 0.62F$$

where M is the molar concentration of cations (Na^+ in this case), XG and XC are the mole fractions of G and C in the DNA, L is the length of the DNA (actually just the shortest strand of DNA in the mix), and F is the concentration of formamide.

The strong negative charge of DNA created by the phosphate backbone (on the outside of the double helix, Figure 1.1) can also be used to sort DNA molecules by size using gel electrophoresis, one of the most common tools in molecular biology. DNA is placed in a well at one end of a gel matrix (commonly an agarose or polyacrylamide gel), and an electrical current is run through the gel. The negatively charged DNA will thus migrate toward the positive electrode, and because the gel acts as a molecular sieve, the rate of DNA migration is dependent upon its size (Figure 1.3a). Smaller DNA fragments run faster than larger ones; graphing the $\log_{10}$ of the molecular weight or fragment length against the distance traveled produces a line (Figure 1.3b).

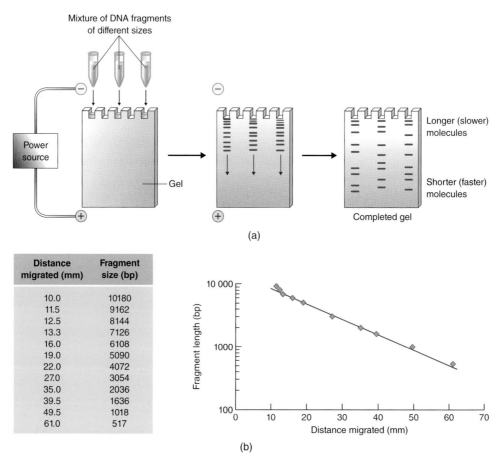

Distance migrated (mm)	Fragment size (bp)
10.0	10180
11.5	9162
12.5	8144
13.3	7126
16.0	6108
19.0	5090
22.0	4072
27.0	3054
35.0	2036
39.5	1636
49.5	1018
61.0	517

Figure 1.3 Gel electrophoresis, a powerful tool for biology. (a) DNA, suspended in an aqueous solution, is taken from a tube and placed in slots in a gel. An electrical current is applied to the gel. Because DNA is negatively charged, it moves toward the positive electrode, with smaller fragments moving more rapidly than larger ones. (b) Table and graph of the size of DNA fragments versus the distance migrated on a representative gel. Note that this relationship is not linear, so that the vertical axis of the graph is logarithmic. (b) Adapted from http://depts.noctrl.edu/biology/resource/handbook.htm

Certain chemicals bind to DNA and fluoresce when illuminated with an appropriate light source. For example, ethidium bromide has been widely used because it will intercalate into the double helix of the DNA; once there it will fluoresce under UV light. Thus, a common method of locating DNA on an agarose gel is to soak the gel in ethidium bromide and then place it on a UV light source. The DNA will then appear as pinkish bands (Figure 1.4a). Unfortunately, ethidium bromide will intercalate into the DNA of anything, including that of the biologist working with it. Because it can absorb light energy, it can damage the DNA; it is thus mutagenic. Its use is becoming less common because of its toxicity.

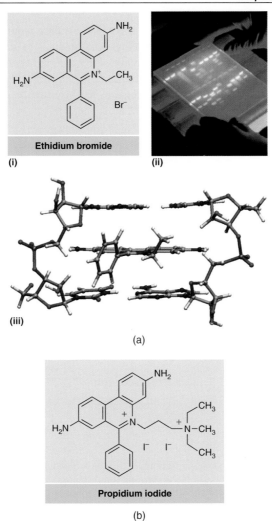

Figure 1.4 Chemicals that can be used to make DNA visible. (a) Ethidium bromide: (i) structure of ethidium bromide; (ii) an agarose gel stained with ethidium bromide and viewed with UV light; and (iii) ethidium bromide inserted into a DNA molecule (ai) http://www.sigmaaldrich.com, (aii) http://www.hamiltoncompany.com/products/syringes/c/893/. Reproduced with permission from Hamilton Company, (aiii)http://en.wikipedia.org/wiki/File:DNA_intercalation2.jpg. Image created by Karol Langner, via Wikimedia Commons. (b) Propidium iodide, from http://www.sigmaaldrich.com

Another fluorescent dye is propidium iodide, which, like ethidium bromide, intercalates into the DNA double helix (Figure 1.4b). Propidium iodide binds to DNA in a quantitative way, with one propidium iodide molecule per 4 or 5 bp; thus more DNA equals more propidium iodide binding, which equals more fluorescence. This direct relationship is used to estimate the genome size, which is the amount of DNA in the nucleus of a cell. Estimates of genome size using propidium iodide fluorescence are relatively rapid, so we have data on the genome size of many organisms, including flowering plants.

Methylated cytosines can be detected by a variety of methods. Restriction enzymes will cut DNA at particular sequences, but some will not cut DNA at 5-methyl cytosine, so that the methyl groups effectively protect the DNA from cleavage at these sites. Such methylation-sensitive restriction enzymes often have the same cut site sequence as methylation-insensitive enzymes. The restriction enzymes can thus be used sequentially. A restriction site that is cut with the methylation-insensitive enzyme and not with the methylation-sensitive one must be methylated. More recently, methods have been developed to find all the methylated cytosines in a genome. One such method is bisulfite sequencing, in which a genome is treated with sodium bisulfite, which converts methylated cytosines to uracils (Figure 1.5a). When the DNA is amplified by PCR the uracils become thymines. By comparing the sequence of the genome (or genomic region) before and after the bisulfite treatment, the cytosines that were methylated can be identified. The non-methylated cytosines are not affected by the bisulfite treatment, and thus remain the same (Figure 1.5b).

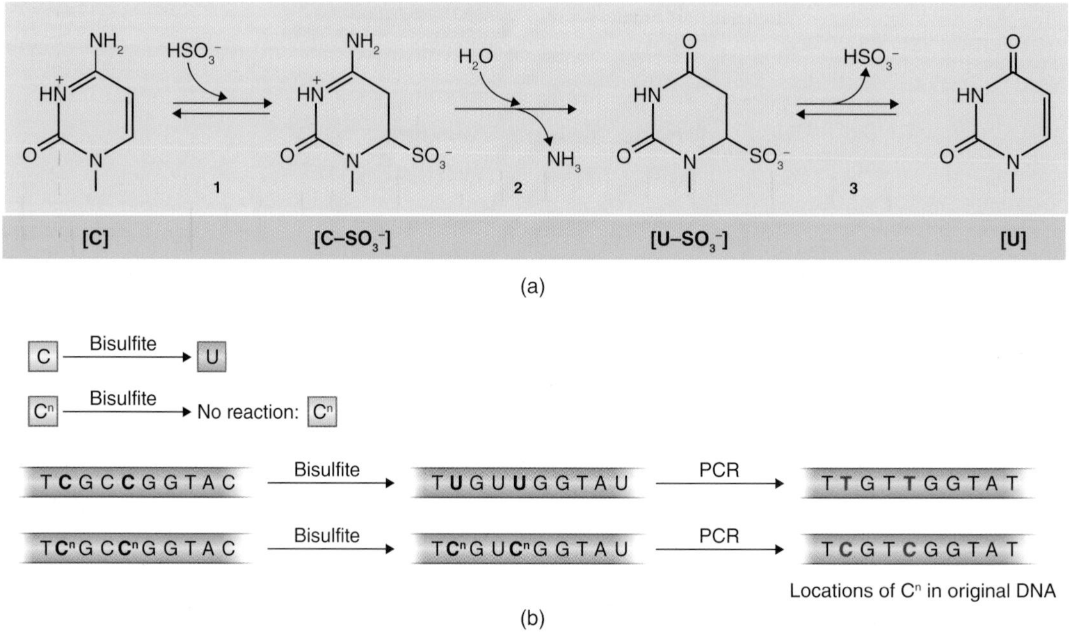

Figure 1.5 Bisulfite sequencing. (a) Chemical reactions necessary to remove the amino group from cytosine and convert it to uracil. This reaction will not occur if the cytosine is methylated. http://www.methylogix.com/genetics/bisulfite.shtml.htm. (b) Inferring the presence of 5-methylcytosine in a sequence of DNA. The DNA is treated with bisulfite and all non-methylated cytosines (C) are converted to uracil (U) (top line of sequences) whereas methylated cytosines are unchanged (bottom line of sequences). DNA is then amplified by PCR, which converts U to thymine (T), and the resulting sequence compared with the original. Any base that is C in the original but T in the amplified product is inferred to be non-methylated; conversely a base that is C in both the original and the amplified product must have been methylated. Redrawn from Hayatsu et al. (2008). Reproduced with permission of John Wiley & Sons, Inc.

Methylation-sensitive restriction enzymes are often used to generate genomic libraries, which are collections of DNA fragments cloned into autonomously replicating vectors such as a plasmid, phage (bacterial virus) or bacterial artificial chromosome (BAC).

1.2   The plant cell contains three independent genomes

The DNA in plant cells is found in the nucleus, the mitochondria and the chloroplasts. The latter two organelles are descendants of bacteria that were captured by a eukaryotic cell and have become endosymbionts; because many of the ancestral bacterial genes were transferred to the nucleus, the organelles can never revert to being free-living bacteria. Such microbial symbioses occur commonly. For example, cyanobacteria occur in the cells of some ferns, and in the stems of cycads. They co-occur with fungi to form lichens, although some lichens form from the symbiosis of a green alga and a fungus. There

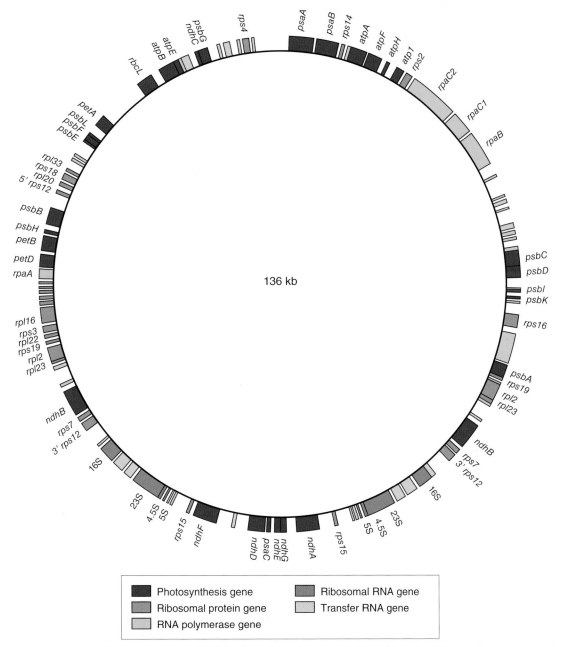

Figure 1.6 The chloroplast genome exemplified by that of rice. Genes are indicated by colored boxes. Note that there is a set of genes, from rps15 to rpl23, that appear twice but in inverse order. These regions are known as the inverted repeat. Redrawn from Brown, 2002

is even one recent example of a slug that has acquired plastids (Rumpho *et al.*, 2008). The extent of gene transfer between members of the symbiosis varies greatly in these examples, and hence the ability of the bacterial endosymbionts to function independently varies accordingly.

Plant mitochondria and chloroplasts have circular genomes, similar to those in Bacteria and Archaea (Figure 1.6). Their translational machinery (ribosomal RNA and proteins) is more similar to that in their bacterial ancestors than to the eukaryotic ribosomes encoded by the nuclear genome. The organellar genomes will be discussed in more detail in Chapter 5.

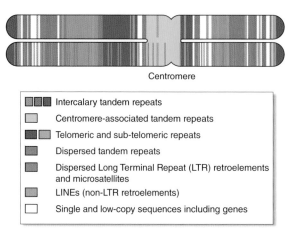

Centromere

▦▦▦	Intercalary tandem repeats
▢	Centromere-associated tandem repeats
▰▢	Telomeric and sub-telomeric repeats
▦	Dispersed tandem repeats
▦	Dispersed Long Terminal Repeat (LTR) retroelements and microsatellites
▦	LINEs (non-LTR retroelements)
▢	Single and low-copy sequences including genes

Figure 1.7 A chromosome, shown at metaphase, with the sister chromatids joined at the centromere. The two chromatids are identical in sequence, but different sorts of sequences are highlighted on each one for clarity. Schmidt and Heslop-Harrison (1998). Reproduced with permission of Elsevier

There is ample evidence from multiple plant species that the broad organization of plant genomes is similar to that of most eukaryotes. In all eukaryotes, including plants, the DNA of the nucleus is organized into linear chromosomes. The ends of the chromosomes are marked by distinctive structures, the telomeres, which have characteristic DNA sequences and have important roles to play in DNA replication. These will be discussed in more detail in Chapter 4 (Figure 1.7).

The other major landmark of eukaryotic chromosomes is the centromere. When chromosomes are viewed through a microscope, centromeres appear as constrictions that divide each chromosome into two segments – the chromosome arms. Centromeres in plants, as in all other eukaryotes, provide the point of attachment for the spindle apparatus during cell division. Centromeres will be discussed in more detail in Chapter 4 (Figure 1.7).

1.3   A gene is a complete set of instructions for building an RNA molecule

DNA is a coded set of instructions for making RNA. The fate of the RNA is hugely varied and is the subject of Part 2 of this book. In general, RNA may function as an independent regulatory molecule (e.g., micro RNA), or as piece of cellular machinery (e.g., transfer RNA, ribosomal RNA), or as a set of instructions for making a protein (e.g., messenger RNA, or mRNA), or some combination of these. In other words, RNA is a central molecule in the life of a cell, and DNA is simply a storage mechanism and blueprint for preserving and expressing the information in the form of RNAs. The genome is thus the full set of RNA-producing instructions, along with the information on when (in developmental or ecological time) and where (in what tissue) to use a particular set of instructions.

The RNAs that make proteins serve two distinct masters. First, and most familiar, are the mRNAs that make proteins to carry out all the cellular functions of the plant, including enzymes, structural proteins, receptors, and transcription factors. The other mRNAs serve the transposable elements, which are mobile pieces of DNA that move around the genome. In any given genome, there are thousands of transposable elements, each of which produces mRNAs that encode proteins that participate in transposon movement (transposition).

With this view of the genome, we may consider a gene as a complete set of instructions for making a particular RNA, including the non-coding DNA sequences that provide information on when and where the coding sequences should be transcribed. Under this very broad definition, the total number

Table 1.1   Genome sizes of selected plants for which genome sequences are available. Note that these are heavily biased toward plants with small genomes, which may affect our ability to generalize about plant genome structure and function. Note also that genome size does not correlate with any clear evolutionary relationships. For instance, Arabidopsis, a common experimental angiosperm species, has about the same genome size as *Chlamydomonas*, a green alga similar to algae that existed hundreds of millions of years prior to Arabidopsis

Species	Classification	Genome size (Mbp)	Estimated number of protein-coding genes
Selaginella moellendorffii (Banks *et al.*, 2011)	Lycophyte	106	22 285
Chlamydomonas reinhardtii (Merchant *et al.*, 2007)	Green alga	121	15 143
Arabidopsis thaliana (Arabidopsis Genome Initiative, 2000)	Angiosperm	125	27 025
Oryza sativa (Goff *et al.*, 2002; Yu *et al.*, 2002)	Angiosperm	420	39 045
Physcomitrella patens (Rensing *et al.*, 2008)	Moss	480	35 938
Populus trichocarpa (Tuskan *et al.*, 2006)	Angiosperm	485	41 335
Vitis vinifera (Velasco *et al.*, 2007)	Angiosperm	505	29 585
Sorghum bicolor (Paterson *et al.*, 2009)	Angiosperm	800	33 032
Zea mays (Schnable *et al.*, 2009)	Angiosperm	2300	39 475

of genes in any given genome is not known with any accuracy. Most commonly, when gene numbers are reported in the literature, the number includes only the mRNA-producing, protein-coding genes, excluding all those produced by the transposable elements. A few of these numbers are shown in Table 1.1, and are about the same order of magnitude ($2–4 \times 10^4$) between different plants. However, these account for only a tiny fraction of the genome; they are vastly outnumbered by the protein-coding genes from the transposable elements and the genes encoding non-messenger RNAs. Even though they constitute only a minority of the genes in the genome, the mRNA producing, protein-coding, non-transposable element sequences are usually the ones simply called "genes" in the literature. We will follow this common usage here unless we specify otherwise.

1.4   Genes include coding sequences and regulatory sequences

Plant nuclear genes are similar in general structure to those of other eukaryotes. The overall architecture of a gene consists of two general components, the regulatory region and the coding or structural region of the

gene (Figure 1.8). The regulatory region is responsible for controlling when a gene is transcribed into RNA. The regulatory region does not appear in the resulting mRNA, but directs the transcription machinery to start RNA biosynthesis (transcription) at a particular position, often but not necessarily, 3′ to the regulatory region. RNA transcription proceeds to the end of the gene generating a large precursor mRNA which will be processed in several ways. The portions of the gene that ultimately end up in the mature mRNA are known as the exons, whereas the portions of the RNA that get spliced out during processing are known as the introns, or intervening RNAs. In protein-coding genes, upstream (5′) of the coding sequence is a region that is transcribed into mRNA, but not translated to protein. This is referred to as the 5′ untranslated region (5′ UTR). A similar untranslated region occurs downstream of the last coding portion of the sequence and is known as the 3′ UTR. UTRs are part of the exons because they are present in the mature mRNA, even though they are not translated into proteins. The UTR regions of mRNAs are known to play roles in initiation of the translation process and stability of the mRNA. After transcription, the introns are spliced out of the messenger RNA, a 5′ cap is added to the 5′ UTR and a polyadenine (polyA) tail is added to the 3′ UTR. These RNA processing steps will be described in more detail in Chapter 13.

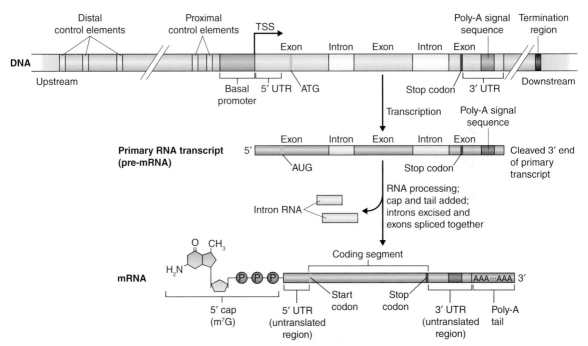

Figure 1.8 Steps in the formation of messenger RNA from DNA. The DNA for a single gene (a protein-coding sequence and all its control elements) is shown at the top. Note that the protein-coding sequence itself is a relatively small portion of the entire gene. The primary RNA transcript (pre-mRNA) also includes sequences upstream and downstream of the part that will be translated. The mature mRNA is formed by removing introns and adding a cap and a tail. Adapted from Campbell *et al.* (2005). While the structure shown is very common, the regulatory elements may occasionally be quite far away from the gene. Also the pre-mRNA may sometimes be spliced in several different ways, a process known as alternative splicing, which is discussed in more detail in Chapter 13

1.5   Nuclear genome size in plants is variable but the numbers of protein-coding, non-transposable element genes are roughly the same

In Bacteria, Archaea, and their mitochondrial and chloroplast descendants, most DNA is made up of sequences that encode RNA or proteins, and the coding sequences are barely separated from each other. In contrast, in the nuclear genome of eukaryotes, DNA encoding RNAs may account for only ~5% of the genome.

Individual plants and plant species differ in the amount of DNA in their genomes. DNA amounts are measured either as picograms (pg) of DNA per cell, or in numbers of base pairs (bp). Genome size is variable, with the largest genomes reported from the monocot *Paris japonica* (Melanthiaceae) (152.23 pg), and the smallest from the eudicot *Genlisea margaretae* (Lentibulariaceae) (0.063 pg) (Table 1.1).

Despite the variation in genome size, the number of protein-coding genes is surprisingly constant; in land plants it varies from approximately 22 300 for the lycophyte *Selaginella* (Banks *et al.*, 2011) to 35 900 for the angiosperm crop, maize (Schnable *et al.*, 2009). The difference in genome size thus cannot be accounted for by the genes themselves, but rather has to do with the size of the space between the protein-coding genes.

As can be seen from Table 1.1, the density of protein-coding, non-TE (non-transposable element) genes in the genome must be remarkably different among species. For example, while *Oryza sativa* (rice)

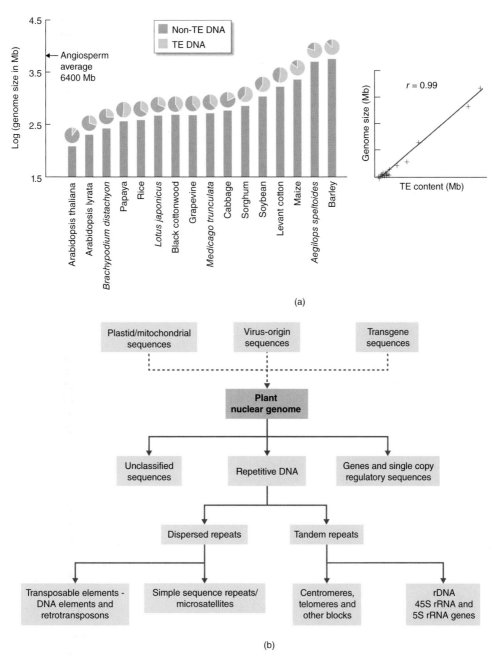

Figure 1.9 Types of sequences in the nuclear genome. (a) Genome sizes for flowering plants for which genome sequences are available. Pie charts show the relative amounts of DNA from transposable elements (TE-DNA) versus non-transposable elements (non-TE DNA). The linear relationship between the log of the genome size and the amount of transposable elements in the graph on the right shows that most of the difference in size is due to the difference in TE content. Tenaillon *et al.* (2010). Reproduced with permission of Elsevier. (b) DNA in the nucleus includes not only nuclear genes but also sequences from other sources such as organelles, viruses or transgenes inserted by humans. Nuclear DNA can then be classified into various categories. (b) Adapted from Heslop-Harrison and Schmidt (2007)

has about 39 000 protein-coding, non-TE genes spread across 400 Mbp of DNA, sorghum has about 33 000 in 800 Mbp of DNA, and maize has about 39 000 genes in a 2300 Mb genome. Clearly the difference must be the size of the "spaces" between the genes.

The word "spaces" is in quotes because in fact there are plenty of genes and regulatory sequences outside the protein-coding genes (Figure 1.7). Over the last decade it has become clear that this DNA (sometimes formerly dismissed as "junk") is a dynamic and active part of the genome and is as important for organismal function as the protein-coding fraction. Many of the discoveries about non-protein-coding genes have been made in plants.

Much of the space between the protein-coding non-TE genes is a complex mixture of repetitive sequences and transposable elements (Figure 1.9a). As noted above, transposable elements are mobile components of the genome with a propensity for creating repetitive bits of DNA sequence. The nature and dynamics of the transposable elements are discussed in the next chapter.

Other repetitive sequences are known as satellite DNA, which falls into several size classes (Figure 1.9b). Satellite DNA, as originally identified, corresponded to a set of repeats between about 150 and 180 bp now known to be located in and around the centromere. Minisatellites were described as repetitive sequences 10–100 bp long. Microsatellites are even smaller, generally consisting of repetitive units 2 or 3 bp long (e.g., ATATATATATAT); these are also known as Simple Sequence Repeats, or SSRs, and are widely used as markers for genetic mapping and for studies of population genetics. Most commonly, plant microsatellites are made up of A's and T's, whereas arrays of G's and C's are more common in animals. In all repetitive sequences, but particularly in microsatellites, the most common mutation is a change in the number of repeats, a process that is caused by replication slippage (Bhargava and Fuentes, 2010) (Figure 1.10). During DNA replication, the DNA polymerase continually dissociates from and reattaches to the DNA strand. Normally this does not create problems, but in a repetitive region the polymerase sometimes reattaches

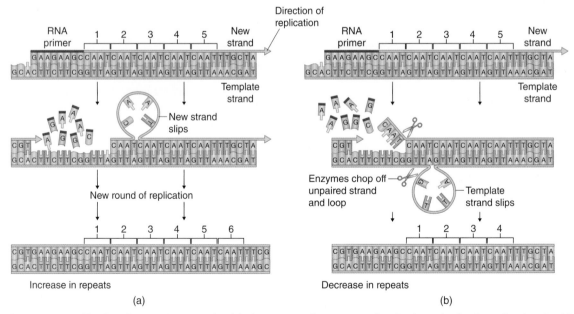

Figure 1.10 Replication slippage. DNA strands with short repeated sequences often lead to mistakes in replication. In this example, the sequence GTTA is repeated 5 times on the template strand. (a) If the new strand dissociates and reanneals such that the copy of repeat 1 anneals with the template for repeat 2, a loop of DNA is created in the copy strand, and repeat 1 on the template strand appears not to be copied. DNA polymerase copies repeat 1 again and the entire set of repeats is lengthened by one. (b) If the new strand slips and reanneals such that the copy of repeat 2 anneals with the template for repeat 1, a loop is formed in the template. This is recognized by DNA repair machinery and is removed; the set of repeats is shortened by one. Modified from Moxon and Willis, 1999

out of register, creating a loop or bubble in the DNA. Repair of this loop results in either loss or gain of one or more repeat units. The mutational process is not simple, and often microsatellites gain or lose several repeat units at once, rather than losses or gains of one at a time.

1.6 Genomic DNA is packaged in chromosomes

Chromosomes have been studied in plants since the nineteenth century. At the time, newly developed stains allowed microscopists to see for the first time colored bodies in the nucleus of some cells. The term chromosome is based on the Greek words *chroma* (color) and *soma* (body) to refer to these newly discovered structures. Once it became clear that the number of colored bodies was constant within an organism and often within a species, the chromosome count became a central piece of biological information. The number of chromosomes can be recorded either for the haploid gametes, in which case the number is referred to as n, or for the diploid (sporophyte) phase, in which case the number is $2n$. Thus for example in rice, $n = 12$ and $2n = 24$.

While the number of chromosomes is generally constant within a species, occasionally "extra" chromosomes appear in some plants but not others. The extra chromosomes are called **B-chromosomes**, or supernumerary chromosomes; they have no discernible effect on growth and development. B-chromosomes are smaller than normal chromosomes, and consist almost entirely of densely packed chromatin (known as heterochromatin). B-chromosomes do not segregate normally at meiosis, but instead accumulate preferentially in the cells that will become the gametes. Studies in maize find that pollen grains from a single plant can have from 0 to 20 B-chromosomes.

1.7 Summary

The structure of plant DNA is similar to that of all living things. The DNA molecule is a double helix, built of two antiparallel strands that are held together by hydrogen bonds. Methylation of cytosine residues is common and is now recognized as a structural feature that influences the function of the genome. The nucleotide composition of the DNA molecule determines its melting temperature, a parameter that is important for many biotechnology applications. The negative charge on the DNA can also be exploited for separation of molecules, and the ability of certain dyes to insert themselves into the double helix can be used to make the molecules visible to the human eye.

Plant DNA is found in the nucleus, the mitochondria and the chloroplasts, the latter two being ancient endosymbionts with genomes similar to those of their bacterial ancestors. Nuclear DNA is organized into linear chromosomes which have internal centromeres and are terminated by telomeres. The number of chromosomes varies between species, and chromosome number and genome size are not necessarily related. DNA provides all the instructions necessary to make RNA, from when the RNA is to be made to what protein will be encoded by the fully processed and mature form of the RNA. We define a gene as the full set of instructions for making a particular type of RNA. While much of the literature focuses on genes that will make protein-coding mRNAs, these are only a small component of the nuclear genome. Much of the rest of the genome is made up of genes that create proteins that are parts of transposable elements.

Genome size is variable among plants, even though the number of non-TE, protein-coding genes is fairly similar. Most of the differences in genome size are caused by differences in the number of transposable elements. Genome size also varies greatly in evolutionary time, so that closely related species may not have similar genome sizes.

1.8 Problems

1.1 You should be able to define the following terms: chromosome, centromere, transposable element, chloroplast, mitochondrion, methylation, bisulfite sequencing, microsatellite.

1.2 Calculate the approximate melting temperature of a DNA molecule with the sequence ATGGGCATAGCCGA.

1.3 Why is the value you calculated in Problem 1.2 an oversimplification?

1.4 What are two ways to determine which sites in a DNA molecule are methylated?

1.5 What is a gene? Show the architecture of a typical protein-coding gene in a diagram.

1.6 Draw the structure of a simple double-stranded DNA molecule with one strand having the nucleotide content of adenine and guanine.

1.7 How might DNA methylation perturb gene expression? And DNA replication?

1.8 Introns are removed from initial RNA transcripts with great fidelity. Can you come up with a design of DNA that could account for this fidelity? What features do you think would be necessary in the DNA to allow for this fidelity?

References

Arabidopsis Genome Initiative. (2000) Analysis of the genome sequence of the flowering plant *Arabidopsis thaliana*. *Nature*, **408**, 796–815.

Banks, J.A., Nishiyama, T., Hasebe, M. *et al.* (2011). The *Selaginella* genome identifies genetic changes associated with the evolution of vascular plants. *Science*, **332**, 960–963.

Bhargava, A. and Fuentes, F.F. (2010) Mutational dynamics of microsatellites. *Molecular Biotechnology*, **44**, 250–266.

Brown, T.A. (2002) *Genomes*, 2nd edn, Wiley-Liss.

Campbell, N.A. and Reece, J.B. (2007) *Biology*, 7th edn, Benjamin Cummings.

Goff, S.A., Ricke, D., Lan, T.-H. *et al.* (2002) A draft sequence of the rice genome (*Oryza sativa* L. ssp. *japonica*). *Science*, **296**, 92–100.

Heslop-Harrison, J.S.P. and Schmidt, T. (2012) *Plant Nuclear Genome Composition*, John Wiley & Sons, Ltd.

Hyatsu, H., Shiraishi, M. and Negishi, K. (2008) Bisulfite modification for analysis of DNA methylation, in *Current Protocols in Nucleic Acid Chemistry*, John Wiley & Sons, Ltd, Unit 6.10.

Merchant, S.S., Prochnik, S.E., Vallon, O. *et al.* (2007) The *Chlamydomonas* genome reveals the evolution of key animal and plant functions. *Science*, **318**, 245–250.

Moxon, E.R. and Wills, C. (1999) DNA microsatellites: agents of evolution? *Scientific American*, **280**, 94–99.

Paterson, A.H., Bowers, J.E., Bruggmann, R. *et al.* (2009). The *Sorghum bicolor* genome and the diversification of grasses. *Nature*, **457**, 551–556.

Raven, P., Johnson, G.B., Mason, K.A. *et al.* (2011) *Biology*, 9th edn, McGraw-Hill.

Reece, J.B., Urry, L.A., Cain, M.L. *et al.* (2011) *Campbell Biology*, Benjamin Cummings.

Rensing, S.A., Lang, D., Zimmer, A.D. *et al.* (2008) The *Physcomitrella* genome reveals evolutionary insights into the conquest of land by plants. *Science*, **319**, 64–69.

Rumpho, M.E., Worful, J.M., Lee, J. *et al.* (2008). Horizontal gene transfer of the algal nuclear gene *Psbo* to the photosynthetic sea slug *Elysia chlorotica*. *Proceedings of the National Academy of Sciences USA*, **105**, 17867–17871.

Schmidt, T. and Heslop-Harrison, J.S. (1998) Genomes, genes and junk: the large-scale organization of plant chromosomes. *Trends in Plant Science*, **3**, 195–199.

Schnable, P.S., Ware, D., Fulton, R.S. *et al.* (2009) The B73 maize genome: complexity, diversity, and dynamics. *Science*, **326**, 1112–1115.

Tenaillon, M.I., Hollister, J.D. and Gaut, B.S. (2010) A triptych of the evolution of plant transposable elements. *Trends in Plant Science*, **15**, 471–478.

Tuskan, G.A., DiFazio, S., Jansson, S. *et al.*, (2006) The genome of black cottonwood, *Populus trichocarpa* (Torr. & Gray). *Science*, **313**, 1596–1604.

Velasco, R., Zharkikh, A., Troggio, M. *et al.* (2007) A high quality draft consensus sequence of the genome of a heterozygous grapevine variety. *PLoS One*, **2**, e1326.

Yu, J., Hu, S., Wang, J. *et al.* (2002). A draft sequence of the rice genome (*Oryza sativa* L. ssp. *indica*). *Science*, **296**, 79–92.

Chapter 2

The shifting genomic landscape

The history of the Earth is recorded in the layers of its crust; the history of all organisms is inscribed in the chromosomes.

H. Kihara, 1946

Genomes change over time. Some changes are rapid and occur between generations of plants, whereas others appear more slowly and accumulate over ecological or evolutionary time. In the category of rapid change are shifts in genome size, gain/loss of repetitive sequences, and epigenetic changes; the latter will be discussed in Chapter 12. Slow – or at least less frequent – change includes chromosomal alterations such as rearrangements and changes in chromosome number. Both sorts of change accumulate at different rates in different parts of the genome, and are influenced by the recombination rate as well as the ability of the plant to tolerate particular mutations.

In this chapter we first outline the kinds of mutations that occur in plant genomes. Then, we move on to a discussion of the amount of DNA in the nucleus, represented by **genome size**. Genome size varies enormously in plants, meaning that DNA must be gained or lost frequently and continually in plant evolution. We discuss the many mechanisms by which DNA can be gained, including duplications of small or large amounts of DNA as well as sharp increases in the numbers of transposable elements. DNA can also be lost, mostly by recombination between repetitive sequences.

All these changes in the genome may have effects on gene expression. Regulatory sequences such as transcription factor binding sites are gained or lost, the three-dimensional structure of the DNA molecule changes, histone and DNA modification patterns change (see Chapter 12), and entire genes may be copied or deleted causing changes in the cell's basic machinery.

2.1 The genomes of individual plants can differ in many ways

Over time, plant genomes, like all other eukaryotic genomes, accumulate mutations. These are caused by many factors, including the simple fact that cells are aqueous environments in which random but infrequent addition of water to DNA can cause damage to a nitrogenous base and then an imperfect repair results in a base substitution. In photosynthetic organisms, daily exposure to light energy also damages the DNA. During mitosis, the process of DNA replication (see Chapter 4 for review) can be error prone, particularly in repetitive regions of the DNA where the polymerase tends to dissociate and reattach out of phase. During meiosis, the double-stranded breaks that occur during recombination also create opportunities for mutations. While DNA repair enzymes continually survey the genome for damage and then correct any errors, some mistakes inevitably occur. When they occur in cells that will become gametes, the mutations can be passed to the next generation.

Plant Genes, Genomes and Genetics, First Edition. Erich Grotewold, Joseph Chappell and Elizabeth A. Kellogg.
© 2015 John Wiley & Sons, Ltd. Published 2015 by John Wiley & Sons, Ltd.
Companion Website: www.wiley.com/go/grotewold/plantgenes.

Mutations may be **base pair substitutions**, **length variation** (insertions/deletions), or **rearrangements**. Base pair substitutions can occur from any nucleotide to any other, so that for example, and adenine (A) can mutate to a guanine (G), cytosine (C), or thymine (T) (Figure 2.1), albeit with distinct frequencies. A substitution that exchanges a purine for a purine or pyrimidine for a pyrimidine is known as a **transition**, whereas a change between a purine and a pyrimidine or vice versa is a **transversion**. There are four possible transversions (A-C, A-T, G-C, G-T) and only two possible transitions (A-G, C-T), so that if mutations were completely random, we would expect to find more transversions than transitions. However, transitions are often more common; this makes sense from a purely structural point of view, since replacing a purine with a pyrimidine or vice versa would affect the width of the double helix, and would likely be detected by the DNA repair machinery and corrected. One particularly common sort of base pair substitution is deamination of 5-methylcytosine to produce uracil. This is then read by the DNA polymerase as thymine. The result is a C to T transition.

The combined effects of mutation and recombination lead to differences in the nucleotide composition of the entire genome. Although we would expect that, on average, about half of the base pairs in the genome would be A-T and about half G-C, the frequencies of the different types of nucleotides vary from place to place in the genome. This frequency variation is particularly obvious at the third position in the codons for amino acids in protein-coding genes. For example, looking just at the third positions in the genome of poppy (Papaveraceae), we find about 36% G-C (and thus 64% A-T), whereas in rye (a cereal grass, Poaceae) the corresponding percentage is about 68% G-C (Serres-Giardi et al., 2012). The reasons for this variation are not clear. The mutation process itself tends to lead to more A-T base pairs. On the other hand, the protein synthesis machinery preferentially uses codons that end in G or C in the most highly expressed genes, a process known as codon bias. However, although the data are most comprehensive within protein-coding genes, the bias toward G-C or A-T often can be found in flanking regions as well, so other phenomena must be occurring in addition to biased codon usage. The most plausible current hypothesis is that the process of recombination has

a bias toward G-C, and indeed the percentage of G-C base pairs correlates well with the frequency of recombination within a gene and along a chromosome. G-C biased genes also have more targets for methylation and often have a higher mutation rate (Tatarinova et al., 2010). Regions of high G-C content are well documented in mammalian genomes, where they are known as C-G islands, and tend to occur in promoters of protein-coding genes (Fenouil et al., 2012). Although they have an important role in gene regulation in mammals, C-G islands seem to be rare or absent in plants.

Corresponding regions of genes and genomes in different plants may be different lengths, which must be caused by either an insertion (i.e., gain of one or more base pairs) or a deletion (i.e., loss of one or more base pairs). When comparing two genomes, it is often unknown whether the length difference represents an insertion in one or a deletion in the other, so the mutations are often simply called **indels** (insertion-deletions). As described in Chapter 1, length variation is particularly common in repetitive regions, which are prone to replication slippage. In addition, rearrangements such as inversions or translocations can occur.

A common way to compare two DNA sequences is to align them so that the 5′ end of one sequence is written above the 5′ end of the other and corresponding bases are arranged in columns until the 3′ end of the sequence is reached (Figure 2.2). Indels and single nucleotide substitutions are easily spotted in such an **alignment**. Because the latter represent different forms of the sequence, they are known as **single nucleotide polymorphisms** or **SNPs**.

The number of matches is expressed as **percent similarity**, and corresponds to the number of matching base pairs divided by the total number of base pairs in the alignment. Percent dissimilarity is of course the percentage of mismatches. In alignments of nucleotides, the terms "similarity" and "identity" are generally used interchangeably. However, some authors make a distinction between the two terms. Two pyrimidines can be considered similar even if they are not identical. Thus, for the alignment in Figure 2.2, the sequences may be 24/42% identical but 27/42% similar. The distinction between similarity and identity is used much more commonly in protein alignments, in which amino acids with similar properties can be

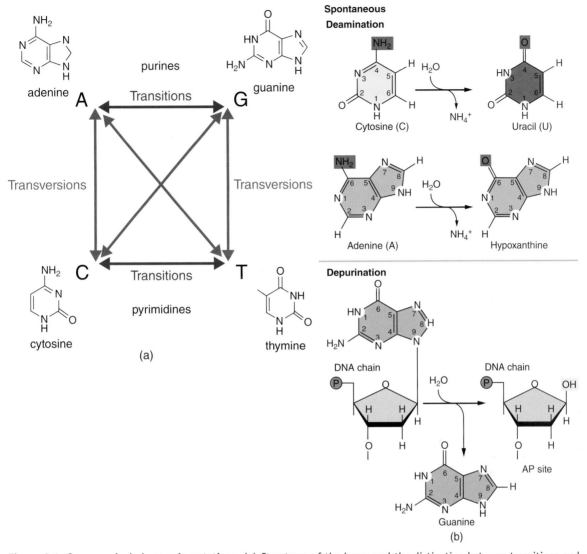

Figure 2.1 Common single base pair mutations. (a) Structures of the bases and the distinction between transitions and transversions. http://commons.wikimedia.org/wiki/File:Transitions-transversions-v3.png. By Petulda (Own work) [Public domain], via Wikimedia Commons. (b) Results of deamination (top) and depurination (bottom), both of which are spontaneous mutations. From Buchanan *et al.* (2000). Reproduced with permission of John Wiley & Sons, Inc.

classified as similar, even if they are not identical. For example, in Figure 2.2, the second amino acid is either aspartate (in sorghum) or glutamate (in maize and Miscanthus). Both residues have R groups that are of similar size and have a negative charge, so they are considered to be functionally similar and are often given the same color in an alignment.

Alignments are also used to find matching sequences in databases. For example, if a gene is

cloned in maize and the investigator wants to determine if a similar gene has been studied in Arabidopsis, the maize gene can be rapidly aligned to all other sequences in Genbank using an algorithm known as **Basic Local Alignment Search Tool**, or **BLAST**. The alignment with the highest similarity is known as the best BLAST hit. Further phylogenetic analysis is then required to determine whether the Arabidopsis gene is really homologous to the one in maize.

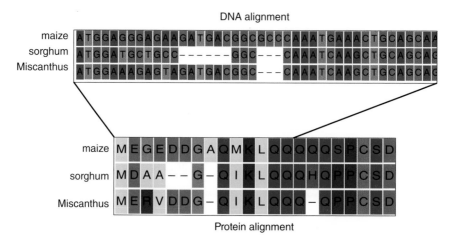

DNA alignment

Protein alignment

Figure 2.2 Alignment of nucleotides and the corresponding amino acids from the beginning portion of the *ramosa1* locus in maize, sorghum, and the ornamental plant *Miscanthus sinensis*. The 42 base pairs of the DNA alignment correspond to the first 14 amino acids in the protein. Two indels are visible in these first 42 base pairs, plus 9 single nucleotide polymorphisms

2.2   Differences in sequences between plants provide clues about gene function

All parts of the genome are assumed to mutate at about the same rate in a random process analogous to the decay of an unstable isotope. However, the effect of the mutation, and hence the likelihood that it will be passed on, differs from one part of the genome to the next. Mutations of some base pairs reduce the ability of the plant to reproduce and are thus eliminated by natural selection. In the most extreme example of this, some mutations are simply lethal and the gamete or plant never survives long enough to pass on the mutation. In such a case, the sequence (often a critical protein-coding gene) will be unaltered in the surviving plants. Thus, sequences in which mutations are immediately lethal are strikingly similar among plants in the same population, species or group of species, and are described as being **conserved**, that is, natural selection has kept them from changing. Other sequences tolerate mutations in particular positions, or in particular combinations of positions, and are less well conserved. For example, many mutations in introns have little or no effect on gene function,

and hence little effect on the reproductive success of the plant. Such mutations tend to accumulate so that comparisons between individuals of a population or closely related species will have different sequences (Figure 2.3).

This basic principle of evolutionary biology can be used to infer the function of different parts of the genome. For example, sequences immediately upstream of a transcription start site (TSS, see Chapter 6) are often more conserved than those farther upstream; these have been called **conserved non-coding sequences** (**CNS**s). These conserved sequences often include the region known as the basal (otherwise known as minimal or core) promoter, where the general transcription factors and components of the RNA transcription machinery bind. DNA sequences that participate in the regulation of gene expression are also known to occur in the introns or other regions of some genes (Chapters 7 and 8).

Plants are quite different from animals in the size and distribution of their CNSs. Mammalian genomes have CNSs that extend for many megabases (mega = million) and are presumed to be regulatory. In contrast, CNSs in plants are often quite short (often less than a few hundred base pairs) and often occur close to genes. This may be one reason that plants can tolerate frequent and extensive genome rearrangement as described below. There are notable

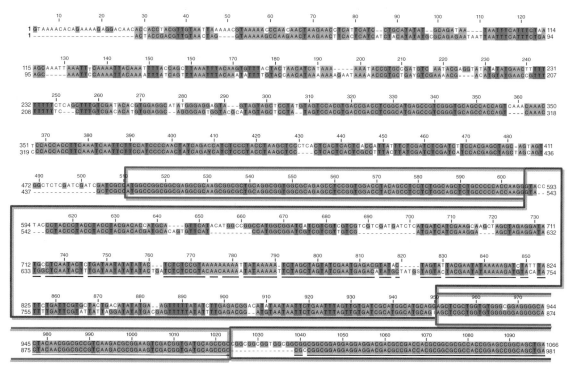

Figure 2.3 Part of the transcribed sequence of the *Rc* gene in two closely related species of *Oryza*. All sequences shown are transcribed, but only the sequences boxed in red are translated. Sequences boxed in blue are spliced out to form the mRNA. The first exon includes the 5′ UTR (not boxed) plus the translation start (red box). This is followed by the first intron (blue box), and the second exon (red box), and a portion of the second intron (blue box). Notice the difference in sequence conservation between the various regions. This particular portion of the 5′ UTR is ca. 70% identical between the two species (358/510 bp), and the first intron is 79% identical. The translated portion of the first exon is 100% identical, whereas the second exon is 99% identical. Data from Gross *et al.*, 2010

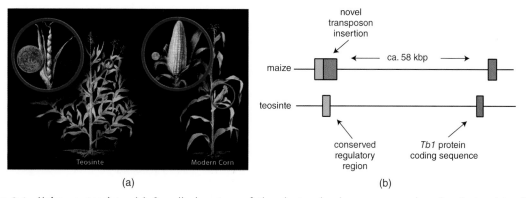

(a)   (b)

Figure 2.4 Maize vs. teosinte. (a) Overall phenotype of the plants, showing many more branches in teosinte; figure from http://nsf.gov/news/mmg/media/images/maize1_f.jpg. Photo Nicolle Rager Fuller, National Science Foundation. (b) Cartoon of the *tb1* locus in maize and teosinte. Orange boxes, *Tb1* coding region; yellow boxes, upstream regulatory regions; brown box, novel insertion of transposon in cultivated maize. Drawing approximately to scale

exceptions, however. Conserved regulatory sequences for the maize gene *Teosinte branched1* occur 58–69 kb upstream of the transcription start site (Clarke *et al.*, 2006). One mutation that causes maize to look different from its wild ancestor teosinte is caused by an insertion of two transposons into this conserved region (Zhou *et al.*, 2011; Studer *et al.*, 2012) (Figure 2.4). (See also http://www.weedtowonder.org/.)

The level of variability in the DNA between any two plants of the same species may be remarkably high. A comparison of two maize inbred lines identified single-copy genes that were unique to one line, as well as hundreds of duplications in which the copy number varied between the lines (Schnable *et al.*, 2009). More than half the sequences outside the protein-coding genes differed in some parts of the genome, reflecting different sets of transposons (Brunner *et al.*, 2005). Similar results have been reported for barley cultivars (Scherrer *et al.*, 2005) and for Arabidopsis (Clark *et al.*, 2007; Zeller *et al.*, 2008), indicating that the pattern found in maize may be common in plants.

2.3   SNPs and length mutations in simple sequence repeats are useful tools for genome mapping and marker assisted selection

A **genetic map** is a useful tool in plant biology, both for basic research and for applications in plant breeding (see Box, Linkage mapping). Like any other DNA-based polymorphism, SNPs and simple sequence repeats (SSRs; see Chapter 1) can be mapped and their positions assessed relative to each other. Once the map is constructed, it can be used to determine the genetic basis of phenotypic variation. It is possible to determine how many loci contribute to a particular phenotype, where those loci are in the genome, and in some cases what the underlying genes are.

Linkage mapping

A linkage map is used to determine which genes are near each other on chromosomes ("linked"), and thus are likely to be inherited together. Even in the current era of genome sequencing, linkage maps are valuable tools. The order of genes and DNA markers over large chromosomal regions provides a valuable constraint and check on genome assembly. In addition, a linkage map allows information from the sequence of a single plant to be extrapolated to other plants in the same species or even the same genus.

Production of a map begins with a cross between two parental plants. Before making the cross, each parental plant is ideally self-pollinated for several generations, a process known as inbreeding. After many generations of inbreeding an organism becomes homozygous throughout the genome, that is, at any given place in the genome, the sequence of DNA on the members of a chromosome pair is the same. The two inbred parents are then chosen to be as different as possible, because only differences (polymorphisms) between parents can be mapped. The differences may be morphological characteristics, such as fruit color or stem branching, but there are usually relatively few of these in any particular pair of parents. More common differences are DNA-level polymorphisms, such as SNPs and SSRs. These sometimes are called markers, because they mark a particular location in the genome.

When the two parents are crossed, they produce a hybrid individual (F1; the F stands for filial) that will be heterozygous, with one allele from the seed parent and one from the pollen parent. The alleles can be identified by particular SNPs. For example, if one parent has an A at a particular position in the genome and the other has a C, the F1 will have both an A and a C.

The F1 is then either self-pollinated (i.e., crossed with itself) or backcrossed to one of the parents. For this example, it is simplest to consider the situation in which the F1 is crossed to itself (selfed, possible in many but not all plants). This will produce an F2 generation, plants that are the "grandchildren" of the original cross. For each marker (SNP, SSR, morphological polymorphism) the F2 plants will fall into three distinct categories – homozygous maternal, homozygous paternal, and heterozygous – following Mendel's First Law (Figure 2.5).

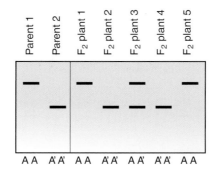

Figure 2.5 Cartoon of an electrophoretic gel for one marker for two parents and a few of their F2 offspring. Genotypes are shown below the gel. The F2s can be either homozygous maternal, in which case they show a single large band on the gel, homozygous paternal, showing a single smaller band, or heterozygous, with two bands

Now consider two markers. If they are completely unlinked, as they would be if they were on different chromosomes, then the maternal alleles of one marker are equally likely to end up with the maternal or paternal alleles of the other one, and vice versa. This is Mendel's Law of Independent Assortment (or Second Law). As an imaginary example, it suggests that if a SNP at one position (say, A or C) is unlinked to a SNP at another position (say, A or G), then in the F2 generation plants with an A at the first position are equally likely to have an A or G at the second position, and plants with a C at the first position are equally likely to have an A or G at the second position. In other words, there is no correlation between the base pair in the first SNP and the one at the second SNP (Figure 2.6a).

If the two markers are linked, however, then the alleles from one parent will occur together more often than we expect. We can check this by counting the numbers in each category. If we find that most plants with an A at the first SNP also have an A at the second SNP, then we infer that the two SNPs must be physically close together so that they are often inherited as a unit (Figure 2.6b).

Markers are thus linked if two alleles are inherited together more than expected. They will become separated only if the chromosomes form a chiasma at meiosis and crossover, and if that crossover happens to fall somewhere between the markers. The percentage of time that this occurs is known as the percent recombination, and is used as a measure of distance between two markers. The closer the two are on the chromosome, the less chance there is for a crossover between them and the lower the percent recombination.

By calculating the percent recombination among a very large number of markers, it is possible to place them in order in linkage groups (Figure 2.7). Markers within a linkage group are inferred to be on the same chromosome, whereas those in different linkage groups are on different chromosomes. The percent recombination is conventionally multiplied by 100 to give a unit of recombination distance known as the centimorgan (cM), honoring T. H. Morgan who first developed the idea of linkage.

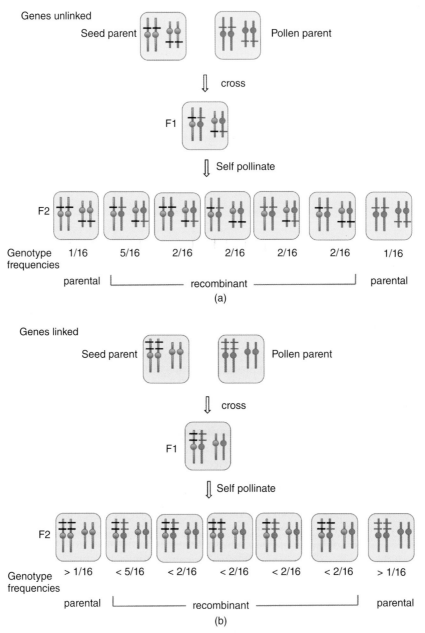

Figure 2.6 Pattern of inheritance of unlinked versus linked genes. (a) If the genes are unlinked, then they sort out randomly in the F2 offspring and the genotype frequencies are as shown. (b) If the genes are linked, then there will be more parental types and fewer recombinants than expected at random. The exact frequencies depend on how closely linked the genes are

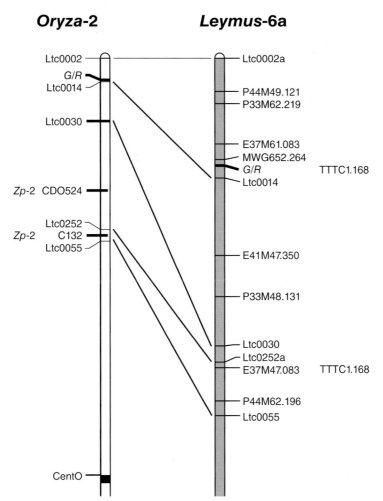

Figure 2.7 Portion of a linkage map of one chromosome of *Leymus*, a forage grass (right), compared with the genome of rice; abbreviations are names of particular DNA markers. Note that in this case the DNA markers, and hence presumably the genes, are in the same order in the two species. From Larson and Kellogg (2009)

In plant breeding, it is often helpful to find a DNA-based polymorphism that is genetically linked to the phenotype because the phenotype itself may be difficult or cumbersome to determine. In the case of tolerance to flooding, for example, rice plants must be grown large enough to test how they respond to several days or weeks of submergence. However, once the gene for submergence tolerance was identified, it was then possible to identify particular sequences that distinguished the tolerant allele, and the allele could be tracked easily even in very young plants using polymerase chain reaction (PCR) methods (Xu *et al.*, 2006).

It was then possible to select for desired progeny plants in a breeding effort to cross the submergence-tolerant allele into a higher yielding variety. This method, referred to as **molecular breeding** or **marker assisted selection**, is far more efficient than having to grow and score every plant for its response to flooding.

A variety of methods exist to correlate variation at a particular locus with **phenotypic variation**. Phenotypic variation can be any aspect of the organism other than the genotype; it might be morphological, such as the size of fruit or the shape of leaves, or physiological, such as the rate of respiration or stomatal conductance, or even biochemical, such as the production of a secondary compound. The most important methods for evaluating the genetic basis of phenotypic variation are **quantitative trait locus (QTL) mapping** and **association analysis**. Although these methods are statistically complex, the underlying logic is similar and reasonably simple. Imagine a group of plants in which some are resistant to a pathogen and some are not; the goal is to find the location in the genome of the resistance gene. Once identified, the locus could be transferred by crossing into a susceptible variety to make it resistant. Now imagine that you have determined the patterns of SNP or SSR variation for all members of this group. Look at the SNP variation at one locus and assign all plants to groups according to their genotype at that SNP, for example, there might be one set of plants with an A and one set with a G. Then determine whether the plants with an A are significantly more resistant than those with a G. If plants with the A SNP (the genotype) are mostly resistant to the pathogen (the phenotype), then there must be a locus near that SNP that controls resistance. Conversely, if the genotype

and phenotype do not correlate, there is no evidence for a resistance gene near that particular SNP. This same sort of analysis is done throughout the genome. (For more details, see From the Experts, QTL analysis and association mapping.)

In the example in the preceding paragraph, we have made several assumptions. First, we have assumed that pathogen resistance is a qualitative trait; that is, the plant is either resistant or not, and no intermediates occur. This is often an unrealistic assumption; many times we are interested in the genetics of **quantitative traits**. Quantitative traits are those that vary continuously, such as plant height, yield, photosynthetic rate, and many others; even some apparently qualitative traits such as color may in fact be quantitative, with continuous variation in the amount of pigment. Secondly, we have assumed that all the plants are completely independent of each other. In fact, this assumption may be violated if some plants are more closely related to each other than to others. The extent of the relationship, also known as population structure, needs to be assessed statistically to make valid comparisons among the plants.

From the Experts

QTL analysis and association mapping

Most traits in plants are quantitative in nature, with a continuous distribution of phenotype. Quantitative traits are usually controlled by multiple genes, the environment, and their interaction. Quantitative trait loci are stretches of DNA containing genes that underlie a quantitative trait. The objectives of QTL mapping are to identify how many QTL control a trait, their genomic locations, and to estimate the genetic effects (how much of the phenotypic variation they control). There are two main approaches for QTL analysis: linkage-based QTL mapping and association-based mapping. Regardless of QTL discovery method, putative QTL must be validated using other methods such as near isogenic lines or transgenic approaches.

Linkage-based QTL mapping relies on recent recombination in a structured population to create linkage between genetically mapped markers (see Box, Linkage mapping) and QTL. A genome scan (e.g., in 2 cM intervals) is conducted to test for a relationship between the trait values and the genotypes. Many types of statistical analyses are used, ranging from simple marker regression to composite interval mapping, where recombination data and background QTL are included to improve mapping precision.

Linkage-based QTL mapping can be a very powerful method because half the alleles in the population come from each parent, creating a balanced statistical test. Statistical power depends on a number of factors, including heritability (the proportion of the trait variation that is due to genetic components versus the environment), the distribution of allelic effect sizes, and population size.

Because there is limited recombination during population development (2 to 6 meioses, depending on the population structure; Figure 2.8a), linkage mapping resolution can be low. Often, QTL are mapped to ~10 cM regions which may contain over a thousand genes. A second drawback of linkage-based QTL

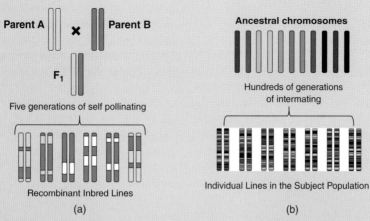

Figure 2.8 (a) Linkage mapping. A bi-parental cross of two inbred lines and the formation of recombinant inbred lines. The linkage blocks along the chromosome are very large as a result of only five meiotic events, resulting in low QTL mapping resolution. (b) Association mapping. An association panel is a collection of lines that can be related or unrelated. The historical recombination that has occurred over hundreds of generations has broken up the linkage blocks along the chromosome, resulting in high QTL mapping resolution

mapping is that only two alleles are sampled, thus limiting the findings to the specific parents used in the study.

Association mapping [also known as linkage disequilibrium (LD) mapping] has been a mainstay in human genetics, but is relatively new to plants. Association analysis can be used to test specific candidate genes or can be conducted as a genome scan. Association mapping is performed at the population level using unrelated or distantly related individuals, and relies on historical recombination events to break down linkage blocks within the genome (Figure 2.8b). Association mapping is highly dependent on the population chosen for analysis due to three issues: LD, population structure, and allele frequencies.

LD is the correlation of polymorphisms within a population usually related to linkage/genetic distance, and is highly variable between species. Generally, LD decays rapidly in outcrossing species and is more extensive in self-pollinating species. In an outcrossing species such as maize, where LD decays within a few kb, high resolution QTL mapping can be achieved, but extremely high marker density is required (i.e., 10–50 million SNPs in diverse maize). In species with extensive LD, many fewer markers are needed, but mapping resolution is more limited.

The presence of population structure within the subject population can lead to spurious associations (i.e., false positives), as alleles can be correlated to the phenotype even if they are not causative. The origin of population structure could be differences in flowering time, plant breeding history, commodity groups, geography, and so on. Populations should be chosen in order to minimize population structure and statistical models are used to account for population structure.

The final factor influencing association mapping is allele frequency. The more genetically diverse the population is, the more rare alleles are present in the population. If the allele frequency is severely imbalanced or all alleles are rare, then the statistical power to detect associations is low.

By Sherry Flint-Garcia and Michael McMullen

2.4   Genome size and chromosome number are variable

The amount of DNA in the nucleus varies between plants, both within and between species. Because the amount of DNA also changes according to the cell cycle, it is common to compare values during the G1 phase, before replication has occurred. This value is known as the **2C value**, meaning the content of DNA in a somatic nucleus; 1C is the gametic amount and is usually determined by simply dividing the somatic value by 2, rather than by measuring gametes directly. One way to measure DNA content is by creating a homogeneous suspension of cells, staining the nuclei with a fluorescent dye, and then measuring the amount of fluorescence in comparison with a standard. The results are reported in picograms of DNA; 1 pg of DNA is roughly equal to 978 Mbp, or 978 x 10^6 bp. For plants, genome size values have been assembled into a large database hosted by the Royal Botanic Gardens at Kew (http://data.kew.org/cvalues/). DNA content and number of chromosomes also vary within a plant, with cells in the epidermis and endosperm particularly undergoing repeated rounds of DNA replication without mitosis. DNA content in these tissues can be assessed by staining the cells with a fluorescent dye and quantifying the amount of fluorescence microscopically, often normalizing to the adjacent guard cells that form the stomatal pore, which in most species have 2C.

As noted in Chapter 1, plant DNA is organized into linear chromosomes, with the numbers in flowering plants varying from $2n = 4$ (e.g., the grass genera *Zingeria* and *Colpodium*) to $2n = 640$ (in the stonecrop *Sedum suaveolens*) (Leitch *et al.*, 2010). The number of chromosomes is often constant within a species, but there are notable exceptions. For example, the spring beauty, *Claytonia virginica* (Figure 2.9a), has chromosome numbers that vary from $2n = 12$ to $2n = 37$, with several cytotypes present even in a single population (Lewis and Semple, 1977). Likewise, different plants of sugar cane (*Saccharum officinarum*), which is a hybrid between two species of *Saccharum*, may have different numbers of chromosomes. While the different chromosome numbers can make plants partially sterile, in the case of sugar cane individual lines are propagated from cuttings (vegetatively) so the chromosome number variation is maintained indefinitely.

Among diploid plants, the number of chromosomes does not correlate with genome size. Thus, two plants could have the same total amount of DNA in the genome, but the DNA might be divided into quite different numbers of chromosomes. For example, barley has seven pairs of chromosomes ($2n = 14$), but a genome size of 5.55 pg of DNA per 1C nucleus, whereas rice has 12 pairs ($2n = 24$) with a genome size of 0.45 pg. Similarly, *Sorghum bicolor* ($2n = 20$; Figure 2.9b) has a genome size of about 0.83 pg, but the closely related *Sarga angusta* has half the number of chromosomes ($2n = 10$) and more than twice as much DNA per 1C nucleus (1.85 pg).

In several well-documented cases of chromosome number reduction, an entire chromosome has been inserted into the centromeric region of another (Luo *et al.*, 2009). Thus, the gene content and order of the inserted chromosome is preserved, whereas the two individual arms of the chromosome into which it is inserted are preserved as the distal ends of the new chromosome. The centromere of the inserted chromosome remains functional, whereas the centromere into which it is inserted is inactivated. For example, the genome of *Aegilops tauschii*, a wild diploid ancestor of bread wheat, has 7 chromosomes whereas that of rice has 12. *Aegilops* chromosome 1 includes many of the same genes as rice chromosomes 5 and 10; the genes found on rice chromosome 10 appear in the middle of the *Aegilops* chromosome, whereas the genes of one arm of rice chromosome 5 appear to have been attached at one end of chromosome 10, with the other arm of rice chromosome 5 attached at the other end of chromosome 10 (Figure 2.10). Similar insertional patterns can be seen in several other *Aegilops* chromosomes. Although such chimeric chromosomes could each have been formed by two successive translocations, it seems more likely that each was formed by a single unique insertion event.

Other sorts of genomic rearrangements have also been observed. In the genomes of Arabidopsis thaliana ($n = 5$) and its close relative Arabidopsis lyrata ($n = 8$), *A. lyrata* chromosomes 1 and 2 appear to have combined end to end to create *A. thaliana* chromosome 1 (Figure 2.11). The genes of *A. lyrata* chromosome 7 appear in inverse order on *thaliana* chromosome 4. This sort of chromosomal rearrangement is known as an **inversion**.

Another common sort of chromosomal rearrangement is a **translocation**. In this case, a portion of one chromosome breaks off and is reattached to a

(a)

(b)

Figure 2.9 (a) *Claytonia virginica*. By User: SB_Johnny (My photo) [GFDL (http://www.gnu.org/copyleft/fdl.html), CC-BY-SA-3.0 (http://creativecommons.org/licenses/by-sa/3.0/) or CC-BY-SA-2.5-2.0-1.0 (http://creativecommons.org /licenses/by-sa/2.5-2.0-1.0)], via Wikimedia Commons. Inset: By Kaldari (Own work) [Public domain], via Wikimedia Commons. (b) *Sorghum bicolor*

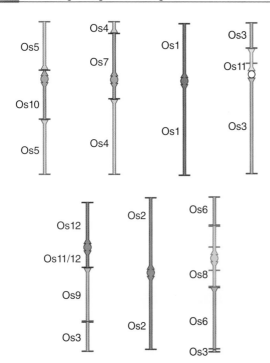

Figure 2.10 Relationship of rice chromosomes (Os for Oryza sativa) to those of *Aegilops tauschii*. *Aegilops* chromosomes 1, 2, 4, and 7 have been formed by insertion of one rice chromosome into another. *Aegilops* chromosome 5 represents a fusion of three rice chromosomes, whereas chromosomes 3 and 6 are largely unrearranged. Adapted from Luo *et al.*, 2009

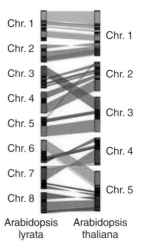

Figure 2.11 Cartoon of karyotypes of *A. thaliana* and *A. lyrata*, showing points of correspondence between the eight chromosomes of *A. lyrata* (left), and the five chromosomes of *A. thaliana* (right). Hu *et al.* (2011). Reproduced with permission of Macmillan Publishers Ltd.

different chromosome via normal mechanisms of double-stranded break repair. This occurs spontaneously in nature but can also be induced artificially by exposing pollen to X-rays. The radiation breaks the chromosomes, which then reanneal in new combinations. Translocations can be detrimental or even lethal if genetic material is lost in the process or if coding or regulatory sequences are disrupted. For example, such highly disruptive translocations are common in many human cancers. However, a translocation may have only modest phenotypic consequences, particularly if chromosomal segments are simply swapped. If no genetic material is lost, the result is a **balanced translocation**.

Balanced translocations are commonly documented in comparisons between species, which often differ in the precise arrangement of blocks of DNA. The many chromosomal rearrangements that have been recognized suggest that translocations may occur fairly frequently in nature and may even lead to

improved fitness in some cases. Translocations have also been used effectively in plant breeding, particularly in wheat, in which crosses between wheat and its wild relatives have been used to select for plants with portions of the wild chromosomes translocated to wheat. Figure 2.12 shows a translocation between wheat and a drought-tolerant relative, *Thinopyrum elongatum*; the resulting wheat plant had higher root biomass and was more tolerant to drought (Placido *et al.*, 2013).

2.5 Segments of DNA are often duplicated and can recombine

Genome sequences are replete with duplicated or repetitive regions, but the dynamics and consequences of duplication differ depending on the function of the DNA in question. The amount of DNA that is duplicated varies from a single base pair to entire genomes and everything in between. Any two identical DNA sequences, whatever their length, can recombine. The resulting breaks in the chromosome are repaired in a way that can cause deletions, insertions, or chimeric sequences that join DNA pieces that have never been joined before. This ability to recombine applies to the

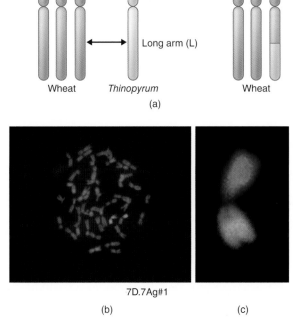

7A 7B 7D   7E   7A 7B 7D

Short arm (S)

Long arm (L)

Wheat   *Thinopyrum*   Wheat

(a)

7D.7Ag#1

(b)   (c)

Figure 2.12 (a) Cartoon of a translocation between the long arm of the wheat chromosome 7D and the long arm of chromosome 7E from the wild plant *Thinopyrum elongatum*. (b) Chromosome squash showing parts of the two *Thinopyrum* chromosomes integrated into the wheat genome by translocation. Placido *et al.* (2013). Copyright 2013 American Society of Plant Biologists. Used with permission. (c) Close-up of one translocated chromosome, with green showing the *Thinopyrum* segment. From Placido *et al.*, 2013

short 1 or 2 bp repeats described in Section 1.6, as well as the various duplications described here.

The so-called "low complexity regions" in genomes may be made up of satellite DNA, minisatellites, or microsatellites (represented by SSRs). Repeat units in these regions are constantly duplicated or deleted via replication slippage or illegitimate recombination. When such changes in copy number occur outside of coding sequences, they are thought to be selectively neutral, with no effect on the phenotype of the organism, although this may reflect the difficulty of measuring the effects of natural selection. Within a coding sequence, SSR variation is more likely to affect the phenotype and its selective value. Duplication and deletion of transposable elements is also common; these will be covered in more detail in Chapter 3. Like the replication of low complexity sequences, changes

in the number of transposable elements often do not affect the phenotype of the organism.

Most of the data on gene and genome duplication come from plants and fungi related to baker's yeast (*Saccharomyces cerevisiae*). Increases, via duplication, have been studied far more than decreases. Data on animals are broadly consistent with those reported for plants and yeast relatives.

Gene duplication can be conveniently divided according to the numbers of genes and portion of the genome involved (Figure 2.13). Duplications that involve one or a few genes are known as small-scale duplications. In some cases, such duplications lead to many copies of a single gene arranged next to each other on the chromosome; these are known as **tandem duplications** or **tandem arrays**. Conversely, those that involve large parts of the genome (e.g., a chromosome arm) are known as **segmental duplications**. The largest scale is the **whole genome duplication**, also called **polyploidy**. Details for each of these duplication events are discussed in the following sections.

2.6 Some genes are copied nearby in the genome

2.6.1 Tandem duplications – genes for ribosomal RNA

Ribosomal RNA genes exist in hundreds to thousands of copies in plant genomes (in contrast to a few hundred copies in animal genomes), but these copies are all arranged head to tail in a long array (Rogers and Bendich, 1987) (see Chapter 7 for discussion of transcription). It is common to have more than one ribosomal array within a plant. These arrays vary in the number of the rRNA genes, with one array apparently being the major source of ribosomal RNA and the other(s) being minor arrays. The numbers of gene copies change over time, sometimes quite rapidly, so that the major array may become a minor one and vice versa; thus the position of the ribosomal RNA array will change its position in the genome over time (Álvarez and Wendel, 2003).

In all land plants and streptophyte algae (the green "algae" most closely related to land plants; see Chapter 1), the genes encoding the 18S, 5.8S, and

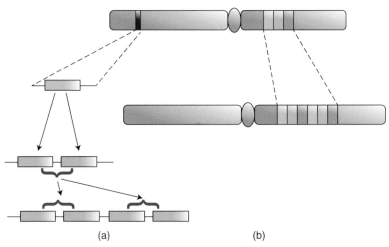

Figure 2.13 (a) Tandem duplication. A gene is copied one or more times in adjacent positions in the genome. (b) Segmental duplication. A set of genes is copied as a block

26S ribosomal subunits are part of a single transcriptional unit, separated by spacers, known as the internal transcribed spacers (ITS). Upstream of the 18S gene and downstream of the 26S gene are relatively short transcribed sequences, known, respectively, as the 5′ and 3′ external transcribed spacers (ETS). The ETS are part of the intergenic spacer (IGS), the central region of which is not transcribed. (For a cartoon of what this looks like, see Figure 7.3.) This basic unit is then repeated many times in a large tandem array. In streptophyte algae, mosses, liverworts, hornworts, lycophytes, and most ferns, the gene for an additional subunit, 5S (not to be confused with 5.8S) occurs just downstream of the 3′ ETS (Wicke *et al.*, 2011). In contrast, in water ferns and seed plants, the 5S genes form their own sizeable tandem array, generally on a separate chromosome from the rest of the rRNA genes.

Long tandem arrays such as the ribosomal arrays develop their own unique dynamics because of the tendency of the gene copies to interact with each other. The RNA-encoding genes of the ribosomal arrays undergo a self-correction process known as **concerted evolution** (Figure 2.14). Concerted evolution is the outcome of frequent illegitimate recombination between the many near-identical gene copies in the array. When the two gene copies interact, mismatched based pairs are treated as DNA errors that must be repaired. As they are repaired the gene copies become similar to each other, more similar than they would be if they were all accumulating mutations independently.

However, the array is so long that the correction process never manages to make all sequences identical, and extensive variation has been found within an array (Kellogg and Appels, 1995). Within any given ribosomal array, some of the copies appear to be non-functional, suggesting that the many copies are redundant and able to compensate when one or more copies are disabled.

2.6.2 Tandem duplications – genes encoding proteins for defense, receptor-like kinases, seed storage proteins

Besides the long, multi-copy ribosomal arrays, smaller arrays occur throughout the genome, generally involving protein-coding, non-transposable element genes. A single protein-coding gene may be duplicated and inserted at a nearby chromosomal location. This process is thought to occur via errors in recombination and replication, generally at meiosis, whereby one chromosome acquires an extra copy of a gene and the other loses its copy. Once there are two adjacent copies of a gene on a chromosome, the possibility of errors in replication increases because of further aberrant replication and/or recombination. Ultimately this leads to the formation of a tandem array in which several copies of the gene are arranged near each other along the chromosome. Analysis of genome sequences

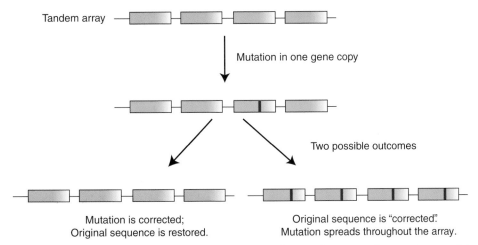

Tandem array

Mutation in one gene copy

Two possible outcomes

Mutation is corrected;
Original sequence is restored.

Original sequence is "corrected".
Mutation spreads throughout the array.

Figure 2.14 Concerted evolution. If a mutation arises in one gene in a tandem array, there are two possible outcomes. Either the mutation is corrected and the gene reverts to its original sequence, or the other gene copies in the array are "corrected" to match the mutation, which then spreads through part or all of the array

shows that such small-scale duplications often involve genes for receptor-like kinases and genes involved in defense responses (*R* genes) (Rogers and Bendich, 1987). Having many copies of a gene or set of genes can lead to higher levels of the respective mRNAs, which are translated to yield much higher levels of the encoded protein. Hence, gene duplication can have a larger effect on the phenotype than expressing a single gene. For example, resistance to nematodes in soybeans is affected by a set of three distinct genes that are close together in the genome. When the set of three genes occurs in 10 copies in the genome, soybean plants are resistant to the soybean cyst nematode (Cook *et al.*, 2012). However, if there is only one copy of each gene, the plant is attacked by the nematode.

While more gene copies may lead to more mRNA, some genes adjust their expression levels so that the amount of mRNA is independent of the number of gene copies. This phenomenon is known as **dosage compensation**. The most familiar example is the X chromosome in mammals; males have one copy of the chromosome whereas females have two. If gene copy number were the only control of mRNA, then females would have twice as much mRNA for every gene on the X chromosome. However, in mammals one entire chromosome is inactivated in females, thus compensating for the increased copy number, or dosage, of the genes. In *Drosophila*, in contrast, expression of all genes on the X chromosome is doubled; thus dosage compensation occurs by very different mechanisms in the two groups of organisms.

Dosage compensation has been demonstrated in plants, particularly in autosomal loci. (Few plants have sex chromosomes.) One particularly clear example is the *alcohol dehydrogenase* (*Adh1*) locus in maize (Birchler, 2007). When plants are bred to have different numbers of *Adh* genes, the amount of ADH protein does not correlate well with gene copy number. It appears that the plant adjusts the amount of gene product so the same amount of ADH is present whether there are 1, 2, or 4 copies of the gene.

Seed storage proteins are also commonly found in tandem arrays (Sabelli and Larkins, 209). In the cereal grasses, the major proteins are known as prolamins, which constitute over half of the protein of the endosperm in crops such as wheat and maize. The prolamins can be divided into groups based on their chemical properties, amino acid sequence and evolutionary history. The maize and sorghum genomes include only one to three genes encoding each of the beta, gamma and delta prolamins. However, the alpha prolamins are in large tandem arrays. Although the alpha prolamins are the most highly expressed of the prolamin genes, most of the alpha prolamin mRNA is the product of only a handful of the genes in the array (Woo *et al.*, 2001). The alpha prolamin arrays appear to have a dynamic history, with arrays expanding and contracting over time; in addition, individual arrays do not have stable positions in the genome (Xu and Messing, 2008).

Recombination among genes within a tandem array is also common. Because the genes each accumulate their own mutations, recombination creates allelic diversity by mixing and matching different components of individual genes. This process in plants is analogous to that in immune system genes in animals. Natural selection favors this kind of shuffling because it permits a response to a wide range of pathogens.

2.6.3 Segmental duplications

Several adjacent genes, or portions of chromosome arms, may be duplicated in a segmental duplication. One such duplication involves rice chromosomes 11 and 12, in which the ends of the short arms have sequences that are unusually similar. This was originally thought to be a recent duplication peculiar to rice, but recently similar duplicated regions have been found in sorghum and foxtail millet (*Setaria*) (Figure 2.15). It is possible that this region is particularly prone to duplication. Another possibility is that the duplication occurred once in the common ancestor of all three species, but then illegitimate base pairing between genes in the duplicated regions caused the sequences to become more similar than they were originally. The latter hypothesis has received some support from close examination of the regions involved (Wang *et al.*, 2011).

2.7 Whole genome duplications are common in plants

In whole genome duplication, the entire chromosome complement of the cell is copied. This can occur by fusion of unreduced gametes, which have the diploid chromosome number due to a failure of meiosis, or by spontaneous doubling in a somatic cell that then becomes part of a floral meristem. In either case, the offspring have twice the number of chromosomes as the parents. The consequences of such duplication depend in part on how distantly related the two parental plants are. At one extreme, if chromosome doubling occurs spontaneously in a single plant or within a single species, it is known as autopolyploidy. This is thought to be rare because it will lead to four identical sets of chromosomes; the

Figure 2.15 *Setaria italica*. http://upload.wikimedia.org /wikipedia/commons/f/ff/Japanese_Foxtail_millet_01.jpg. By STRONGlk7 (Own work) [CC-BY-SA-3.0 (http://creative commons.org/licenses/by-sa/3.0)], via Wikimedia Commons

formation of tetravalents (instead of bivalents) will lead to abnormalities in meiosis. Autopolyploidy is also difficult to detect however, and hence it may be more common than currently supposed.

Allopolyploidy is the opposite of autopolyploidy; the parents may be members of different species or (rarely) different genera. In this case, polyploidy is necessarily accompanied by hybridization. It is common for two distantly related plants to cross and to produce offspring that are partially or wholly sterile because the dissimilar chromosomes cannot pair at meiosis. (A similar problem in the cross between a horse and donkey causes the offspring, a mule, to be

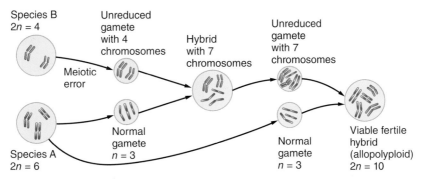

Figure 2.16 Allopolyploidy. From Reece *et al.*, 2010

sterile.) Although such plants cannot reproduce, they often persist long enough for spontaneous mutations to occur in tissues that will give rise to gametes. One common mutation is the production of non-reduced (i.e., diploid) gametes. When diploid gametes create a zygote, the result is a plant that has twice the number of chromosomes as the original. Because every chromosome is duplicated, pairing at meiosis is restored and with it, fertility (Figure 2.16).

In **allopolyploids**, the two chromosomes from a single parent are known as **homologues**; the sets of chromosomes from different parents are **homoeologues**. While homologues will pair with each other, as normal, homoeologues will not pair. Hence the plant behaves cytogenetically as a diploid, and is sometimes called an amphidiploid.

It is not clear what prevents homoeologues from pairing with each other. After all, the species are often closely enough related that the chromosomes are generally fully collinear and retain many very similar sequences. Some insight has been provided by some of the best known of the allopolyploids, durum wheat (*Triticum durum*, used for making pasta) and bread wheat (*Triticum aestivum*, used for most baking other than pasta products) (Figure 2.17). *T. durum* is an allotetraploid, formed by the cross of Einkorn wheat (*T. monococcum*) or its close relative *T. urartu* with a common weed in wheat fields, *Aegilops speltoides* or a very similar plant. Although Einkorn is still cultivated in a few parts of Turkey, durum is much more widespread. Within the last 10 000 years, *T. durum* again crossed with a diploid weed, this time *Aegilops tauschii*; the combination of a diploid and a tetraploid led to the hexaploid *T. aestivum*.

In *T. aestivum*, pairing of homoeologues is under the control of a locus known as *Ph1* for *Pairing*

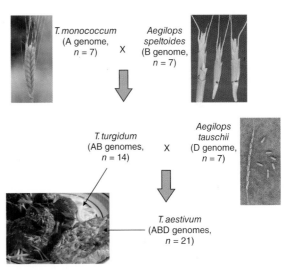

Figure 2.17 History of wheat. Diploid species of wheat were domesticated from wild ancestors similar to *T. monococcum*, which is still cultivated a few parts of the world today and is known as Einkorn. *T. monococcum* crossed spontaneously with a weed similar to *Aegilops speltoides*, growing in the same fields, and formed the tetraploid *T. turgidum*. The ancient domesticated form of *T. turgidum* is known as Emmer wheat and the modern commercial domesticate is durum, which is the source of flour used to make pasta. *T. turgidum* crossed spontaneously with another species of goatgrass, *A. tauschii*, to produce the hexaploid bread wheat http://upload.wikimedia.org/wikipedia/commons/5/53 /Aegilops_tauschii_ARS-1.jpg. By Marknesbitt at en. wikipedia (January 2006). Later uploaded by Jeannot12 at fr.wikipedia (December 2006) [Public domain], from Wikimedia Commons

homeologous1. When *Ph1* is mutated, homoeologues pair and cross over. Over time, chromosomes become rearranged and the plants lose fertility. *Ph1* has been identified as a tandem array of cyclin-dependent

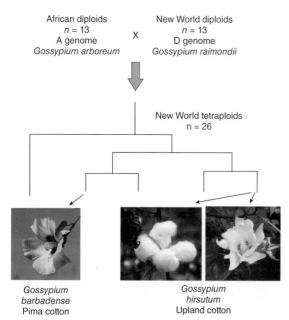

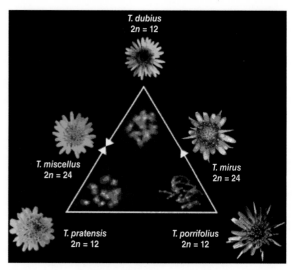

Figure 2.18 History of cotton. Evolutionary history of the cotton genus, *Gossypium*, based on the work of Cronn and Wendel (2004) and Flagel and Wendel (2010). A diploid species most similar to modern *G. arboreum*, from Africa, crossed with a diploid species most similar to modern *G. raimondii*, from the New World. It is not clear how the African plant arrived in the New World since the hybridization event was 1–2 million years ago, well before the origin of *Homo sapiens*. The tetraploids then underwent speciation in Central and South America and produced five modern species; two of these tetraploid species, *G. hirsutum* and *G. barbadense*, produce commercial cotton, which is used for clothing and also paper and money. Not all of the tetraploid species are labeled. Left figure: http://upload.wikimedia.org/wikipedia/commons/2/2c /Gossypium_barbadense.jpg. By Brian Gratwicke [CC BY 2.0 (http://creativecommons.org/licenses/by/2.0)], via Wikimedia Commons. Middle figure: http://upload.wikimedia .org/wikipedia/commons/thumb/6/68/CottonPlant.JPG /1280px-CottonPlant.JPG Right figure: http://commons .wikimedia.org/wiki/File:Gossypium_hirsutum_BotGardBln 1105FlowerLeaves.JPG. CC BY SA 3.0 (http://creative commons.org/licenses/by-sa/3.0/deed.en)

Figure 2.19 Recent polyploids in the genus *Tragopogon* (salsify or goatsbeard). The species of salsify shown at the corners of the triangle are diploid, with *n* = 6; they were introduced into the USA in the early years of the twentieth century. Since then they have crossed and formed two new polyploid species, *T. miscellus* and *T. mirus*, with *n* = 12. These species are morphologically distinct from their parents. From Pires *et al.*, 2004

kinases, which appear to affect chromosome pairing by influencing the process of DNA replication and phosphorylation of histones (Giffths *et al.*, 2006; Greer *et al.*, 2012).

Another familiar allopolyploid is cotton, *Gossypium hirsutum*, and its close relative *G. barbadense*, sold under the name of Pima cotton (Figure 2.18). Both species are allotetraploids, formed after a cross between a diploid similar to the Peruvian species *G. raimondii* and an African diploid species similar to *G. arboreum* or *G. herbaceum*. The polyploidy event that led to *G. hirsutum* probably occurred several million years ago, so is older than that leading to bread wheat.

The genus *Tragopogon* (salsify) includes several spontaneous polyploids that have formed within the last century (Figure 2.19). Diploid *Tragopogon* species were introduced to North America from Europe and have become weedy. In parts of the Pacific Northwest, the diploids have crossed repeatedly and formed allotetraploids. These were first discovered in a narrow area outside Pullman, Washington, but their range has been increasing (Soltis *et al.*, 2004).

Cytogenetic and genomic studies show that the flowering plants have undergone repeated rounds of polyploidy and gradual diploidization. The pattern of whole genome duplication has been found coinciding with the origin of the monocots, the eudicots, and several major families such as the Brassicaceae and Poaceae (Jiao *et al.*, 2011) (Figure 2.20). Polyploidy is also common in ferns (Wood *et al.*, 2009) and mosses

(Såstad, 2005), but is surprisingly rare in gymnosperms (Khoshoo, 1959; Murray, 1998).

2.8 Whole genome duplication has many effects on the genome and on gene function

Immediately after whole genome duplication, every cell in the plant has a duplicate copy of every piece of DNA, including every gene, transposable element, and regulatory sequence. However, this fully duplicated state does not last, and within one or a few generations DNA is lost, such that the genome size of a polyploid is usually slightly less than the sum of the sizes of the parental genomes. Although this result has been reported repeatedly (Leitch and Bennett, 2004), in a few cases allopolyploid genomes are larger than the sum of the parental genomes (Anssour *et al.*, 2009).

Polyploidy also affects gene expression. Gene expression in a newly formed polyploid is not simply the sum of that in the ancestral diploids; although all the genes in the genome have been duplicated, they do not retain the same pattern of expression that they had in their diploid ancestors. Studies in newly synthesized polyploids in wheat, *Gossypium*, *Brassica*, and *Tragopogon* have shown that gene expression changes immediately. Some genes are upregulated, while others are downregulated. Often, one parental genome appears to dominate the other, in that the gene copies from one parent are preferentially upregulated.

Another immediate effect of whole genome duplication is that every gene copy is now redundant (Haldane, 1933). Because every cellular function is now performed by two genes instead of one, one gene copy becomes dispensable. If a mutation disables one gene copy, natural selection will not eliminate the plant with the mutation because the other, non-mutated copy is still functioning. The mutated copy often becomes a **pseudogene**, a sequence that looks like a gene but that cannot be transcribed or translated properly. Over time, the pseudogene will acquire more and more mutations, and will eventually decay to look like a random sequence. Thus, the number of genes in the polyploid will gradually drop back to the approximate number that was in the diploid parents; genes that were duplicated will return to being single copy.

Not all extra genes become pseudogenes, however. Some are retained in duplicate and the copies then begin down their own independent evolutionary paths accumulating mutations and responding to selection. As more genome sequences become available, it has become possible to ask whether there are particular patterns of gene retention and loss that follow whole genome duplication. In other words, are there rules governing which genes are returned to single copy and which are retained in duplicate? Certain broad classes of genes are often retained in duplicate, whereas other classes are generally returned to single copy. In general, transcription factors, protein kinases, and ribosomal genes are often retained in duplicate, whereas many genes of metabolism are single copy, suggesting that their duplicates were eliminated (or at least, were permitted to be lost) soon after genome duplication.

Various hypotheses have been developed to account for the retention of duplicate genes (Figure 2.21); Innan and Kondrashov (2010) assign these to four major categories. Category I includes models in which the initial duplication has no effect on the fitness of the organism. Over time, one copy may mutate to acquire a new expression domain or may change in such a way that the protein product acquires different binding partners or substrates; this is called **neofunctionalization**, and was first suggested by Ohno (1970). Another possibility is that the two duplicates will partition the functions of the ancestral gene, so that the two genes together cover the same biochemical functions and developmental roles as covered by the ancestral gene; this is known as **subfunctionalization**. Various mechanisms have been proposed for the occurrence of subfunctionalization, including the duplication-degeneration-complementation (DDC) model and the Escape from Adaptive Conflict (EAC) model; these differ in the presumed role of natural selection on the ancestral, preduplication gene. Category II includes models in which the duplication confers an immediate benefit on the plant, and Category III includes models that assume ancestral polymorphism in the duplicated gene (Innan and Kondrashov, 2010). For a well-developed example of the fate of duplicate genes, see the "From the Experts" section on duplication in anthocyanin genes.

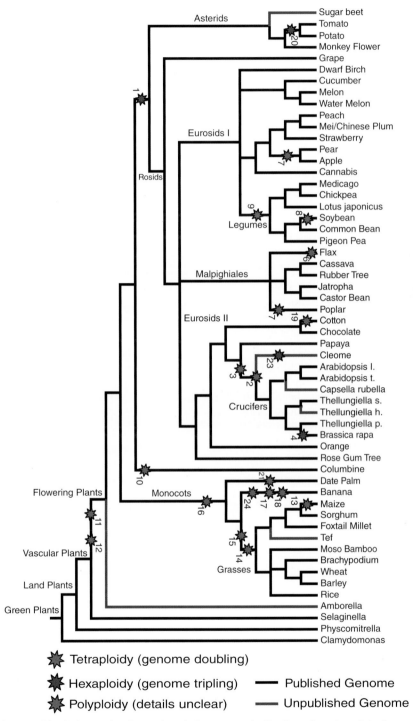

Figure 2.20 Phylogeny of land plants, showing major whole genome duplications, based on data from currently available genome sequences. http://genomevolution.org/wiki/index.php/Plant_paleopolyploidy. Reproduced with permission

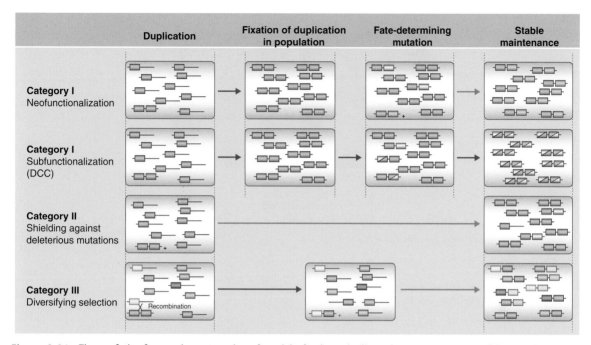

Figure 2.21 Three of the four major categories of models for how duplicated genes are preserved in populations. In Category I models, it is assumed that the duplication occurs spontaneously and spreads through the population without any selection occurring. After all the plants in the population have two copies of the gene, one or more mutations occur. In neofunctionalization, the mutated copy (green) is advantageous because it provides a new gene function; natural selection causes it to spread through the population. In subfunctionalization, mutations occur in both gene copies (white sectors); these cause the two copies to divide up the ancestral gene functions. In Category II models, the duplication is advantageous immediately; in the example shown, it may provide a back-up copy of the gene that permits the organism to function even if a mutation (white) knocks out one copy. In Category III models, it is assumed that different plants in the population already have different forms of the gene. In this example, recombination between the two duplicate copies and the other forms in the population provide many more different gene forms. Adapted from Innan and Kondrashov (2010). Reproduced with permission of Macmillan Publishers Ltd.

From the Experts

Duplication in anthocyanin genes

Clarkia gracilis is a wildflower that grows naturally in California, Oregon and Washington. The petals of some populations of this species have a large petal spot that is dark red, whereas the remaining petal background is light violet. These different colors result from the deposition of different anthocyanin pigments in the spot and in the petal background (Figure 2.22a). Related species lack petal spots, which is believed to be the ancestral condition (Figure 2.22b). Gene duplication has played an important role in the evolution of the spot (Martins *et al.*, 2013).

    C. gracilis is an allotetraploid. In both parental species, which are diploid, the gene coding for the enzyme dihydroflavonol reductase (*Dfr*) is duplicated. This enzyme catalyzes an intermediate reaction in the anthocyanin biosynthetic pathway. These copies may be called Dfr_1 and Dfr_2. This means that in *C. gracilis* there are two copies of each of these genes. The two copies of Dfr_1 have a different pattern of expression from the two copies of Dfr_2: both copies of Dfr_2 are expressed only in the region of the petal corresponding to the spot, whereas both copies of Dfr_1 are expressed throughout the petal (Figure 2.22c).

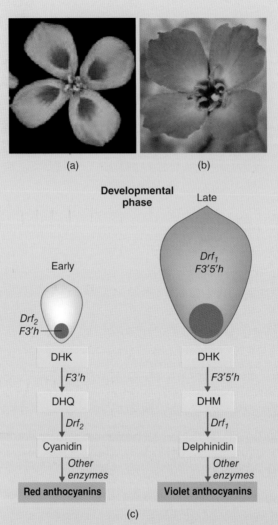

Figure 2.22 (a) *Clarkia gracilis* with petal spots. (b) *Clarkia dudleyana* lacking spots (ancestral condition). http://www.rahul.net/raithel/Plumas/html/clarkiaDudleyana_FarewellToSpring00015.html. Reproduced with permission of John Raithel and Linda Herbert. (c) Schematic of development of spots and background color. Spots appear early in development. Below each petal is the portion of the anthocyanin biosynthetic pathway that contributes to spot and background characteristics. In the spots, red coloration is caused by conversion of dihydrokaempferol (DHK) to dihydromyricetin (DHQ) by the enzyme F3'h. DHQ is then converted by the enzyme Dfr to cyanidin, which is then converted by other enzymes into red anthocyanins. Later in development, activation of Dfr1 and F3'5'h causes production of violet pigments in the background. F3'5'h converts DHK to dihydromyricetin (DHM). Subsequently, Dfr2 converts DHM to delphinidin. Finally, other anthocyanin enzymes convert delphinidin to violet anthocyanins

The Dfr_1 copies are also expressed at different times from the Dfr_2 copies. Dfr_2 is expressed early in petal development, while Dfr_1 is expressed later. These times correspond to two different waves of expression of the other anthocyanin enzyme-coding genes, which means that the petal spot appears earlier in development than the background coloration (Figure 2.22c).

Some individuals of *C. gracilis* lack the petal spot and the entire petal is violet. In these individuals, Dfr_2 is not expressed in the region of the petal spot, but Dfr_1 remains expressed throughout the petal. This observation demonstrates that expression of Dfr_2 is required for spot production. However, expression of this gene does not explain why the spot is a different pigment from the background. This occurs because the gene *F3'h* is expressed during spot development, but not later, while the gene *F3'5'h* is not expressed during spot development, but is expressed later. *F3'h* converts the precursor dihydrokaempferol (DHK) to dihydroquercitin (DHQ), which is eventually transformed into red anthocyanin pigments. By contrast, *F3'5'h* transforms DHK into dihydromyricetin (DHM), which is eventually converted into violet anthocyanin pigments (Figure 2.23c).

The differential involvement of the two copies of *Dfr* in spot and background formation indicates that duplication of this gene facilitated the evolution of petal spots. This work provides one example of how gene duplication can contribute to the evolution of new characters.

By Mark Rausher

The models in Category I are well documented in the literature. For example, the MADS box transcription factor AGAMOUS (see Chapter 10) directs floral organ primordia to become stamens and carpels in most angiosperms. (MADS is an acronym for *Mcm*1, *Agamous*, *Deficiens*, and *Srf*, which were the first genes in this transcription factor family to be identified.) However, AGAMOUS has been duplicated in the grasses, and now one copy is primarily involved in specifying stamen development and the other copy specifies carpel development.

All the models in Categories I–III require the accumulation of mutations followed by the action of natural selection to shape the functions of duplicate genes over time. Although it is unclear exactly how much time is involved, the process is not likely to be instantaneous and it may happen too slowly to account for the initial retention of duplicates. In other words, something must permit the duplicates to persist long enough for mutation and selection to modify their functions. The Category IV model, the **dosage balance hypothesis (DBH)**, addresses this problem and is receiving increasing support. In this model, genes are retained in duplicate if their stoichiometry is particularly critical for the organism. This is most easily understood in terms of genes whose products function in physical complexes or closely linked components of metabolic networks. If the complex or network requires a particular ratio of subunits, then either all components of the complex must be single copy, or all must be duplicated, but a mix will be selected against. This hypothesis, of course, requires that the amount of gene product is related directly to gene copy number.

The DBH itself may predict different outcomes. It is possible that gene products must be available in the cell in the same relative proportions, and that the cell will function correctly as long as all gene products increase or decrease together. Alternatively, there may be selection for the absolute amount of gene product, independent of its relationship to other gene products. Obviously, both relative and absolute dosage may be important. Investigation of duplications 2 and 3 (Fig. 2.20) in the history of Arabidopsis suggests that the immediate effect of a whole genome duplication is to maintain relative dosage, but over longer periods of time the requirements for absolute dosage become more important (Bekaert *et al.*, 2011).

Consistent with the DBH, genes controlling the circadian clock are generally retained in multiple copies after triplication of the genome in *Brassica rapa* (Lou *et al.*, 2012). Work in Arabidopsis had previously shown that gene dosage is important for clock genes, and the pattern of gene retention after the polyploidy event is completely consistent with that observation.

The chromosomal position of genes that are retained in duplicate after genome duplication tends to be stable over evolutionary time. In contrast, members of gene families that tend to appear as tandem arrays of duplicates (e.g., genes encoding F-box proteins and some disease resistance proteins) often change their chromosomal location following genome duplication, presumably via recombination (Woodhouse *et al.*, 2011).

2.9   Summary

Far from being a static entity, the plant genome changes constantly. DNA is modified by replacement of nucleotides. Insertions and deletions occur in sizes as small as 1 bp and as large as arrays of transposable elements, causing frequent shifts in genome size. Chromosomes divide and fuse. Polyploidy, also known as whole genome duplication, is common. As soon as polyploidy occurs, the genome begins to be remodeled and reverts gradually to a diploid state. Polyploidy is cyclical, and generates novel combinations of genes. This constant activity leads to changes in gene transcription, although we are only beginning to understand the many ways in which this happens. Genome variation can be exploited for many practical applications, most obviously for production of linkage maps, associating genotype with phenotype, and accelerating the process of crop breeding.

2.10   Problems

2.1   What is the difference between a transition and a transversion?

2.2   Define (a) indel, (b) SNP, and (c) percent similarity.

2.3   Search Genbank for the sequence with the accession number AY260164.1. What organism is it from? Use BLAST to determine the name of the most similar gene from rice.

2.4   Mutation is a random process and happens continually in all parts of the genome. However, when protein-coding sequences are compared between species (e.g., between Arabidopsis and Brassica, or rice and sorghum), they are often very similar and are described as being "conserved". How can sequences be mutating all the time and yet still appear to be conserved?

2.5   In comparing protein sequences, the methionine (M) in the first position is almost always a conserved residue. Why is the M conserved? In other words, what would happen to the protein if there were a mutation in the M codon?

2.6   What is meant by the term molecular breeding and how does it increase the efficiency of the breeding process?

2.7   What is a linkage map and what do the distances on the map represent?

2.8   What is genome size and how is it measured?

2.9   What is (a) a translocation and (b) an inversion?

2.10   Distinguish between tandem, segmental, and whole genome duplications.

2.11   Which familiar plants are allopolyloids?

2.12   What happens to genes after duplication?

Further reading

Conant, G.C. and Wolfe, K. (2008) Turning a hobby into a job: how duplicated genes find new functions. *Nature Reviews Genetics*, **9**, 938–950.

References

Álvarez, I. and Wendel, J.F. (2003) Ribosomal ITS sequences and plant phylogenetic inference. *Molecular Phylogenetics and Evolution*, **29**, 417–434.

Anssour, S., Krügel, T., Sharbel, T.F. *et al.* (2009) Phenotypic, genetic and genomic consequences of natural and synthetic polyploidization of *Nicotiana attenuata* and *Nicotiana obtusifolia*. *Annals of Botany*, **103**, 1207–1217.

Bekaert, M., Edger, P.P., Pires, J.C. and Conant, G.C. (2011) Two-phase resolution of polyploidy in the *Arabidopsis* metabolic network gives rise to relative and absolute dosage constraints. *Plant Cell*, **23**, 1719–1728.

Birchler, J.A. (2007) The gene balance hypothesis: from classical genetics to modern genomics. *Plant Cell*, **19**, 395–402.

Brunner, S., Fengler, K., Morgante, M. *et al.* (2005) Evolution of DNA sequence nonhomologies among maize ibreds. *Plant Cell*, **17**, 343–360.

Buchanan, B.B., Gruissem, W., and Jones, R.L. (2002) Biochemistry and Molecular Biology of Plants. (Rockville, MD: American Society of Plant Biologists).

Clark, R.M., Schweikert, G., Toomajian, C. *et al.* (2007) Common sequence polymorphisms shaping genetic diversity in *Arabidopsis thaliana*. *Science*, **317**, 338–342.

Clark, R.M., Wagler, T.N., Quijada, P. and Doebley, J. (2006). A distant upstream enhancer at the maize domestication gene *Tb1* has pleiotropic effects on plant and inflorescence architecture. *Nature Genetics*, **38**, 594–597.

Cook, D.E., Lee, T.G., Guo, X. *et al.* (2012) Copy number variation of multiple genes at *Rhg1* mediates nematode resistance in soybean. *Science*, **338**, 1206–1209.

Cronn, R. and Wendel, J.F. (2004) Cryptic trysts, genomic mergers, and plant speciation. *New Phytologist*, **161**, 133–142.

Fenouil, R., Cauchy, P., Koch, F. *et al.*, (2012). CpG islands and GC content dictate nucleosome depletion in a transcription-independent manner at mammalian promoters. *Genome Research*, **22**, 2399–2408.

Flagel, L.E. and Wendel, J.F. (2010) Evolutionary rate variation, genomic dominance and duplicate gene expression evolution during allotetraploid cotton speciation. *New Phytologist*, **186**, 184–193.

Greer, E., Martin, A.C., Pendle, A. *et al.*, (2012). The *Ph1* locus suppresses Cdk2-type activity during premeiosis and meiosis in wheat. *Plant Cell*, **24**, 152–162.

Griffiths, S., Sharp, R., Foote, T.N. *et al.*, (2006) Molecular characterization of *Ph1* as a major chromosome pairing locus in polyploid wheat. *Nature*, **439**, 749–752.

Gross, B.L., Steffen, F.T. and Olsen, K.M. (2010) The molecular basis of white pericarps in African domesticated rice: novel mutations at the *Rc* gene. *Journal of Evolutionary Biology*, **23**, 2747–2753.

Haldane, J.B.S. (1933) The part played by recurrent mutation in evolution. *American Naturalist*, **69**, 446–455.

Hu, T.T., Pattyn, P., Bakker, E.G. *et al.* (2011) The *Arabidopsis lyrata* genome sequence and the basis of rapid genome size change. *Nature Genetics*, **43**, 476–481.

Innan, H. and Kondrashov, F. (2010) The evolution of gene duplications: classifying and distinguishing between models. *Nature Reviews Genetics*, **11**, 97–108.

Jiao, Y., Wickett, N., Ayyampalayam, S. *et al.* (2011) Ancestral polyploidy in seed plants and angiosperms. *Nature*, **473**, 97–102.

Kellogg, E.A. and Appels, R. (1995) Intraspecific and interspecific variation in 5S RNA genes are decoupled in diploid wheat relatives. *Genetics*, **140**, 325–343.

Khoshoo, T.N. (1959) Polyploidy in gymnosperms. *Evolution*, **13**, 24–39.

Larson, S.R. and Kellogg, E.A. (2009) Genetic dissection of seed production traits and identification of a major-effect seed retention QTL in hybrid *Leymus* (Triticeae) wildryes. *Crop Science*, **49**, 29–40.

Leitch, I.J., Beaulieu, J.M., Chase, M.W. *et al.* (2010) Genome size dynamics and evolution in monocots. *Journal of Botany*, Article ID 862516, 1–18.

Leitch, I.J. and Bennett, M.D. (2004) Genome downsizing in polyploid plants. *Biological Journal of the Linnean Society*, **82**, 651–663.

Lewis, W.H. and Semple, J.C. (1977) Geography of *Claytonia virginica* cytotypes. *American Journal of Botany*, **64**, 1078–1082.

Lou, P., Wu, J., Cheng, F. *et al.* (2012) Preferential retention of circadian clock genes during diploidization following whole genome triplication in *Brassica rapa*. *Plant Cell*, **24**, 2415–2426.

Luo, M.C., Deal, K.R., Akhunov, E.D. *et al.* (2009). Genome comparisons reveal a dominant mechanism of chromosome number reduction in grasses and accelerated genome evolution in Triticeae. *Proceedings of the National Academy of Sciences USA*, **106**, 15780–15785.

Martins, T.R., Berg, J.J., Blinka, S. *et al.* (2013) Precise spatio-temporal regulation of the anthocyanin biosynthetic pathway leads to petal spot formation in *Clarkia gracilis* (Onagraceae). *New Phytologist*, **197**, 958–969.

Murray, B.G. (1998) Nuclear DNA amounts in gymnosperms. *Annals of Botany*, **82**, 3–15.

Ohno, S. (1970) *Evolution by Gene Duplication*, Springer.

Pires, J.C., Lim, K.Y., Kovarik, A. *et al.* (2004). Molecular cytogenetic analysis of recently evolved *Tragopogon* (Asteraceae) allopolyploids reveal a karyotype that is additive of the diploid progenitors. *American Journal of Botany*, **9**, 1022–1035.

Placido, D.F., Campbell, M.T., Folsom, J.J. *et al.* (2013) Introgression of novel traits from a wild wheat relative improves drought adaptation in wheat. *Plant Physiology*, **161**, 1806–1819.

Reece, J.B., Urry, L.A., Cain, M.L. *et al.* (2011) *Campbell Biology*, Benjamin Cummings.

Rodgers-Melnick, E., Mane, S.P., Dharmawardhana, P. *et al.* (2012) Contrasting patterns of evolution following whole genome versus tandem duplication events in *Populus*. *Genome Research*, **22**, 95–105.

Rogers, S.O. and Bendich, A.J. (1987) Ribosomal RNA genes in plants: variability in copy number and in the intergenic spacer. *Plant Molecular Biology*, **9**, 509–520.

Sabelli, P.A. and Larkins, B.A. (2009) The development of endosperm in grasses. *Plant Physiology*, **149**, 14–26.

Såstad, S.M. (2005) Patterns and mechanisms of polyploid formation in bryophytes. *Regnum Vegetabile*, **143**, 317–334.

Scherrer, B., Isidore, E., Klein, P. *et al.* (2005). Large intraspecific haplotype variability at the *Rph7* locus results from rapid and recent divergence in the barley genome. *Plant Cell*, **17**, 361–374.

Schnable, P.S., Ware, D., Fulton, R.S. *et al.* (2009). The B73 maize genome: complexity, diversity, and dynamics. *Science*, **326**, 1112–1115.

Serres-Giardi, L., Belkhir, K., David, J. and Glémin, S. (2012) Patterns and evolution of nucleotide landscapes in seed plants. *Plant Cell*, **24**, 1379–1397.

Soltis, D.E., Soltis, P.S., Pires, J.C. *et al.* (2004) Recent and recurrent polyploidy in *Tragopogon* (Asteraceae): cytogenetic, genomic and genetic comparisons. *Biological Journal of the Linnean Society*, **82**, 485–501.

Studer, A., Zhao, Q., Ross-Ibarra, J. and Doebley, J. (2012) Identification of a functional transposon insertion in the maize domestication gene *Tb1*. *Nature Genetics*, **43**, 1160–1163.

Tatarinova, T.V., Alexandrov, N.N., Bouck, J.B. and Feldman, K.A. (2010) GC_3 biology in corn, rice, sorghum and other grasses. *BMC Genomics*, **11**, 308.

Wang, X., Tang, H. and Paterson, A.H. (2011) Seventy million years of concerted evolution of a homoeologous chromosome pair, in parallel, in major Poaceae lineages. *Plant Cell*, **23**, 27–37.

Wicke, S., Costa, A., Muñoz, J. and Quandt, D. (2011). Restless *5S*: the re-arrangement(s) and evolution of the nuclear ribosomal DNA in land plants. *Molecular Phylogenetics and Evolution*, **61**, 321–332.

Woo, Y.-M., Hu, D.W.-N., Larkins, B.A. and Jung, R. (2001) Genomics analysis of genes expressed in maize endosperm identifies novel seed proteins and clarifies patterns of zein gene expression. *Plant Cell*, **13**, 2297–2317.

Wood, T.E., Takebayashi, N., Barker, M.S. *et al.* (2009) The frequency of polyploid speciation in vascular plants. *Proceedings of the National Academy of Sciences USA*, **106**, 13875–13879.

Woodhouse, M.R., Tang, H. and Freeling, M. (2011) Different gene families in *Arabidopsis thaliana* transposed in different epochs and at different frequencies throughout the rosids. *Plant Cell*, **23**, 4241–4253.

Xu, J.-H. and Messing, J. (2008) Organization of the prolamin gene family provides insight into the evolution of the maize genome and gene duplications in grass species. *Proceedings of the National Academy of Sciences USA*, **105**, 14330–14335.

Xu, K., Xu, X., Fukao, T. *et al.* (2006) *Sub1A* is an ethylene-response-factor-like gene that confers submergence tolerance to rice. *Nature*, **442**, 705–708.

Zeller, G., Clark, R.M., Schneeberger, K. *et al.* (2008) Detecting polymorphic regions in *Arabidopsis thaliana* with resequencing microarrays. *Genome Research*, **18**, 918–929.

Zhou, L., Zhang, J., Yan, J. and Song, R. (2011) Two transposable element insertions are causative mutations for the major domestication gene *Teosinte branched 1* in modern maize. *Cell Research*, **21**, 1267–1270.

Chapter 3
Transposable elements

3.1 Transposable elements are common in genomes of all organisms

Anyone who has taken introductory biology has probably learned of Mendel's experiments on peas, in which he outlined the basic laws of inheritance. One phenotype he investigated was whether the peas were smooth or wrinkled. Since the time of Mendel, we have learned that the wrinkled peas have a defect in starch production; a particular starch branching enzyme has been disabled by the insertion of a piece of DNA that comes from another part of the genome, a transposable element.

As described in Chapters 1 and 2, transposable elements, also known as **transposons**, are major components of plant genomes. These segments of DNA are capable of moving from one spot in the genome to another. As such, they behave as semi-independent entities, existing, reproducing, and going extinct within the ecosystem of the plant genome and the cell that contains it. In this sense, transposons are analogous to parasites, for which the host is simply the means by which the parasite carries out its life cycle. A transposon is successful if it is able to make copies of itself and if these copies are maintained through time. Because a transposon follows its own evolutionary trajectory that may or may not have any benefit to the cell that contains it, it is sometimes described as "selfish," although this is a very human characteristic

to ascribe to a short piece of DNA. Like a successful parasite, a transposon will persist if the host evolves a way to tolerate it; if the host cell dies as a result of transposon activity, then the transposon itself is eliminated from the population. Thus natural selection acts on the transposon to persist and reproduce, but at the same time not to disrupt the genome and kill the cell that contains it. Plant genomes, like those of all other eukaryotes, tolerate enormous variation in the number of transposable elements, and in most cases the elements do not have a detectable effect on the fitness of the plant. However, with the large numbers of transposable elements moving around the genome, it is not surprising that, from time to time, they will land in a gene and have an obvious phenotypic effect like that seen by Mendel in the wrinkled peas.

Transposable elements affect plant genomes in four major ways: first they can account for most of the DNA in the genome (e.g., in maize, humans) and thus directly affect genome size. Secondly, they create mutations when they insert and excise. Thirdly, transposable elements may affect the arrangement of genes by moving genes or gene fragments from one genomic location to another. Fourthly, they may affect gene expression. While these general effects can be caused by any transposon, different types of transposons are often associated with particular effects.

In considering the effects of transposons as outlined in this chapter, it will be important to distinguish **somatic** (soma = body) from **germinal mutations** in plants. A somatic mutation is one that happens in a vegetative (plant body) tissue and is not transmitted to the

Plant Genes, Genomes and Genetics, First Edition. Erich Grotewold, Joseph Chappell and Elizabeth A. Kellogg.
© 2015 John Wiley & Sons, Ltd. Published 2015 by John Wiley & Sons, Ltd.
Companion Website: www.wiley.com/go/grotewold/plantgenes.

progeny. In contrast, a germinal mutation is one that is transmitted to the next generation. Somatic mutations, such as the insertion or excision of transposons, often happen late in development. In animals, because of the way in which the gamete-producing tissues form, all mutations that happen late in development and that do not involve the reproductive organs will be somatic. For example, it is easy to understand that a skin mole in our face will not be transmitted to our children.

The situation is slightly more complex in plants. The body plan of a plant, unlike that of a vertebrate, is not determined in the embryo. Rather, groups of pluripotent cells (in animals referred to as stem cells) make most plant organs. One such group of pluripotent cells is the **apical meristem**. The apical meristem will produce leaves, stems, and – once the switch to the reproductive phase happens – flowers. If a mutation happens in the meristem in floral precursor cells, then the mutation will be represented in some or all of the ovules or pollen grains and can be passed on to the next generation. Mutations caused by transposons are no different from other mutations in this respect, and may be somatic or germinal.

Transposable elements fall into three broad groups – RNA elements, DNA elements, and *Helitrons* – depending on whether they execute their transposition via a DNA or RNA intermediate and the type of repeat signature sequence they leave behind. **RNA elements** are also known as **retrotransposons**, because they copy themselves into RNA, then are transcribed back to DNA and are inserted (Figure 3.1). In contrast, **DNA elements** operate by cutting themselves out of their original position and moving to another point in the genome. Different types of elements vary in terms of where they prefer to insert in the genome, how the DNA strand is modified during the insertion process, and how the elements themselves tend to vary over time. All groups of transposable elements contain some that are able to cause their own movement, and some that need another element to activate them. The former are known as **autonomous elements** and the latter as **non-autonomous**. In most cases, the non-autonomous elements are derived from autonomous ones by loss of part or all of the internal sequences, and thus depend on an autonomous element for the machinery necessary to move around the genome. This combination of autonomous and non-autonomous elements can be manipulated by geneticists to control transposon activity, as will be described below.

Increases in genome size are most influenced by the retrotransposons because of their "copy and paste"

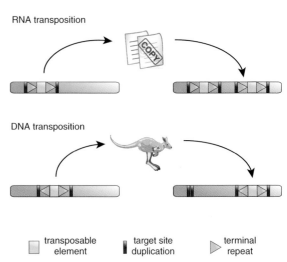

Figure 3.1 DNA elements differ from RNA elements in their mode of transposition. DNA elements excise themselves from the DNA and hop to a new position, leaving behind only a few base pairs of duplicated sequence (black bars). RNA elements copy themselves via an RNA intermediate and then reinsert into the DNA. The original element thus stays in place. Note also that the terminal repeats (blue triangles) are generally in inverse orientation in the DNA elements but in the same (direct) orientation in RNA elements. Clip art 1: http://all-free-download.com/free-vector/vector-clip-art /kangaroo_clip_art_5892.html. Courtesy of Clker.com. Clip art 2: http://www.dreamstime.com/royalty-free-stock -photo-copy-icon-image6655585ID 6655585 © Dimensionsdesign.

method of RNA-based transposition, resulting in a duplication of the element each time it transposes. These elements tend to accumulate between genes so they affect the overall architecture of the genome (Figure 3.2a). DNA elements, in contrast, transpose their DNA directly through a "cut and and paste" mechanism, creating mutations in and near genes (Figure 3.2b). Incorporation of genes and gene fragments, and movement of these around the genome is particularly well documented for DNA elements and for the recently discovered *Helitrons* (Figure 3.2c).

3.2 Retrotransposons are mainly responsible for increases in genome size

As noted in Chapter 1, genome size in plants is highly variable and does not correlate with any obvious aspect of the whole-plant phenotype (e.g., number of protein-coding genes). Instead, genome size

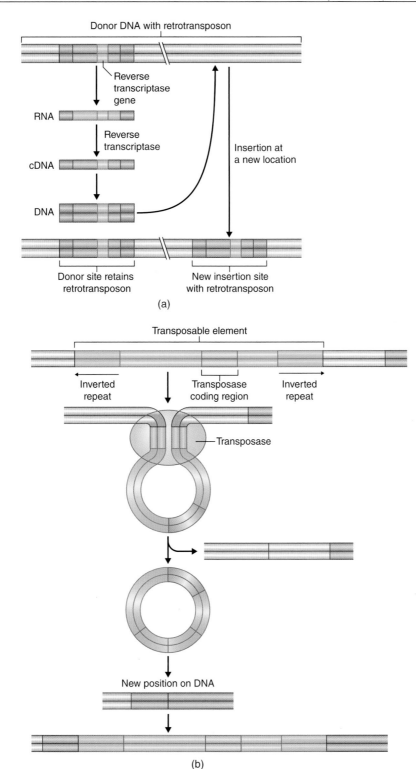

Figure 3.2 A more detailed look at the mechanisms of transposition of three major sorts of transposable elements. (a) RNA elements. RNA polymerase II transcribes the element to RNA. Reverse transcriptase then creates a DNA product that is integrated into the genome at a new position, creating a small target site duplication. (b) DNA elements. The transposase encoded by the element cuts the transposon out of its original position and pastes it into a new position. As with the RNA element, a target site duplication is produced. The break at the original position is repaired. (a–c) Adapted from Buchanan *et al.* (2002)

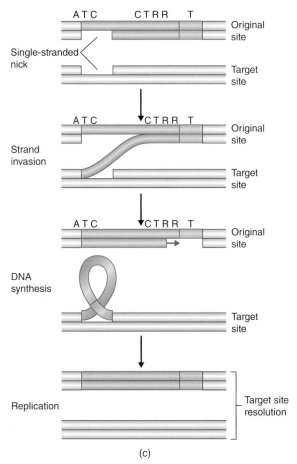

(c)

Figure 3.2 (c) *Helitrons*. The tranposase creates a single-stranded nick at the 5′ end of the *Helitron* (red line) between the A in the adjacent DNA and the TC at the end of the helitron. A corresponding nick is created in the DNA at the target site. This is followed by invasion of one strand of DNA from the original site into the target site. DNA is then synthesized via a rolling circle mechanism, repairing the original site. The target site is repaired via DNA replication. It is common for DNA synthesis to proceed beyond the terminus of the original element so that a small piece of adjoining DNA (green) may be transposed along with the element (red). CTRR is the sequence at the 3′ end of the helitron, where R is either purine base (A or G). T is in the adjoining DNA. Redrawn from Lisch, 2013

correlates with the percentage of the genome made up of transposons, particularly the retrotransposons. Retrotransposons are transcribed to RNA and then reverse transcribed back to DNA and reinserted into the genome (Figure 3.2a). Because the original transposon remains in the genome even as it is replicated, copy number steadily increases.

Retrotransposons are structurally similar to retroviruses, to which they are closely related. Like retroviruses, retrotransposons contain sequences that encode reverse transcriptase (RT) and an enzyme required for replication (RNase H), both of which are necessary for transposition. Also like the retroviruses, they encode a protein known as *gag* (short for "glycosamineglycan"), which encapsulates the transposon's RNA along with the proteins necessary for insertion into the genome, creating a viral-like particle. Unlike their viral cousins, however, the retrotransposons are unable to leave the cell in which they occur.

Retrotransposons are classified into two major groups, according to the DNA sequences at their ends. In one major group, the 5′ and 3′ ends of the transposon have identical DNA sequences of about 100 to several thousand base pairs; these sequences are known as **long terminal repeats**, or **LTRs**. The other major group of retrotransposons lacks the repeats.

3.2.1   *LTR retrotransposons*

LTR retrotransposons are particularly common in plants, much more so than in animals, and have been well characterized. Each LTR retrotransposon is a single transcription unit, with transcription beginning somewhat downstream of the 5′ end of the first LTR and ending upstream of the 3′ end of the second LTR. Within this unit are two multi-functional genes, *gag* and *pol. pol* encodes the Pol precursor protein, which is composed of several different functional proteins. First, within Pol is a protease (PR) that cleaves the precursor protein into functional units. Also encoded by Pol are the RT, RNase H, and an integrase (INT). These allow the entire LTR transcript to be replicated as DNA (the role of RT and RNase H) and then to reinsert into the genome (role of INT) (Figure 3.3). In some elements, all the genes are in the same reading frame, whereas in others the reading frames overlap so that translation must be stopped and restarted in the new frame.

Transcription of the element occurs in the nucleus to create a message RNA that is exported to the cytoplasm, where it is translated. The gag protein then forms the virus-like particle within which reverse transcription takes place. The cDNA is then released along with the integrase, returned to the nucleus, and integrated into the genome (Siomi *et al.*, 2011).

Each LTR contains three regions or domains – U3, R, and U5 – that are necessary for both transcription and reverse transcription. The DNA is transcribed to produce an mRNA that extends from one R sequence through the other one. This mRNA is then

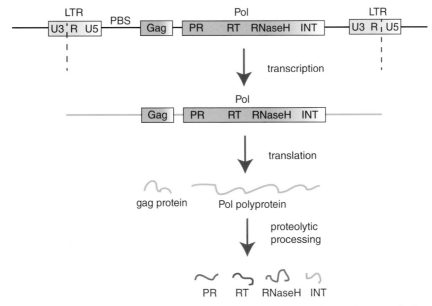

Figure 3.3 Diagram of the structure of a *copia*-type retrotransposon. The entire element is transcribed as a unit and then translated to produce two proteins, gag and the pol polyprotein. The latter is then cleaved into its four component proteins. LTR, long terminal repeat; R, transcription start site; U3, U5, characteristic sequences in the LTR; Gag, gene for the Gag protein; PR, protease; RT, reverse transcriptase; RNAseH, RNAse; INT, integrase; Pol, polymerase; PBS, primer binding site. Adapted from Kumar and Bennetzen (1999)

reverse-transcribed to DNA that will be reinserted into the genome. The mechanism behind this reverse transcription is intricate. RT requires a double-stranded template for initial binding, so the first step must be to bind a primer to the mRNA (Figure 3.4a). The primer is not encoded by the transposon but rather is supplied by the cell itself, generally the 3′ end of a transfer RNA (tRNA). This binds to the primer-binding site (PBS) and reverse transcription then proceeds through the 5′ U5 and R sequences, until it reaches the 5′ end of the mRNA (Figure 3.4b). At this point, the RNase H (encoded by the transposon) removes the 5′ single-stranded RNA, leaving a single strand of DNA ending in the R sequence (Figure 3.4c). This DNA is complementary to the R sequence at the other end of the mRNA; the two R sequences thus pair and the entire molecule forms a circle (Figure 3.4d). Once the mRNA is circularized, reverse transcription can proceed until complementary DNA (cDNA) is synthesized for the entire molecule (Figure 3.4e). RNase H then removes the RNA and the second strand of DNA is synthesized (Figure 3.4f, g). The new double-stranded DNA molecule is then inserted into the genome with the help of the integrase encoded by the transposon.

LTR retrotransposons appear to insert preferentially into themselves, thus destroying the original transposon, pushing apart the LTRs, and creating tandemly repeated sequences (Figure 3.5). After this insertion event, there is no selection to retain function of the original pair of LTRs so these gradually begin to accumulate mutations such that the original 5′ LTR will have a sequence that is slightly different from that of the 3′ LTR. After multiple rounds of such retrotransposon insertions, the LTRs from the original retrotransposon can often be found far apart in the genome, with LTRs from the second oldest retrotransposon inside them, and so on until the most recently inserted retrotransposon is located. Because the original retrotransposon is the oldest, its LTRs will have the most mutations and thus be more different from each other than the LTRs of the second oldest retrotransposon, and so forth. By tracking the numbers of mutations in the pairs of LTRs it is possible to obtain a rough time estimate for the age of arrays of retrotransposons.

Amplification of retrotransposons is clearly a mechanism whereby genome size can increase. However, the retrotransposons also provide the basis of genome size decreases. Recombination can occur between the LTRs of retrotransposons, causing excision of large pieces of

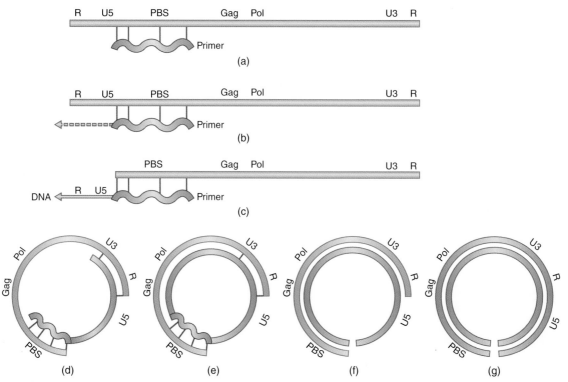

Figure 3.4 Reverse transcription of a retrotransposon. (a) Primer (wavy line) binds to the RNA (red). (b) DNA (blue) is synthesized to complement the RNA. (c) RNAse removes the R and U5 sequence of RNA. (d) newly synthesized DNA base pairs at U5 and R with the RNA from the other end of the message. (e) DNA is synthesized complementary to the RNA. (f) RNA is digested. (g) Second strand of DNA is synthesized, readying the element for reinsertion

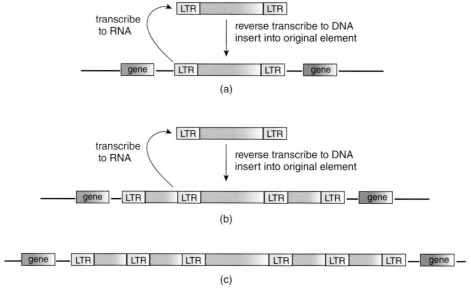

Figure 3.5 Pattern of nested retrotransposons. (a) Element is transcribed to RNA and then reverse transcribed and inserted into itself. (b) The newly inserted element is transcribed, reverse transcribed, and again inserted into itself. (c) Structure of the DNA after two rounds of retrotransposition. At each step, the original LTRs are pushed farther apart. In this figure, the original LTRs are the ones farthest to the left and to the right

DNA. All that is left after such recombination is a single LTR (Figure 3.6).

Among the LTR retrotransposons, the two major classes are Ty1-*copia* and Ty3-*gypsy*; these differ from each other in sequence and also in the order of the genes that they encode. In Ty3-*gypsy*, the integrase gene (INT) is 3′ of RNAseH (RH); this is the config-uration shown in Figure 3.3. However, in Ty1-*copia* elements, INT is between PR and RT. Like most other LTR retransposons, Ty3-*gypsy* elements are found

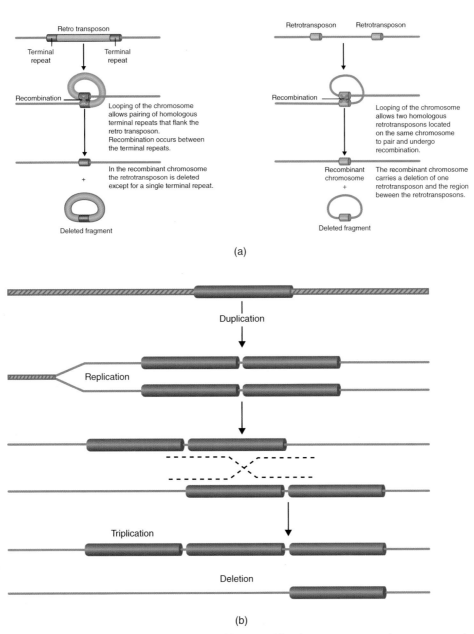

(a)

(b)

Figure 3.6 Mechanisms of genome rearrangement caused by recombination among repeated sequences, such as provided by transposons. (a) Recombination between terminal repeats (left) or between entire elements (right) can lead to loss of fragments of DNA. (b) Unequal crossing over between repetitive sequences such as transposons will lead simultaneously to gene duplication on one strand and deletion on the other. From Fedoroff, 2012

in regions away from genes such as pericentromeric regions. In contrast, Ty1-*copia* elements are found preferentially near genes.

3.2.2 Non-LTR retrotransposons

Retrotransposons that lack LTRs are known as **long interspersed nuclear elements (LINEs)** and **short interspersed nuclear elements (SINEs)**. These are particularly common and well characterized in vertebrate genomes, but also occur in plants. As their name implies, LINEs are long enough to contain a gene for RT. The non-autonomous counterparts of the LINEs are the SINEs; the latter are less than 500 bp long. Both LINEs and SINEs end in either polyA, polyT, or other simple repeats. When they insert into the DNA, they create small duplications. The SINEs have some resemblance to tRNAs (Yoshioka *et al.*, 1993), and like tRNAs are transcribed by RNA polymerase III (Sun *et al.*, 2007) (see Chapter 7). Copy number of LINEs and SINEs is high. Because the sequences of SINEs vary considerably, they have been hard to detect, but new computational tools are being developed that show how common they are (Wenke *et al.*, 2011).

Most of our knowledge about the mechanism of action of LINE elements comes from studies in mammalian genomes; for example, the only known active transposon in humans is a LINE element, LINE-1 or L-1 (Beck *et al.*, 2011). Thus the function of LINE elements in plant genomes is based on extrapolation. In general, a LINE element contains a 5′ UTR followed by two open reading frames (ORFs), ORF1 and ORF2, and a 3′ UTR. ORF2 encodes an RT. The element is transcribed in the nucleus and then exported to the cytoplasm for translation. The proteins produced from ORFs 1 and 2 associate with their own mRNA to form a unit containing both RNA and protein. The RNA-protein complex then returns to the nucleus, where reverse transcription occurs to produce cDNA. ORF2 encodes an endonuclease that causes a break in one strand of the target DNA, and then the RT produces a cDNA that is integrated into the genome. Sequences of LINE elements from both mammalian and plant genomes show that the vast majority of elements have been truncated at the 5′ end, rendering them inactive (Kumar and Bennetzen, 1999; Beck *et al.*, 2011).

3.3 DNA transposons create small mutations when they insert and excise

DNA transposons excise themselves from their position in the double helix and reinsert in another location. As part of this process they leave a tiny signature in the form of a duplication of two or more base pairs (Figure 3.1b). This signature is known as a **target-site duplication**, the sequence of which is specific for a particular transposon. Such duplications, which often occur in or near protein-coding, non-transposable-element genes, are mutagenic; they have the potential to modify gene structure and function, for better (occasionally) or worse (more often). The DNA repair mechanisms of the cell can also insert a copy of the element back into the site from which it has excised, thus increasing the copy number.

The cut-and-paste elements appear in all eukaryotes, suggesting an ancient origin. All of these consist of a gene encoding a **transposase**, with a **terminal inverted repeat (TIR)** on either side. (Note that the repeats are inverted, rather than the direct repeats of the retrotransposons.) The transposase proteins of all cut-and-paste elements share a conserved set of amino acids, the DDE/D domain, which is central for catalysis of the transposition reaction (Yuan and Wessler, 2011). Class II elements are classified largely by similarities in the sequences of their transposases, but each superfamily also has characteristic target sites and modes of transposition. In plants, the major superfamilies of these elements are *Tc1/mariner*, *PIF/Harbinger*, *Mutator*-like elements (MULEs), *hATs*, and *CACTA* elements.

3.3.1 hAT elements

The first transposable elements discovered in any organism are now called *hAT* elements. These were identified by Barbara McClintock in maize in the 1940s. McClintock had been investigating chromosomal breakage and loss of chromosomal fragments when she discovered a group of plants in which one part of a particular chromosome consistently "dissociated" from the rest of the chromosome (Comfort, 2001). McClintock concluded that dissociation was caused by a gene (which she initially called *D* and later, *Ds*) at the site of the breakage. By comparing several maize stocks with the *Ds* locus, she concluded that *Ds* was necessary but not sufficient for dissociation; a second locus was required. The second locus was named *Ac*, or *Activator*. As long as *Activator* was present, *Ds* caused chromosomal breakage.

As was common in genetics in the 1940s, McClintock proceeded to try to map their locations relative to each other and to other genes known at the time. She found contradictory results, and eventually, in early 1948, concluded that both *Ds* and *Ac* could move around in the genome. From this earliest discovery it became clear that (a) certain genes were able to move, (b) some could move on their own, whereas (c) some needed other genes to permit them to move. She had discovered the autonomous element *Ac*, and its non-autonomous derivative, *Ds*. Although she had discovered what are now called transposons, McClintock focused instead on their apparent role in affecting plant development and called them "controlling elements" (Comfort, 2001).

Since the discovery of *Ac/Ds* in maize, a similar element, *Tam3*, was found in *Antirrhinum majus* (snapdragon) (Figure 3.7), and another element, *hobo*, in *Drosophila*. The name *h*AT element is thus an acronym for *hobo, Ac*, and *Tam*. The terminal inverted repeats of these elements are short – 11 bp in *Ac/Ds* and 12 bp in *Tam3*. Both *Ac* and *Tam3* encode a single

gene, the transposase; although the sequences of the two transposases differ, as would be expected for proteins encoded by such phylogenetically distant plants, the resemblance of their amino acid sequence is high enough to show that they are part of the same general family of transposable elements.

When a *h*AT element is cut out of its original position in the genome, the transposase binds to the terminal inverted repeats and to parts of the subterminal sequence. Mutations in the terminal inverted repeats abolish transposition, so transposase binding is clearly sequence-specific. The transposase brings the two ends of the element together and the repeat sequences synapse, creating a hairpin structure. The element is then cut out of the DNA strand. Binding of the transposase to the DNA generally will not occur if the DNA is methylated; this is thought to be one mechanism for preventing transposition and keeping transposable elements under control.

Another of McClintock's discoveries was that *Ac* in maize is negatively regulated by dose – the more copies of *Ac* in the genome, the less transposition is observed. She was able to track the movement of *Ac/Ds* by observing pale kernels of maize, in which the transposon was in a gene that controlled pigment (anthocyanin) formation. When one copy of *Ac* was present, the transposon moved out of the pigment gene in some cells, causing the formation of colored spots on the pale background. With two copies of *Ac*, the number of spots increased, although they were smaller. With three copies, the spots failed to appear. McClintock concluded that the dose of *Ac* determined the time of development at which *Ac* left the pigment locus. Two copies caused transposition and hence spot formation to occur later, resulting in more but smaller spots. Three copies caused transposition to occur so late that most of kernel development had already occurred and little color phenotype was observable. Because of this dosage dependence, copy number of *Ac* cannot get very high in the maize genome. Curiously, when *Ac* is introduced into tobacco or Arabidopsis, this negative dosage regulation is lost and more doses of *Ac* lead to more transposition (Kunze and Weil, 2002). *Tam3* does not exhibit such marked dosage sensitivity. *Ac/Ds* preferentially transposes to closely linked sites, generally within 5 cM (a cM or centimorgan is a measure of the recombination frequency between two genetic loci; see Chapter 2); physical transposition distances may be within 15 kb and often much less.

Figure 3.7 *Antirrhinum majus* (snapdragon). The *h*AT class of elements has been studied extensively in snapdragon, where they are known as Tam (Transposon of *Antirrhinum majus*) elements. From http://commons.wikimedia .org/wiki/File:Antirrhinum_aka_Snap_dragon_at_lalbagh _7118.JPG

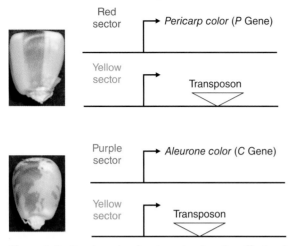

Figure 3.8 Two kernels of maize, showing the effects of transposon insertion into two different genes. In both cases the gene controls production of pigment, so transposon insertion results in a yellowish set of cells. In the top kernel the transposon is inserted into a gene, *P*, that controls color in the outer layer of the kernel, the pericarp. In the bottom kernel the transposon is inserted into a different gene, *C*, that controls color in the outer layer of the endosperm, the aleurone. The difference in pattern is due to the different patterns of cell division during development of the two tissues. Adapted from Athma *et al.* (1992)

The exact phenotype caused by *Ac/Ds* transposition depends on the particular gene it disrupts. For example, the gene required for pigment production in the outer layer of the kernel of maize (pericarp) is known as *P*. If a transposon inserts into this gene early in development, all cells containing the element will lack pigment, whereas cells lacking the element will have their normal color (Figure 3.8, top diagram and kernel). Because the cells in the pericarp divide in long rows, the differently colored cells will form stripes. A second gene, known as *C*, is required for pigment production in the outer layer of the endosperm (called aleurone in cereals). If the transposon inserts into *C*, pigment production will again be prevented, but this time the sectors lacking color will be irregular (Figure 3.8 bottom diagram and kernel). This is because the pattern of cell division is less directional in the aleurone.

3.3.2   Mutator-like elements

In 1978, D. S. Robertson published a paper in which he described a set of maize plants that had an abnormally high rate of mutations; he concluded that there must be a genetic element (which he called a "controlling element" following McClintock) that was causing these plants to have a mutation rate about 30 times higher than that of normal maize (Robertson, 1978). Later it was discovered that this was a novel transposable element, which came to be known as *Mutator*, or *Mu* for short. In maize, *Mu* elements may be autonomous or non-autonomous, and the autonomous ones are now called *MuDR*, for Don Robertson. *MuDR* contains two genes, *mudrA* and *mudrB*, and can itself occur in many copies (up to 20) in the maize genome. While the levels of *mudrA* and *mudrB* mRNA are proportional to copy number, the levels of MURA and MUDB protein are maintained no matter how many copies of *MuDR* are present in the genome, indicating some sort of post-transcriptional control, and also hinting at one way that the plant can tolerate such high copy number of the element (Walbot and Rudenko, 2002).

Non-autonomous elements can be very common in the maize genome. Because of this high copy number, a plant with active *MuDR* can have many transpositions leading to many new mutations, with each new insertion creating a 9 bp duplication. In somatic tissues, *Mu* elements often excise and insert during the last rounds of cell division, so that tiny revertant sectors are formed. In gametophytes however, *Mu* can insert but cannot excise. The factors that affect this differential regulation are not known.

The internal sequences of non-autonomous *Mu* elements are not particularly similar to each other nor to *MuDR*, and often appear to contain fragments of host genes. *Mu* has a number of properties that make it different from *Ac/Ds*. First, *Mu* will insert in unlinked sites, affecting distant regions of the genome. Secondly, unlike *Ac/Ds* elements, which tend to leave the host duplication site intact, *Mu* often does considerable damage to the DNA when it excises, and can result in insertions or deletions of as many as 500 bp. However, *Mu* excision generally occurs somatically (i.e., in the body of the plant), but only very rarely in cells that will give rise to gametes. This means that the heritable mutations that *Mu* causes are largely a consequence of the transposon being present, and not due to the mutations characteristic of the footprints that transposons usually leave behind. Thirdly, there are generally many more *Mu* elements than *Ac/Ds* elements in the genome of a single plant.

While *Mu* elements are best characterized in maize, they have been discovered throughout angiosperms and bryophytes, but have not been found in lycophytes (Yuan and Wessler, 2011). Outside of maize they are known as ***Mu*-like elements** (**MULEs**). They can be recognized by their TIRs, which are approximately 200 bp long as well as the sequence of *mudrA*. *mudrB* does not occur in species other than *Zea mays*.

3.3.3 CACTA elements

The *En-Spm* (*CACTA*) elements are part of a group with a characteristic 5 bp sequence, CACTA, which occurs in the terminal repeats. These elements are similar to *Mirage* and *Chapaev* elements, and so are sometimes placed in the CMC (*CACTA-Mirage-Chapaev*) superfamily (Yuan and Wessler, 2011). The terminal inverted repeats are 13 bp long, and when the element inserts, it produces a 3 bp duplication. Unlike *Mu* or *hAT* elements, the CACTA elements encode several genes. The transcript produced by these elements can be alternatively spliced (see Chapter 13) to generate several different mRNAs and hence several different proteins. The major proteins are TNPA, which is transcribed from a 2.4 kb transcript, and TNPD, from an overlapping 6 kb transcript. Like the transposase in *Mu* elements, TNPA binds to the terminal repeats of the element and brings them together; such binding is reduced by DNA methylation. TNPD then binds to TNPA and appears to be responsible for cleavage of the DNA.

3.3.4 MITES

Hundreds to tens of thousands of miniature inverted-repeat transposable elements (MITEs) are found in the genomes of plants, animals, and fungi (Feschotte *et al.*, 2002). As their name implies, they are small stretches of DNA, generally less than 600 bp long, marked by inverted repeats at the ends. They do not encode proteins, and for some years after their initial discovery, they were not known to move about the genome. It was thus a mystery as to how they achieved the observed high numbers. By analysis of their target sites and their terminal inverted repeats, it eventually was possible to link the MITEs with particular classes of DNA transposons. For example, a large class of MITES found in plants, known as *Stowaway* elements, is related to transposons of the *Tc1/mariner* class based on similarity of the target site and the terminal inverted repeats. In this case, the *Mariner* element encodes the transposase, which then activates not only *Mariner* elements themselves but also the related MITEs.

Another well-characterized class of MITES is known as *Tourist*, a group of elements that appear to be activated by elements known variously as *Harbinger* or *PIF*. *PIF/Harbinger* elements encode two genes, ORF1 and a transposase, and produce a 3 bp duplication upon insertion. Unlike other Class II elements, however, they remove one copy of the duplication when they excise, thus leaving no evidence of having been present and creating no permanent change in the DNA. Two *PIF/Harbinger* elements have been particularly well studied in rice, where they are known as *Ping* and *Pong*. Both are capable of causing active transposition of *mPing*, a Tourist-like MITE that is derived from *Ping* by deletion.

DNA transposons in forward and reverse genetics

Genetic studies involve manipulating genes and then correlating changes in the gene with changes in some aspect of phenotype. One way to manipulate genes is simply to introduce mutations. This can be done by mutagenizing pollen or seeds, growing up the mutant progeny and then assessing the sorts of mutant phenotypes that result. This is known as forward genetics (population → phenotype → gene identification); it provides immense opportunity for gene discovery and is ideal for identifying loci that all participate in a particular pathway or that affect a particular phenotype. On the other hand, it is generally not directed, so it can be an inefficient way to answer some questions. Conversely, it is possible to choose a particular gene, and then select from a population the mutation in that particular gene and then assess the phenotype, a process known as reverse genetics (gene → population → gene mutation). This approach is obviously a highly targeted way to answer questions about a locus already known to be involved in a process, but makes it less likely that other unexpected players in a process will be discovered.

Transposons are powerful tools for both forward and reverse genetics. Because they generate mutations when they insert and excise, they are widely used for forward genetics, particularly in maize. Maize lines carrying *MuDR/Mu* have particular advantages in this regard in that the transposition rate is high, elements can move almost anywhere in the genome, and insertions frequently occur in exons. Several collections of maize lines have been developed in which each line has one or more *Mu* insertions in a different gene or part of the genome. These lines are then an effective source of mutant alleles. Cloning the disrupted gene then requires identifying the site of the relevant *Mu* insertion, a process that was time-consuming in the past. However, with the advent of cheap and high-throughput sequencing it is now possible to locate where all the insertions are in a given population of plants, and then to generate sequence-indexed collections that can be easily queried for the gene of interest.

Maize lines carrying *Ac/Ds* elements are used in slightly different ways. Because *Ac/Ds* will transpose with a high frequency to sites close by, an element can be used to create alleles of a single gene or nearby genes. For example, if an *Ac/Ds* insertion has been identified in or near a gene of interest, the plant carrying the element can be crossed with one with an active transposase. This will cause the element to move into various positions, but always in or near the target gene. It is then possible to correlate the different insertion positions with different phenotypes. For example, an insertion into the first exon may block transcription entirely and create a non-functional gene (a null mutant), whereas an insertion near the stop codon may simply create a truncated protein with slightly impaired function. By comparing the phenotypes of each of these mutants with wild type plants, it is often possible to gain considerable insight into the function of the protein. A null mutant in a gene controlling a major aspect of plant morphology will often create a plant whose development is so severely disrupted that it is impossible to determine exactly which processes have been affected, or the plant may simply die. On the other hand, a milder mutation will allow some aspects of development to proceed.

An example of an allelic series created by *Ac* transposition is shown in Figure 3.9. In this example, an *Ac* element inserted into the *P* locus in maize, similar to that in Figure 3.8, was allowed to move into many

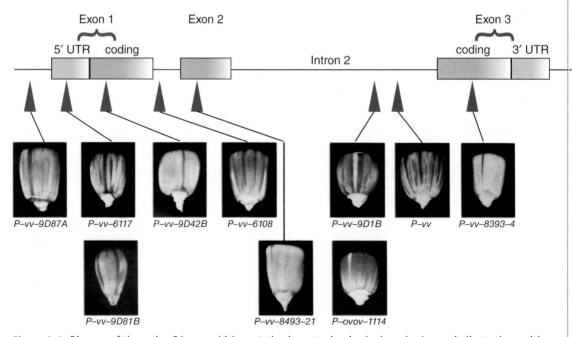

Figure 3.9 Diagram of the maize *P* locus, which controls pigmentation in the kernels. Arrows indicate the position of the *Ac* element in corresponding kernels. Insertions of *Ac* elements at different positions in the gene lead to different patterns of color in the kernels. In the kernels in the lower row, *Ac* is inserted in an orientation opposite to that of the kernels of the upper row. Adapted from Athma *et al.* (1992). Reproduced with permission of the Genetics Society of America; permission conveyed through Copyright Clearance Center, Inc.

nearby locations. The kernels show the effect of the element moving into each different location. Notice that insertions into the exons prevent production of the protein, which in turn blocks pigment production and leads to largely white kernels (Athma *et al.*, 1992).

Transposable elements are the workhorses of maize genetics, where they are used constantly as tools for genetic research. Other systems in which naturally occurring transposons are used include petunia, rice, and snapdragon. Transposons have also been introduced into different species. Maize *Ac/Ds* and *En/Spm* have been introduced into Arabidopsis, petunia, and tobacco. The frequency with which they transpose and the distance they move are often different in these foreign hosts. This appears to reflect the ability of the host's silencing mechanisms (generally DNA methylation) to recognize the element and to control its movement.

3.4 Transposable elements move genes and change their regulation

Nearly all transposons have been characterized and studied in domesticated plants; there is little information on the effects of transposons in nature. Thus the long-term effects of transposons in nature are based on extrapolation from observations and anecdotes from cultivated plants. It is virtually certain that transposons have played an important role in evolution because they are mutagens, and mutation is the raw material for evolution. Nonetheless, the importance of transposons relative to other mutagens, and the conditions under which they are important, remain largely unknown. Transposon activity can be triggered by environmental stress, which presumably faces most natural plant populations much of the time. In addition, a cross between two very different parent plants (e.g., ones from different species) can stimulate transposons to move, presumably by altering the methylation state of the DNA. In the lab, transposons can be activated by mutagenesis, tissue culture, and plant transformation. While they can be a nuisance in this context, they can also be harnessed for genetic research (see Box). The ability of transposons to cause movement of genes or gene fragments, to modify gene expression, and to create novel genes can thus be viewed as a utility for genetic improvement as well as a tool for unraveling basic physiological, biochemical, and genetic properties of plants.

3.4.1 Disruption of genes and regulatory regions

A common effect of transposons is simply to inactivate the gene into which it inserts; as shown in Figure 3.7, this can occur in various ways depending on exactly where the insertion is in the locus. We have already mentioned Mendel's wrinkled peas, in which a transposon inserts into an enzyme that helps in starch production. Another starch production enzyme, granule-bound starch synthase I (GBSSI), is important in the endosperm of cereals where it helps produce amylose. GBSSI is often disabled by transposable elements, blocking amylose production, and causing the grain to become sticky when cooked. Humans have selected for this particular mutation because they like sticky rice and sticky millet; these are cultivated in Asia, and used for particular specialty dishes.

In another example, insertion of a retrotransposon into a pigment gene in deep purple Cabernet grapes, used to make red wine, led to the green grapes of Chardonnay, used to make white wine. In some of the descendants of the Chardonnay grape, the element excised but the grape color did not revert back to deep purple; instead it reverted only partially, to a reddish color (Figure 3.10). Why didn't the color return to deep purple? Recall that all elements create small duplications when they insert and these are left behind when the element excises; in addition, retrotransposons may leave behind a single LTR if they are removed by recombination. Thus even when the element is gone it leaves extra base pairs behind, and these extra base pairs can disrupt transcription and

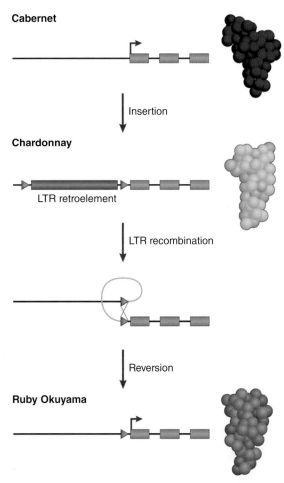

Cabernet

Insertion

Chardonnay

LTR retroelement

LTR recombination

Reversion

Ruby Okuyama

Figure 3.10 Control of fruit color in grapes by a retrotransposon. Cabernet grapes have a fully functional pigment gene (exons indicated by gray boxes). Insertion of a retrotransposon just upstream of the gene blocks pigment production and leads to green Chardonnay grapes. The LTRs of the element can recombine and remove most of the transposon, but one LTR is left, causing reduced transcription of the locus in Ruby Okoyama grapes. Lisch (2013). Reproduced with permission of Macmillan Publishing Ltd.

alter the amount or nature of the transcript. In other words, disruptions may persist even if the element is no longer present.

3.4.2 Movement of genes or gene fragments

In some cases, when a transposon excises from the DNA, another piece of DNA is copied from elsewhere in the genome and inserted into the broken strand. If the piece of DNA being moved contains a gene, then the process results in a duplicate gene at some position in the genome other than the original one (Puchata, 2005; Wicker *et al.*, 2010). In general, plants with more transposable elements, such as wheat, have more genes "out of place" than plants with fewer (Wicker *et al.*, 2011). As might be expected, most of these "out-of-place" genes do not function; either they are truncated or have lost their regulatory sequences or have acquired mutations in their coding sequence. However, some continue to work in their new genomic environment. One well documented example concerns fruit shape in tomato. While wild and many domesticated tomatoes have fruits that are roughly spherical, some domesticated tomatoes have elongated fruits, a shape change that is caused by increased expression of a gene known as *SUN* (Xiao *et al.*, 2008). The increased expression in elongated fruits results from a complex duplication of *SUN*, which in turn results from an abnormal transcript initiating from a retrotransposon (Figure 3.11). The abnormal transcript including *SUN* and several other genes is inserted into an intron of the gene DEFL1, so that *SUN* becomes regulated by the promoter of DEFL1. This leads to elevated levels of SUN.

MULEs (see Section 3.3.2) often transport genes or gene fragments, in which case they are known as **Pack-MULEs**. In rice, 22% of the genes carried by the Pack-MULEs are transcribed and some of these are translated. This occurs most frequently when the Pack-MULEs are carrying several genes. There is also evidence that natural selection is removing mutations (i.e., they are undergoing purifying selection), which would occur only if the genes were performing a cellular function (Hanada *et al.*, 2009).

The recently discovered *Helitrons* are particularly important in transporting non-transposon DNA from one part of the genome to another (Figure 3.2c). *Helitrons* were only identified once whole-genome sequences became available, and could be spotted because their ends are conserved and because they occur in high copy numbers. As with other classes of transposable elements, both autonomous and non-autonomous *Helitrons* have been identified. Compared with other kinds of elements, autonomous *Helitrons* are unusually large, 5–15 kb long; they encode a protein similar to a DNA helicase and one similar to a replicator initiator protein that functions in rolling circle replication in some plasmids and single-stranded

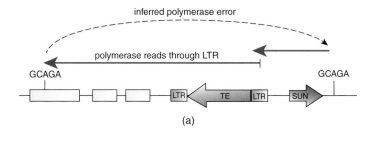

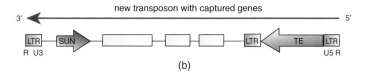

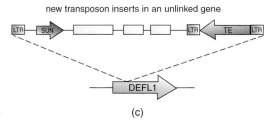

Figure 3.11 Complex structure of the SUN locus in tomato. (a) Transcription begins from a transposable element (TE) 5′ LTR, reads through the 3′ LTR and continues through the sequence GCAGA in a downstream gene. The polymerase appears to have dissociated and then picked up transcription from the GCAGA upstream from where transcription began. (b) Structure of the new retrotransposon with "captured" genes, including SUN. (c) Insertion of the new composite transposon into the gene DEFL1; whenever DEFL1 is transcribed, SUN will also now be transcribed. (a–c) Adapted from Xiao *et al.* (2008)

DNA viruses. They lack terminal repeats and do not produce target site duplications.

Non-autonomous *Helitrons* move genes or pieces of genes around the genome; about 60% of the elements in the maize genome are carrying genes. They clearly have the potential to cause changes in gene position, gene expression, and even gene structure by moving parts of genes into positions where they can create chimeras with other genes (Barbaglia *et al.*, 2012).

3.4.3 Modification of gene expression

In some cases, insertion of a transposon can increase gene expression rather than prevent it. This is because transposable elements contain promoter sequences required for their own transposition, and these promoters may occasionally be co-opted as promoters for adjacent genes. For example, the LTRs of retrotransposons encode promoters; the 3′ LTR of a retrotransposon can initiate transcription of host sequences immediately downstream of the element. Thus, movement of a retrotransposon also results in movement of promoter sequences.

An example of retrotransposon influence on gene expression is provided by a study of aluminum tolerance in wheat. Wheat cultivars vary in their tolerance for aluminum (Al^{3+} ions) in the soil. High concentrations of Al^{3+} in the soil make some parts of the world unsuitable for wheat production even if the climate is otherwise favorable. One way that wheat can tolerate positively charged ions such as Al^{3+} in the soil is to excrete small organic acids such as malate or citrate from their roots; these acids form complexes with positively charged ions in the soil and thus effectively immobilize them. Consequently, wheat breeders seek to increase aluminum tolerance by increasing the ability of the plant to excrete anions. A wheat cultivar has recently been identified with high activity of a citrate transporter in the roots (Tovkach *et al.*, 2013). A transposable element has inserted immediately upstream of the gene in this cultivar and has caused citrate export to be 20 times higher than in other cultivars. In other words, there is a direct connection between movement of a transposable element and improved agronomic value of a particular cultivar.

3.5 How are transposable elements controlled?

Transposable elements occur in all living organisms, but only in eukaryotes do they accumulate and exert the many effects described above. In bacteria, transposons are efficiently removed by recombination, a process that also occurs in eukaryotes. Thus if all organisms have the ability to get rid of them, why do they persist? Federoff (2012) has recently suggested that transposons accumulate because of the effective mechanisms that eukaryotes have developed to silence genes, including methylation of DNA, modification of histone tails, and changes in the structure of chromatin, which together are known as epigenetic phenomena (see Chapter 12) (Slotkin and Martienssen, 2007); she argues that these epigenetic processes originally developed to suppress homologous recombination between repeated components of the genome. Once transposons began to accumulate in eukaryotic genomes, the potential for evolution itself was enhanced, or in other words, transposons increase "evolvability". The concept of evolvability is a slippery one, in that organisms will not accumulate variation just in case an appropriate environment appears somewhere down the line; variation must be selectively advantageous or at least not disadvantageous in the moment when it appears. However, it is certainly true that there is a tight link between transposon activity, homologous recombination, and the mechanisms by which both are controlled.

While these mechanisms effectively silence most transposons in the genome, transposons periodically escape silencing and bursts of amplification occur, particularly among retrotransposons. In other words, transposons sometimes begin to take over the genome. The events that trigger such a release of epigenetic control are not well understood. McClintock suggested half a century ago that transposable elements might be mobilized in response to stressful environments and might thus provide diversity on which natural (or artificial) selection can act. There is some evidence that environmental stress such as high temperature or drought may reverse the epigenetic controls on the elements and allow them to explode in copy number. Polyploidy, which brings two entire genomes together in a single cell, also may release transposon activity (Slotkin and Martienssen, 2007). While reduced epigenetic control may release all transposons, more often it applies to a single class of elements, so that an individual plant or plant species will be characterized by a massive number of copies of one particular transposon. In general, epigenetic silencing mechanisms are specific for particular DNA sequences. Thus when an element first appears in the genome (e.g., by hybridization), it is not controlled, but as it replicates and copies begin to accumulate, the sheer number of copies or the formation of defective copies triggers silencing of all of the new elements.

After inactivation, transposons are gradually lost because of recombination and deletion. Since loss of DNA is driven in part by illegitimate recombination, it follows that transposon loss is slower in regions of low recombination such as the ends of chromosomes and the centromeric region. These regions are sometimes known as "safe havens", in which transposable elements can persist, often in an inactive state, without being removed for long periods of time.

Removal of transposons is also subject to natural selection. When transposons are silenced by DNA methylation or by changing the state of the chromatin, nearby genes are also affected. Thus, a gene near a set of silenced transposons may itself be silenced. If silencing the gene makes the plant less fit, then there is a selective pressure for removing the transposon. The result of this selection is a general depletion of transposons close to genes.

Epigenetic mechanisms will be covered in much more detail in Chapter 12. Whether they evolved as a mechanism to control transposons, or whether transposons accumulate in response to epigenetic controls (or both), epigenetics and transposons are intimately related in the history of the plant cell.

3.6 Summary

Transposable elements are found in all plant genomes and have enormous influence on genome architecture. They are powerful mutagens; they are a disruptive force when they insert and often leave characteristic footprints when they excise. In addition, they can cause movement of genes around the genome, can create novel regulatory environments, and can even sometimes create new proteins. All classes of elements contain both autonomous elements, which encode a transposase and can direct their own movement, and non-autonomous elements, which are derived from the autonomous elements by deletion and can only move when an autonomous element is present. Major classes of transposable elements are the retrotransposons (RNA elements), the cut-and-paste transposons (DNA elements), and the *Helitrons*. Because the retrotransposons transpose by replication, they can accumulate rapidly; their numbers correlate with genome size. The DNA elements are varied in their structure, mechanism of transposition, and their effects on the genome. Many of these have become valuable tools for genetic research. The *Helitrons* are not as well understood as the others, but appear to be replicate by a rolling circle mechanism; unlike the other classes they do not create a target site duplication when they insert and do not have characteristic terminal repeats. Transposon-induced mutations have been identified in starch biosynthetic pathways and in regulators of pigment production. In addition, transposons have moved genes from their original position to a position inside another gene where they come under the control of new regulatory sequences. While genes embedded in tranposons are often expressed, the effects on the phenotype are unknown.

3.7 Problems

3.1 What are the major kinds of transposable elements and how are they distinguished?

3.2 In which organisms have transposable elements been best characterized?

3.3 Explain the difference between a somatic and a germinal mutation.

3.4 Draw a diagram of a situation in which one transposon lands inside another. What do you think would be the consequence of such an event?

3.5 What makes transposons such effective tools for determining gene function?

3.6 Do you expect most transposons in a plant to be moving at any given time? Why or why not?

3.7 What is the difference between a somatic and a germinal transposon excision?

3.8 It is easy to find examples in which transposon insertions are detrimental to the plant, but occasionally they are beneficial. Provide at least one example in which the insertion of a plant transposon resulted in a favorable mutation.

3.9 State whether the following are true or false.
(a) All plant transposons have LTR at the end.
(b) All transposons replicate through an RNA intermediate.
(c) The plant can control the number of transposons, for example through DNA methylation.
(d) All transposons are equally useful for forward and reverse genetics.

References

Athma, P., Grotewold, E. and Peterson, T. (1992) Insertional mutagenesis of the maize *P* gene by intragenic transposition of *Ac*. *Genetics*, **131**, 199–209.

Barbaglia, A.M., Klusman, K.M., Higgins, J. *et al.* (2012) Gene capture by *Helitron* transposons reshuffles the transcriptome of maize. *Genetics*, **190**, 965–975.

Beck, C.R., Garcia-Perez, J.L., Badge, R.M. and Moran, J.V. (2011) LINE-1 elements in structural variation and disease. *Annual Review of Genomics and Human Genetics*, **12**, 187–215.

Buchanan, B.B., Gruissem, W. and Jones, R.L. (2002) Biochemistry and Molecular Biology of Plants, American Society of Plant Biologists.

Comfort, N.C. (2001) *The Tangled Field: Barbara McClintock's Search for the Patterns of Genetic Control*, Harvard University Press.

Fedoroff, N.V. (2012) Transposable elements, epigenetics, and genome evolution. *Science*, **338**, 758–767.

Feschotte, C., Zhang, X. and Wessler, S.R. (2002) Miniature inverted-repeat transposable elements and their relationship to established DNA transposons, in *Mobile DNA II* (eds N.L. Craig, R. Craigie, M. Gellert and A.M. Lambowitz), ASM Press, pp. 1147–1158.

Hanada, K., Vallejo, V., Nobuta, K. *et al.* (2009) The functional role of Pack-MULEs in rice inferred from purifying selection and expression profile. *Plant Cell*, **21**, 25–38.

Kumar, A. and Bennetzen, J.L. (1999) Plant retrotransposons. *Annual Review of Genetics*, **33**, 479–532.

Kunze, R. and Weil, C.F. (2002) The *h*AT and CACTA superfamilies of plant transposons, in *Mobile DNA II* (eds N.L. Craig, R. Craigie, M. Gellert and A.M. Lambowitz), ASM Press, pp. 565–610.

Lisch, D. (2013) How important are transposons for plant evolution? *Nature Reviews Genetics*, **14**, 49–61.

Puchta, H. (2005) The repair of double-stranded breaks in plants: mechanisms and consequences for genome evolution. *Journal of Experimental Botany*, **56**, 1–14.

Robertson, D.R. (1978) Characterization of a mutator system in maize. *Mutation Research*, **51**, 21–28.

Siomi, M.C., Sato, K., Pezic, D. and Aravin, A.A. (2011) PIWI-interacting small RNAs: the vanguard of genome defence. *Nature Reviews Molecular Cell Biology*, **12**, 246–258.

Slotkin, R.K. and Martienssen, R. (2007) Transposable elements and the epigenetic regulation of the genome. *Nature Reviews Genetics*, **8**, 272–285.

Sun, F.J., Fleurdépine, S., Bousquet-Antonelli, C. *et al.* (2007) Common evolutionary trends for SINE RNA structures. *Trends in Genetics*, **23**, 26–33.

Tovkach, A., Ryan, P.R., Richardson, A.E. *et al.* (2013). Transposon-mediated alteration of *Tamate1b* expression in wheat confers constitutive citrate efflux from root apices. *Plant Physiology*, **161**, 880–892.

Walbot, V. and Rudenko, G.N. (2002) *Mudr/Mu* transposable elements of maize, in *Mobile DNA II* (eds N.L. Craig, R. Craigie, M. Gellert and A.M. Lambowitz), ASM Press, pp. 533–564.

Wenke, T., Döbel, T., Sörenson, T.R. *et al.* (2011) Targeted identification of short interspersed nuclear element families shows their widespread existence and extreme heterogeneity in plant genomes. *Plant Cell*, **23**, 3117–3128.

Wicker, T., Buchmann, J.P. and Keller, B. (2010) Patching gaps in plant genomes results in gene movement and erosion of collinearity. *Genome Research*, **20**, 1229–1237.

Wicker, T., Mayer, K.F.X., Gundlach, H. *et al.* (2011) Frequent gene movement and pseudogene evolution is common to the large and complex genomes of wheat, barley, and their relatives. *Plant Cell*, **23**, 1706–1718.

Xiao, H., Jiang, N., Schaffner, E. *et al.* (2008) A retrotransposon-mediated gene duplication underlies morphological variation in tomato fruit. *Science*, **319**, 1527–1530.

Yoshioka, Y., Matsumoto, S., Kojima, S. *et al.* (1993) Molecular characterization of a short interspersed repetitive element from tobacco that exhibits sequence homology to specific tRNAs. *Proceedings of the National Academy of Sciences USA*, **90**, 6562–6566.

Yuan, Y.-W. and Wessler, S.R. (2011) The catalytic domain of all eukaryotic cut-and-paste transposase superfamilies. *Proceedings of the National Academy of Sciences USA*, **108**, 7884–7889.

Chapter 4

Chromatin, centromeres and telomeres

4.1 Chromosomes are made up of chromatin, a complex of DNA and protein

In eukaryotic cells, DNA is packed into linear structures known as chromosomes, the structure of which is related to their function, which includes the need for preservation of genetic information, faithful replication of that information, and precise regulation of transcription. Key in the processes of replication and chromosome segregation during cell division (mitosis) are the chromosome centromeres, often but not always located near or at the center of the chromosome, and the telomeres at the end of the chromosomes. In this chapter we discuss the architecture of chromosomes, centromeres and telomeres.

In a cell, DNA always associates with a large number of proteins. DNA and associated proteins are known as **chromatin**. When viewed with a light microscope, the chromosomes can only be seen when they are condensed at meiosis or mitosis; during interphase they are elongate and too slender to see. Chromosomes are commonly viewed at mitosis, when the sister chromatids are still attached at the centromere; this produces the familiar X-shaped structure. After the sister chromatids separate, however, the chromosome consists of only one side of the X.

Our knowledge of chromosomal behavior comes primarily from studies using light microscopy. Actively dividing cells are placed in a solution containing a stain, and then cells are placed on a microscope slide and squashed under a cover slip in such a way that the chromosomes spread out. (This is done simply by hand; a skilled cytogeneticist learns exactly how hard to press with the thumb to get the desired result.) Using this simple technique, individual chromosomes can be tracked based on their shapes, and in particular the position of the centromere. A chromosome with the centromere at or near the middle is known as metacentric, whereas if the centromere is near or apparently at one end, the chromosome is acrocentric or telocentric (Figure 4.1a). While the position of the centromere is one reliable way to identify individual chromosomes, it is rarely sufficient to distinguish all the chromosomes of a plant. However, different chromosomes also react differentially to stains. In particular some parts of each chromosome stain more darkly than others, producing a characteristic set of bands (Figure 4.1b). The dark staining bands are known as **heterochromatin** and the lighter areas are known as **euchromatin**.

As the molecular structure of chromosomes has become better understood, a method known as **fluorescent in situ hybridization (FISH)** has been developed to label individual sequences, particularly repetitive sequences. An individual sequence is isolated and a fluorescent label is attached to it.

Plant Genes, Genomes and Genetics, First Edition. Erich Grotewold, Joseph Chappell and Elizabeth A. Kellogg.
© 2015 John Wiley & Sons, Ltd. Published 2015 by John Wiley & Sons, Ltd.
Companion Website: www.wiley.com/go/grotewold/plantgenes.

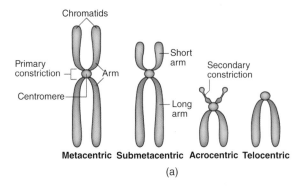

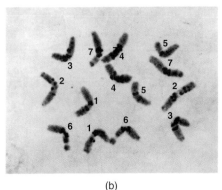

(b)

Figure 4.1 (a) Metacentric, submetacentric, acrocentric, and telocentric chromosomes. Reproduced from Jones *et al.*, 2012. (b) Chromosomes stained to show heterochromatin and euchromatin. From Barley Genetics Newsletter, Colorado State University, 1982, front cover; reproduced there from Singh and Tsuchiya, 1981

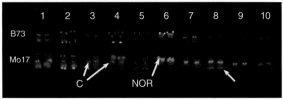

Figure 4.2 Karyotypes of two of the most commonly used inbred lines of maize, B73 and Mo17. Numbers across the top indicate the chromosome number. The chromosomes have been stained with a mixture of fluorescent probes that bind to distinct repetitive sequences. For example, most centromeres (C) are colored green except for chromosome 4 in which the centromeric sequence is different and is colored red. The nucleolar organizing region (NOR) encoding the ribosomal RNAs is on chromosome 6 and is also colored green, whereas the *5S* RNA array is in yellow on chromosome 2. Despite the overall similarity between the karyotypes, each inbred line is slightly different from the others reflecting differences in the number and location of particular repetitive sequences. Unlabeled arrow notes a set of repetitive sequences that are present in Mo17 but not B73. Adapted from Kato *et al.* (2004). Reproduced with permission. Copyright (2004) National Academy of Sciences, U.S.A

This labeled piece of DNA is then used exactly like a traditional cytogenetic stain. It pairs with the matching DNA on the chromosome, and the chromosome will fluoresce wherever the DNA binds. Using an appropriate combination of sequences and differently colored fluorescent labels, individual chromosomes can be identified (Figure 4.2). Chromosome visualization and FISH are at the core of many human genetic tests, for example to determine if a fetus carries any genetic abnormalities.

As described in Chapter 1, chromatin, the material that makes up chromosomes, consists of DNA and associated proteins. Each DNA molecule is long and is negatively charged because of the phosphate groups along the backbone. This charge is important for the association with positively charged proteins. In addition, the DNA needs to be folded carefully so that it fits inside the nucleus but remains accessible to proteins needed for replication and for transcription. Take, for example, the genome of *Escherichia coli*: the 4.6×10^6 bp genome has a length of about 1.5 mm, which needs to fit in a cylindrical cell with a diameter of 0.5 μm and a length of 1 μm. This involves at least a 1000-fold compression. The challenge is significantly more complex in eukaryotes, where longer genomes need to pack into a nucleus that occupies only a fraction of the cellular space.

The first level of DNA packaging is accomplished by wrapping the DNA around **histone** proteins. Histones are very conserved, small (~11–21 kDa) basic proteins. The histones contain many arginine and lysine residues, which are positively charged; the positive charge helps strengthen the interactions with DNA and also helps to neutralize the negative charge on the DNA. Chromatin contains five distinct sorts of histone proteins in the cell, numbered H1, H2A, H2B, H3, and H4, as well as several less common histone variants that serve more specialized functions (e.g., H2X, H2Z, CENH3). H2A, H2B, H3, and H4 bind together to form a tetrameric protein and two of these tetramers join to create an octamer that is roughly cubical (Figure 4.3). Slightly less than two turns of DNA (~146 bp) are wrapped around the histone octamer to form a structure known as a **nucleosome**. Histone H1 then associates with ~10 bp on the DNA

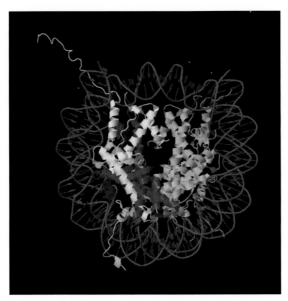

Figure 4.3 Structure of the nucleosome. Histones form the core of the nucleosome, and are arranged as two tetramers, each of which contains one molecule of H2A , H2B, H3, and H4; the α-helices of the histones are visible as colored ribbons and the N- and C-terminal tails as lines. DNA is shown in purple. View the original 3D image at http://www.rcsb.org/pdb/explore/jmol.do?structureId=1AOI&bionumber=1

at the entry/exit of the nucleosome (in a structure that is often known as a chromatosome), providing a "clip" that stabilizes the DNA wrapped on the nucleosome. The structure of the core nucleosome was established by X-ray crystallography in 1997 by Timothy J. Richmond and his colleagues (Swiss Federal Institute of Technology, Zurich).

On either side of a nucleosome is a stretch of 20–200 bp of DNA that is not bound to histones. The resulting structure is a string of nucleosomes separated by short uncoiled regions; under the transmission electron microscope this looks like beads on a string. The "beads on a string" structure is about 10 nm in diameter (Figure 4.4). This level of chromatin structure is often known as the 10 nm fiber or nucleosome array. A more detailed perspective on how histones and DNA interact is provided in Chapter 12.

The position of nucleosomes in the genome is not random, and is controlled both by the primary sequence of the DNA and by the structure of the histones. Nucleosome positioning can be studied by cutting isolated chromatin with a nuclease (such as micrococcal nuclease [MNase], an endo-exonuclease

obtained from the bacterium *Staphylococcus aureus*), which cuts between nucleosomes but cannot touch the DNA that is coiled around histones. Cutting chromatin with MNase thus produces isolated nucleosomes. The DNA can then be separated from the histones and the DNA sequence determined. By assessing thousands of such nucleosome "footprints" it is then possible to determine the pattern of nucleosome binding.

In all organisms studied so far, nucleosomes appear to be absent for several hundred base pairs upstream of the transcription start site of protein-coding genes, creating a short nucleosome-free region. This seems to be necessary to give the transcription initiation machinery access to the DNA. There is another nucleosome-free region at the 3′ end of the gene. In addition, all organisms have a nucleosome just downstream of the transcription start site. This downstream nucleosome, called the +1 nucleosome (the +1 refers to nucleosome number, not base pairs), is usually modified and may be needed to position the transcription initiation complex (Jiang and Pugh, 2009). In animals and fungi, there is also a nucleosome at −300 to −150 bp upstream of the transcription start site (Jiang and Pugh, 2009). Surprisingly, this upstream nucleosome appears not to be present in plants (Fincher *et al.*, 2013). It is not clear why the kingdoms differ in this way. Different classes of transposable elements also often have their own particular pattern of nucleosome positioning, with nucleosomes often in the repetitive elements at the ends of the transposon (Fincher *et al.*, 2013).

Nucleosomes get in the way of the protein machinery that acts on DNA; they create a mechanical problem for gene transcription and DNA replication. During transcription, nucleosomes may slide so that their position shifts by a few base pairs, or they may be modified so that they bind DNA less tightly, or they may be evicted entirely from the DNA strand. During DNA replication, nucleosomes must also be removed and then replaced afterwards. To control access to the DNA, histones are frequently modified by covalently attaching functional groups that strengthen or weaken the DNA–histone interaction. Histone modifications are discussed more extensively in Chapter 12.

While the 10 nm fiber is shorter than unbound DNA, it is still too long to fit into the nucleus. To shorten the structure even more, the DNA is looped and folded into an even more compact structure consisting of coils of six nucleosomes held together by the histone H1 (also known as the linker histone). The exact structure of the 30 nm fiber is not known, with

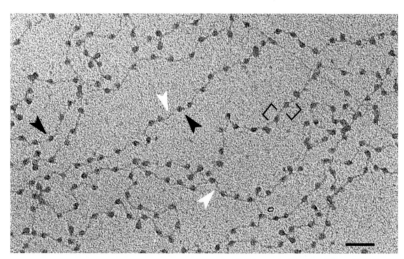

Figure 4.4 Electron micrograph showing the 10 nm fiber, consisting of DNA wound around nucleosomes ("beads on a string"). Black brackets and black arrowheads indicate individual nucleosome core particles; white arrowheads point to linker DNA. http://commons.wikimedia.org/wiki/File:Chromatin_nucleofilaments_%28detail%29.png. By Chromatin_nucleofilaments.png: Chris Woodcock derivative work: Gouttegd (Chromatin-nucleofilaments.png) [CC-BY-SA-3.0 (http://creativecommons.org/licenses/by-sa/3.0)], via Wikimedia Commons

zigzag and solenoid models having been proposed (Figure 4.5). The coiled structure of many solenoids held together is 30 nm wide and so is called the 30 nm fiber. The 30 nm fiber is the configuration of DNA at interphase, when it is stretched too thin to be visible under a light microscope.

Although chromosomes are often drawn as though they are floating free inside the nucleus, in fact the DNA is attached to the nuclear matrix, which is a network of fibers that connects with the nuclear envelope. Proteins bind to the matrix and to specific matrix attachment regions (MARs) in the DNA. The MARs can be separated by tens to hundreds of kilobases, so that the DNA forms long loops extending from the matrix (Figure 4.6). It is unclear exactly how the attachment proteins recognize a MAR; MARs have no specific shared DNA sequence, although they do have a few general characteristics (e.g., they have an excess of A-T base pairs, tend to contain repetitive sequences, and are in regions where the DNA is bent or kinked) (Ottaviani *et al.*, 2008). MARs appear to isolate the loops of DNA from each other, so that genes within one loop may be regulated differently from genes in an adjacent loop, and the MAR itself may serve as a dock for transcriptional regulators.

The isolating capacity of MARs might be useful in biotechnology by solving the problem of random insertion of transgenes (Singer *et al.*, 2011). In genetic engineering in general, a foreign gene is usually introduced into a plant along with a strong promoter that causes a high level of gene transcription. The gene plus promoter (the construct) inserts at a random spot in the genome. Depending on where it lands, the introduced promoter may affect expression of nearby genes, or the abundance of mRNA may lead to gene silencing. Such positional effects are discussed in Chapter 11. It was hoped that perhaps by placing MARs on either side of the construct, the transgene could be isolated from the rest of the genome and the detrimental effects of random insertion could be mitigated. However, the results of these experiments have been mixed, with the insulating effect of the MAR depending on the specific MAR. This suggests that MARs could be as heterogenous in function as they are in DNA sequence.

4.2 Telomeres make up the ends of chromosomes

The very end of each eukaryotic chromosome does not contain genes, but rather consists of a structure known as a telomere (Figure 4.7a). The telomere contains repetitive arrays of short sequences, which in most plants are 7 bp long, TTTAGGG, repeated

Nucleosome structure

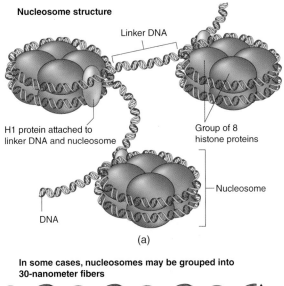

(a)

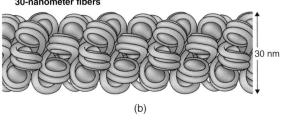

(b)

Figure 4.5   Structure of the 10 nm and 30 nm fiber. (a) The 10 nm fiber, showing DNA wrapped about the histone ocatmer; histone H1 is attached on the outside of each nucleosome. (b) The solenoid, or 30 nm fiber. Coils of six nucleosomes are stabilized by histone H1. Jones *et al.* (2012). Reproduced with permission of John Wiley & Sons, Ltd.

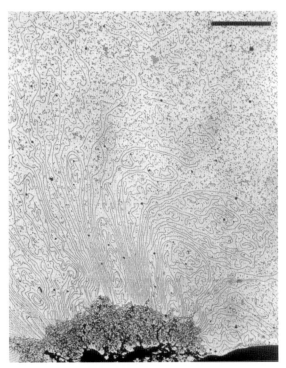

Figure 4.6   Electron micrograph showing loops of DNA attached to the periphery of the nucleus. The attachment is created by interactions between matrix attachment regions (MARs) of the DNA with nuclear scaffold proteins. Paulson and Laemmli (1997). Reproduced with permission of Elsevier

many hundreds of times. The number of repeats varies among cell types and among organs. In some plant groups, such as members of Asparagales, the telomere repeat sequences are only 6 bp, TTAGGG, similar to but independently derived from those in vertebrates.

The DNA molecules making up the chromosomes are not the same length. The 3′ ends are longer than the 5′ ends, creating an overhanging sequence in which almost half the nucleotides are G. This G overhang is folded back and tucked into the double-stranded DNA so that the raw 3′ end is protected from nucleases (Figure 4.7b). The telomere is then surrounded by a set of proteins, forming a complex known as shelterin. Although most studies of the shelterin complex have been done in animals and fungi, a few of the proteins have been identified in plants (Watson and Riha, 2010). These include PROTECTION OF TELOMERES1 protein (POT1), for which there are three

copies in Arabidopsis (POT1A, POT1B, and POT1C). POT1C may bind the telomere and effectively mark it as a chromosomal end, thus preventing modification or degradation by enzymes that might treat it as damaged DNA, whereas POT1A appears to bind to and function with telomerase (Watson and Riha, 2010). Proteins similar to mammalian TRF proteins are also found in plants, although it is not yet clear whether they function the same way they do in mammals and fungi. G-strand-specific single-stranded telomere binding proteins (GTBPs) bind to the G overhang, regulate its length, and control the process of folding into upstream sequences (Lee and Kim, 2013).

Understanding the important role of telomeres requires a review of DNA replication, which occurs during S phase of the cell cycle (Figure 4.8). Replication begins at specific points along the DNA molecule, known as origins of replication (ori) (Figure 4.9). DNA helicase binds and unwinds the double helix, which is then held in an open configuration by single-stranded binding proteins; the point at which the strands separate is known as the replication fork. Formation

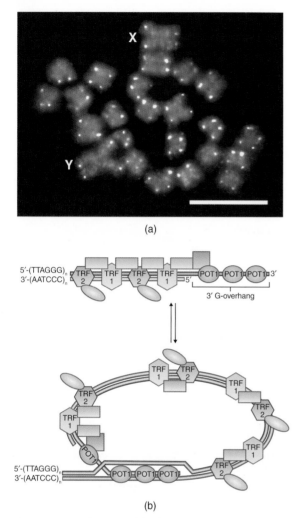

(a)

(b)

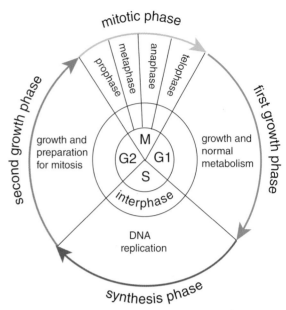

Figure 4.7 Telomeres. (a) Chromosomes of *Melandrium album* (= *Silene latifolia*) stained with a fluorescent probe designed to bind to the telomeric repeat TTTAGGG. *M. album* is one of the few plants with sex chromosomes; the X and the Y chromosome are labeled. Riha *et al.* (1998). Copyright 1998 American Society of Plant Biologists. Used with permission. (b) Model of a telomere, showing the G overhang and the T-loop (foldback loop). Adapted from Lamarche *et al.* (2010). Reproduced with permission of Elsevier

Figure 4.8 The cell cycle. Cell growth occurs primarily during G1 or Gap1 phase. During S phase, DNA is replicated, and the cell synthesizes the machinery necessary for division during G2 or Gap 2 phase. Division of the chromosomes occurs during mitosis, which may or may not be followed by cell division, or cytokinesis

To create a double-stranded structure for the polymerase to bind, the enzyme primase creates a small RNA primer that is complementary to the DNA. DNA Polymerase III then can bind and begin inserting nucleotides complementary to each strand in a 5′ to 3′ direction.

DNA replication runs into a problem because the strands of DNA are antiparallel. This means that DNA synthesis on one strand, known as the **leading strand**, will proceed toward the replication fork. As helicase continues to unwind the DNA, synthesis will continue smoothly and continuously all the way to the end of the chromosome. However, DNA replication on the other strand, known as the **lagging strand**, proceeds away from the replication fork. This means that DNA Polymerase III is moving in one direction while single-stranded DNA is being exposed behind it. DNA Polymerase III must therefore bind, produce a DNA strand, dissociate from the DNA and then bind again closer to the fork. Primase must make a new primer each time this happens. The result is that DNA is synthesized in pieces, known as Okazaki fragments, on the lagging strand. After the Okazaki fragments are built, DNA Polymerase I binds to

of the replication fork creates strain on the helix (think of what happens when you try to untwist a telephone cord), which is relieved by topoisomerase, an enzyme that cuts the DNA strands and rejoins them so that they are untwisted. Once the replication fork is created, DNA Polymerase III must begin synthesis of complementary strands of DNA. However, DNA Polymerase III is unable to bind single-stranded DNA.

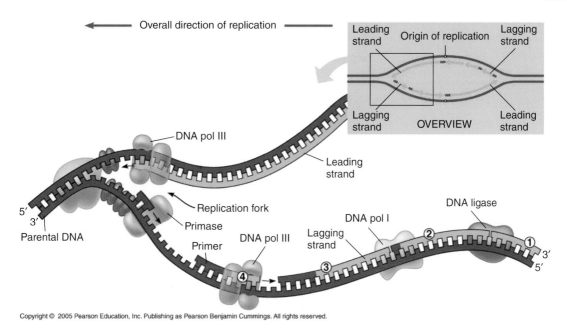

Figure 4.9 DNA replication. Replication begins at one of many origins of replication on the chromosome and proceeds in two directions at once (inset). To copy the two strands of DNA, it is unwound by a helicase (light green) and held open by single-stranded DNA binding proteins (gray). Primase (pink) creates an RNA primer (red), permitting DNA polymerase III (orange) to bind and synthesize the complementary strand. DNA polymerase I (lighter orange) removes the primer and replaces it with DNA, and DNA ligase (dark green) ligates the ends of the Okazaki fragments on the lagging strand. From Reece *et al.,* 2011

the new double-stranded DNA just upstream of the RNA primers, removes the RNA, and replaces it with DNA. Finally, a ligase stitches the fragments together into a single continuous molecule. Nucleosomes are reassembled on the new daughter strands of DNA, with the H3/H4 components being brought in first, followed by H2A/H2B (Annunziato, 2005).

The very end of the lagging strand cannot be copied. (This is the 3′ end of the lagging strand, which corresponds to the 5′ end of the newly synthesized copy.) Primase will put the RNA primer at the terminus of the strand, DNA Polymerase III will synthesize DNA away from the end, and this last fragment will be joined to the preceding fragment as usual. However, there is no upstream DNA for DNA Polymerase III to bind to and the DNA complementary to the primer site cannot be filled in (Figure 4.10).

Because of the processes described above, eukaryotic chromosomes get shorter every time they are replicated. However, the long repetitive telomere sequences mitigate this problem because shortening of the chromosome results simply in loss of a few repeats, usually totaling about 50–100 bp. Genes are not lost. Animal cells that have divided many times will eventually lose too much telomere sequence and further cell division is prohibited; this has been particularly well documented in human-derived cells. In plant cells, however, cessation of division does not appear to occur so abruptly, but is preceded by increasing chromosomal abnormalities. For continual division, a cell requires that telomeres be replaced. This is accomplished by the enzyme telomerase, an enzyme that was discovered in the early 1980s by Elizabeth Blackburn and Carol Greider; they were awarded the Nobel Prize in 2009 for this discovery, along with another telomere researcher, Jack Szostak.

Telomerase is an enzyme with both protein and RNA components, so is known as a ribonucleoprotein. It combines a reverse transcriptase (telomerase reverse transcriptase, or TERT) with an RNA molecule that includes a template complementary to the telomeric repeat (5′-CUAAACCUUA-3′, in Arabidopsis). TERT binds to the overhanging end of the DNA and adds new repeats to replace those that were lost during replication. The RNA molecule is much longer than the repeat sequence, and also includes sequences

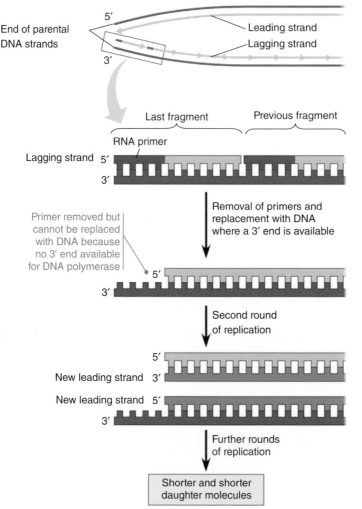

Figure 4.10 Shortening of telomeres during replication. From Reece *et al.*, 2011

that bind to other proteins that locate the telomerase complex to the chromosome end and position it properly. Although we might expect that the telomerase RNA would be highly conserved among plants, it appears to change very rapidly over evolutionary time; this raises the possibility that what we know about telomere maintenance in Arabidopsis may have to be modified when we learn more about other species (Belistein *et al.*, 2012).

In plants, telomerase appears to be active in many cell types but primarily functions in meristematic and germ cells. This is different from the situation in animals, in which telomerase is only active in stem cells. In plants, it is unclear whether telomeres shorten during development, with conflicting reports from different plant species. When plant cells are placed into culture, dedifferentiated, and then re-differentiated, telomeres increase in length.

When telomerase is knocked out in Arabidopsis thaliana, the plants survive and reproduce for a couple of generations, even though their telomeres lose about 500 bp each generation (Riha *et al.*, 2001). However, after about 5 generations the plants begin to exhibit developmental defects. Chromosome ends lose their caps and chromosomes fuse in ways that led to abnormal cell division. After 10 generations, the plants are unable to produce normal organs and appear to be permanently vegetative.

4.3   The chromosome middles – centromeres

The centromere is the part of the chromosome that forms the kinetochore, the structure to which microtubules attach at mitosis and meiosis (Figure 4.11). It is visible in mitotic or meiotic chromosomes as the familiar constriction in the middle of the X-like pattern of sister chromatids. In most organisms investigated, centromeres are found in regions of heterochromatin. Heterochromatic regions replicate later than euchromatic regions, contain relatively few transcribed genes, generally do not recombine at meiosis, and are associated with specific proteins and modifications of histones. Heterochromatic regions are also comparatively stiff, which may be necessary to withstand the forces of

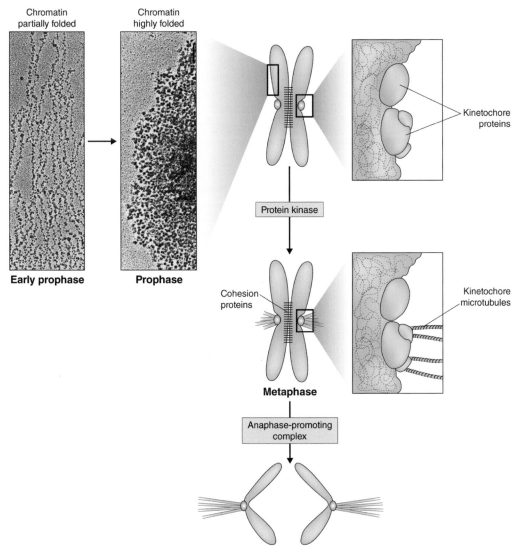

Figure 4.11 Structure and function of the kinetochore. As the cell prepares to divide, the chromosomes must condense so that they can be separated without becoming entangled. This condensation occurs during prophase. Kinetochore proteins assemble at the centromere but the chromatids do not separate because they are held together by cohesion proteins. During metaphase microtubules attach to the kinetochore proteins. The anaphase promoting complex then breaks down the cohesion proteins and the chromatids are pulled to opposite poles. Jones *et al.* (2012). Reproduced with permission of John Wiley & Sons, Ltd.

kinetochores pulling the chromatids apart during cell division. In addition, the proteins that hold the centromeres together during meiosis I are attached to heterochromatin.

The centromere paradox

Comparative studies of plants, animals, and fungi have shown that the functions of centromeres are highly conserved; in contrast, the DNA sequence is variable and the sequence of a number of the associated proteins is highly variable. This is known as the "centromere paradox" (Henikoff *et al.*, 2001): how is it possible to have an essential and presumably strongly selected structure (the centromere) made up of DNA and proteins that differ widely among species?

One answer comes from the pattern of meiosis in the ovule (female parent). Meiosis in the ovule (in the megasporocyte), as in the anther (the microsporocyte), produces four haploid cells. In the pollen, each of these cells will go on to form a pollen grain. However, in most angiosperm ovules, three of the four haploid cells degenerate; only one proceeds to form the megagametophyte. Any bias in distribution of cellular components to the single surviving haploid cell will be passed on to the megagametophyte and thence to the seed. Furthermore, the bias will increase each generation. This general phenomenon is known as meiotic drive because it drives accumulation of particular sequences rapidly in one direction.

In the context of centromeres, any component of the kinetochore that is more efficient at attracting microtubule binding will tend to accumulate by meiotic drive (Figure 4.12). Because of the drive mechanism, each generation will accumulate more of the "strong" centromere sequences, causing the novel sequence to spread rapidly through a population. This has been documented in monkey flowers

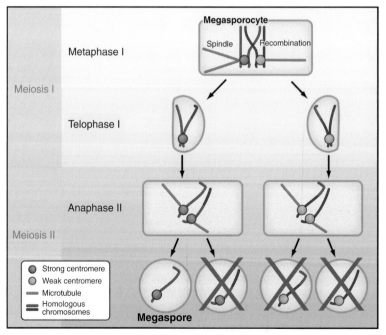

Figure 4.12 Model for meiotic drive; a hypothesis. During meiosis I, homologous chromosomes pair, recombine, and microtubules attach to the kinetochore. If one centromere has undergone a mutation such that the attachment to the microtubules is more effective, it will be preferentially pulled to the position in the megasporocyte that will form the surviving gamete. In most plants this is the end away from the micropyle. The other megaspores degenerate so only the "strong" centromere is passed on. Malik and Henikoff (2009). Reproduced with permission of Elsevier

(*Mimulus guttatus*), where a repeat associated with a centromere is transmitted at high frequency to off-spring of a cross (Fishman and Saunders, 2008). In this particular case, the centromeric repeat is prevented from spreading rapidly through the population because plants homozygous for the repeat have lower pollen fertility than heterozygotes or plants lacking that particular repeat sequence. A similar mechanism leads to accumulation of B chromosomes.

A second answer to the centromere paradox comes from data on gene conversion, a process by which sequences from one centromere of a chromosome pair are copied on the other member of the pair. While this process is both rare and hard to detect, it has been documented in maize (Shi *et al.*, 2010). It thus appears that both meiotic drive and gene conversion can lead to rapid evolution of novel centromere repeats.

A major evolutionary transition occurred between the circular chromosomes of bacteria and the linear ones of eukaryotes. Whereas segregation of bacterial chromosomes is determined by the sequence of the DNA (i.e., genetic determination), in most eukaryotic cells, the determination of the point of segregation is independent of sequence (i.e., epigenetic determination) (Malik and Henikoff, 2009). The exception is budding yeast, *Saccharomyces cerevisiae*, and its close relatives, which have reverted to a genetically determined centromere that lacks many components of the standard eukaryotic centromeric apparatus.

4.3.1 Centromere nucleosomes contain a centromere-specific histone

The major proteins in the centromere, as everywhere in the chromosome, are **histones**. In a subset of centromeric nucleosomes, however, histone3 (H3) is replaced by the centromere-specific histone, CENH3. In eukaryotes, CENH3 is the marker for the centromere; wherever CENH3 is incorporated into the nucleosome, a kinetochore can form. Curiously, DNA appears to wrap around CENH3 nucleosomes in a right-handed coil, rather than the left-handed coil in non-centromeric nucleosomes; the direction of the coil affects the configuration of the DNA. Blocks of CENH3 nucleosomes alternate with blocks of H3 histones; the latter are transcriptionally active. This suggests a model in which the blocks of nucleosomes may coil such that the CENH3 regions are on the external face of the chromatids at mitosis and meiosis; in this position they can interact with other kinetochore proteins (Wu *et al.*, 2012). In contrast, the H3 nucleosomes are in an internal position on the chromatids (Figure 4.13). Whether CENH3 nucleosomes include centromere-specific variants of H2A, H2B and H4 is unclear, although some data from yeast suggest that this may be the case (Henikoff and Furuyama, 2012).

Both H3 and CENH3 have an N-terminal "tail", which is required for appropriate chromosome segregation during mitosis (Ravi *et al.*, 2010). Whereas the sequence of the histone folds of CENH3 is highly conserved among eukaryotes, the N-terminal tail is strikingly different (Figure 4.14). Even among flowering plants, the sequence of CENH3 is nearly species-specific; CENH3 from Arabidopsis arenosa can complement a *cenh3* mutant of *A. thaliana*, but CENH3 from *Brassica rapa* (in the same family as Arabidopsis) cannot (Ravi *et al.*, 2010). The *Brassica* CENH3 is localized appropriately in the centromere, but does not allow mitosis to proceed properly.

Differences in the sequence of CENH3 control chromosome segregation in mitosis and meiosis, presumably by affecting formation of the kinetochore complex (Ravi and Chan, 2010). When an Arabidopsis plant with a null mutant CENH3 is crossed to a wild type plant, fertilization occurs normally but in the first few mitotic divisions in the zygote, all centromeres containing nucleosomes with the mutant CENH3 are eliminated (Figure 4.15). This means that the plants have only one set of chromosomes, that is, they are haploid. Unlike animals, in which haploidy outside of the gametes is most likely lethal, plants can tolerate haploidy; the haploid plants are smaller than their diploid parents, but are otherwise healthy. Meiosis is impossible, however, so the plants are sterile. Occasionally the haploid plants produce normal gametes because the single set of chromosomes all segregate to one daughter cell during meiosis. Self-fertilization

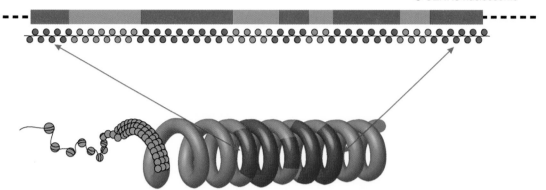

Figure 4.13 Hypothesized three-dimensional structure of a rice centromere. Blocks of chromatin with normal (H3, in blue) nucleosomes alternate with blocks with centromeric (CENH3, in red) nucleosomes. When the chromatin coils the CENH3 regions line up on one side where they can bind the kinetochore, whereas the H3 regions are on the interior. From Wu *et al.* (2012). Copyright 2012 American Society of Plant Biologists. Used with permission

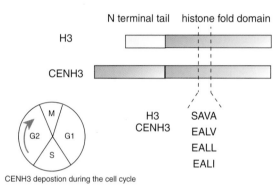

CENH3 deposition during the cell cycle

Figure 4.14 Comparison of histones H3 and CENH3 in plants. All histones have a conserved histone fold domain (green). In histone H3 the N-terminal tail (cream) is also conserved among eukaryotes, while in CENH3 histones, the N-terminal tail is different in length and sequence (orange). Four residues in the histone fold domain distinguish H3 from CENH3; in the latter the fourth residue can vary. CENH3 histones are deposited during G2 phase of the cell cycle, while normal histones H3 are always placed during S phase. Adapted from Malik and Henikoff (2009)

then will produce a fully homozygous plant, a doubled haploid (Ravi and Chan, 2010).

The specificity of the CENH3 sequence has practical applications in plant breeding. Many crop plants (e.g., wheat) are polyploid; because each gene in a polyploid has more than two copies, conventional breeding and genetic analysis are slow and difficult. However, if the polyploid genome can be returned to diploid in one generation, genetics becomes simpler. This process has

been used in some cereal crops, but has depended on identifying specific haploid-producing parental lines. It now appears that modification of CENH3 sequences may permit production of haploid lines.

A few plants have diffuse centromeres, in which kinetochores form along the entire length of the chromosome rather than just in the middle; chromosomes with diffuse centromeres are known as **holocentric.** (Diffuse centromeres also occur in *Caenorhabditis elegans*.) The best-documented group with diffuse centromeres is in the Cyperaceae/Juncaceae clade. Structure and function of diffuse centromeres have been studied in *Luzula nivea* (Juncaceae) (Figure 4.16). In these plants, CENH3 marks a central "starting point;" as chromosome condensation proceeds, CENH3 appears at more places along the chromosome (Nagaki *et al.*, 2005).

4.3.2 Centromeric DNA generally contains repeated sequences and few genes

The DNA sequence in *most* centromeres in *most* eukaryotes is dominated by a set of repeat units that are about 150–180 bp long; these repeats appear only in the centromeres and nowhere else in the genome. The repeat sequence is often the same for all centromeres in the cell, pointing to some sort of mechanism for propagating a single sequence or for correcting mutant

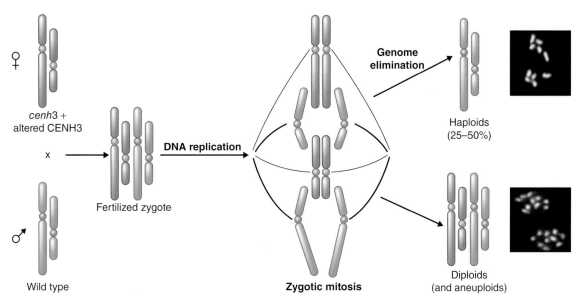

Figure 4.15 Production of haploids by chromosome elimination. Plants can be engineered to produce an altered form of CENH3 (purple chromosomes). These are then crossed to wild type plants (green). The chromosomes with the altered CENH3 form less effective kinetochores and thus fail to segregate efficiently during anaphase. There are many possible outcomes but about 25–50% of the time, the resulting plants lose the purple (altered) chromosomes entirely and develop as haploid. Adapted from Chan (2010). Reproduced with permission of Elsevier

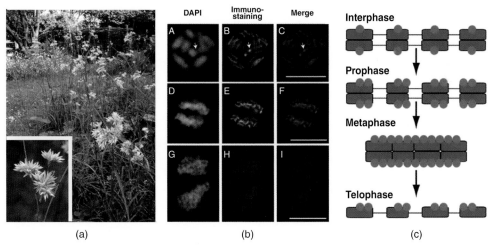

Figure 4.16 (a) *Luzula nivea*. (b) *Luzula nivea* chromosomes stained with DAPI, which binds to all DNA (left column), and with antibodies to CENH3 (middle column); images overlaid on each other (right column). A–C, metaphase; D–F, anaphase, G–I, telophase. Arrow indicates a chromosome seen end-on from the telomere. (c) Model for recruitment of CENH3 to the chromosome at various stages of cell division, forming a holocentric chromosome; heterochromatic regions shown as blue boxes, CENH3 as red circles. (a) http://upload.wikimedia.org/wikipedia/commons/9/9b/Luzula-nivea -total.JPG. By Sten Porse (Self-published work by Sten Porse) [GFDL (http://www.gnu.org/copyleft/fdl.html) or CC-BY-SA-3.0 (http://creativecommons.org/licenses/by-sa/3.0/)], via Wikimedia Commons Inset: http://www.florealpes .com/fiche_luzuleneige.php. Reproduced with permission of Franck Le Driant. (b) Nagaki *et al.* (2005). Copyright 2005 American Society of Plant Biologists. Used with permission. (c) Nagaki *et al.* (2005). Copyright 2005 American Society of Plant Biologists. Used with permission

sequences to match the others. These repeat sequences are often still called **satellite DNA**, a historical term based on early experiments to characterize DNA by creating a gradient that separated all the DNA from an organism by density. Most of the DNA of the cell would gravitate to a single spot in the gradient, whereas highly repetitive sequences would form a separate, or "satellite" band.

In addition to the centromeric repeats, centromeres often have characteristic transposable elements (see Chapter 3). Retrotransposons and degenerate viruses tend to accumulate in regions of low recombination. Because recombination is infrequent in the region around the centromere, repetitive elements, including transposons, tend to accumulate in centromeric and pericentromeric regions.

As in the centromeric histones, the DNA sequence of the centromeric repeats is species-specific. This is surprising because the function of centromeres is assumed to be under very strong natural selection. This raises the question of how repeats evolve and how centromeres arise. (See Box.) In most plant species that have been studied to date, the repetitive sequences are a mix of centromere-specific sequences plus many copies of centromere-specific retrotransposons. Like all repetitive sequences, the centromeric arrays expand and contract because of recombination between repeat copies. In addition, large duplications and inversions can occur among the arrays. Identification of sequences in the functional centromere has taken advantage of the technique known as chromatin immunoprecipitation (ChIP), described in Chapter 9. By precipitating and sequencing pieces of DNA bound directly to the histone CENH3, it is possible to know which DNA sequences are part of the centromere (Nagaki et al., 2004).

Because heterochromatin is often associated with lack of transcription, it was long assumed that the centromeric region would contain no genes, and would be transcriptionally inactive. However, the centromere of chromosome 8 in rice harbors 16 transcribed genes (Nagaki et al., 2004). Surprisingly, seven of these are also found in other species of *Oryza*. If the genes were generally non-functional in one or more species, then comparisons between the species would show accumulation of many mutations. However, there are many fewer mutations in these genes than expected and none obviously disrupts the protein, which thus indicates that they are functional (Fan et al., 2011). The core centromere of chromosome 8 has about

the same density of genes and retrotransposons, and about the same histone modifications as the flanking regions, suggesting that the DNA sequence of the centromere has nothing particularly distinctive about it (Yan et al., 2005). In contrast to rice, none of the five centromeres of Arabidopsis appears to contain transcribed genes.

For many years it was assumed that long stretches of repetitive sequences were necessary for centromere function. However, new centromeres (called **neocentromeres**) can arise in non-centromeric parts of the genome. Neocentromeres are fully functional but generally do not contain repeat sequences. This raises a question: how do neocentromeres form? Some of the centromeres in chicken and in potato lack repeats and yet appear to function well (Gong et al., 2013). Since centromeres have been sequenced in relatively few organisms, it seems likely that these two domesticated species are not the only ones with non-repetitive centromeres. The centromere on chromosome 8 in rice, with its relatively short repetitive portion and included genes, is interpreted as being intermediate between a neocentromere and a fully developed repetitive centromere.

The variation in centromeric structure suggests a model for centromere origin and evolution (Nagaki et al., 2004; Malik and Henikoff, 2009; Gong et al., 2013). Short neocentromeres could appear in otherwise normal stretches of DNA perhaps by incorporation of CENH3 into a set of nucleosomes. Presumably such neocentromeres would be eliminated

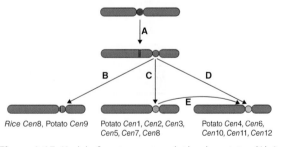

Rice Cen8, Potato Cen9   Potato Cen1, Cen2, Cen3,   Potato Cen4, Cen6,
        Cen5, Cen7, Cen8   Cen10, Cen11, Cen12

Figure 4.17 Model of centromere evolution in potato. (A) A neocentromere is activated in a novel site. (B–D), The new centromere acquires repeat sequences. (B) Repeat sequence derived from other centromeres, as in rice Cen8 and potato Cen9. (C) Repeat sequence expands and persists for several million years, accumulating mutations and new repeat sequences. (D and E) New repeat may invade and spread, creating a repeat-based centromere with a different repeat sequence. Gong et al. (2012). Copyright 2012 American Society of Plant Biologists. Used with permission

from the population rapidly (they would lead to sterile plants), but rarely they might persist. With the suppression of recombination at the new centromere, blocks of repetitive DNA would accumulate, as would retrotransposons. With the continued suppression of recombination plus the effect of meiotic drive, the repetitive region would expand to produce the massive megabase-sized centromeres found in many plants. This scenario is currently speculative, but is supported by observations of centromeric sequences representing all stages of the evolutionary process (Figure 4.17).

4.4   Summary

The architecture of the plant genome is marked by telomeres at the ends of chromosomes and centromeres in internal regions. In between, regions are demarcated by MARs. These major landmarks of the genome are only partially determined by DNA sequence, with other non-genetic signals also involved. For example, telomeres are composed of many copies of very short repeated sequences (usually about 7 bp) but the specific sequence does not appear to be especially important; it must only be G-rich. MARs have no sequences in common and it is unclear how they are recognized by nuclear attachment proteins. Centromeres are usually in regions of heterochromatin and generally contain repeated sequences, but the repetitive sequences are not necessary for function and do not appear in all centromeres in all species. The mark of a centromere is instead the presence of a novel histone in the nucleosome, CENH3.

4.5   Problems

4.1 What would happen to a chromosome if it lacked telomeres?

4.2 Are telomeres and centromeres needed in circular chromosomes? Why or why not?

4.3 Why do nucleosomes have to be removed before DNA replication or transcription? How do they know where to reassemble?

4.4 What structures are required for centromere function? What aspects of the DNA are common but not required?

4.5 An artificial chromosome would be a useful tool for plant breeding because it could be introduced and removed by crossing. If you were going to build an artificial chromosome:

(a) What components would you need to include to insure faithful transmission of the chromosome from one generation to the next?

(b) What challenges might you run in to in constructing those components?

(c) To what extent could you rely on components from other species, and how would you address any problems that might arise? For example, what data, sequences or proteins from Arabidopsis chromosome structure would be helpful in creating an artificial chromosome in maize?

Further reading

Jiang, C. and Pugh, B.F. (2009) Nucleosome positioning and gene regulation: advances through genomics. *Nature Reviews Genetics*, **10**, 161–172.

Jiang, J., Birchler, J.A., Parrott, W.A. and Dawe, R.K. (2003) A molecular view of plant centromeres. *Trends in Plant Science*, **8**, 570–575.

References

Annunziato, A.T. (2005) Split decision: what happens to nucleosomes during DNA replication? *Journal of Biological Chemistry*, **280**, 12065–12068.

Belistein, M.A., Brinegar, A.E. and Shippen, D.E. (2012) Evolution of the Arabidopsis telomerase RNA. *Frontiers in Genetics*, **3**, Article 188.

Chan, S.W.L. (2010). Chromosome engineering: power tools for plant genetics. *Trends in Biotechnology*, **28**, 605–610.

Fan, C., Walling, J.G., Zhang, J. *et al.* (2011) Conservation and purifying selection of transcribed genes located in a rice centromere. *Plant Cell*, **23**, 2821–2830.

Fincher, J.A., Vera, D.L., Hughes, D.D. *et al.* (2013) Genome-wide prediction of nucleosome occupancy in maize reveals plant chromatin structural features at genes and other elements at multiple scales. *Plant Physiology*, **162**, 1127–1141.

Fishman, L. and Saunders, A. (2008) Centromere-associated female meiotic drive entails male fitness costs in monkeyflowers. *Science*, **322**, 1559–1562.

Gong, Z., Wu, Y., Koblizková, A. *et al.* (2013) Repeatless and repeat-based centromeres in potato: implications for centromere evolution. *Plant Cell*, **24**, 3559–3574.

Henikoff, S., Ahmad, K. and Malik, H.S. (2001) The centromere paradox: stable inheritance with rapidly evolving DNA. *Science*, **293**, 1098–1102.

Henikoff, S. and Furuyama, T. (2012) The unconventional structure of centromeric nucleosomes. *Chromosoma*, **121**, 341–352.

Jiang, C. and Pugh, B.F. (2009) Nucleosome positioning and gene regulation: advances through genomics. *Nature Reviews Genetics*, **10**, 161–172.

Jones, R., Ougham, H., Thomas, H. and Waaland, S. (2012) *The Molecular Life of Plants*, Wiley-Blackwell.

Kato, A., Lamb, J.C. and Birchler, J.A. (2004) Chromosome painting using repetitive DNA sequences as probes for somatic chromosome identification in maize. *Proceedings of the National Academy of Sciences USA*, **101**, 13554–13559.

Lamarche, B.J., Orazio, N.I. and Weitzman, M.D. (2010) The MRN complex in double-strand break repair and telomere maintenance. *FEBS Letters*, **584**, 3682–3695.

Lee, Y.W. and Kim, W.T. (2013). Telomerase-dependent 3′ G-strand overhang maintenance facilitates GTBP1-mediated telomere protection from misplaced homologous recombination. *Plant Cell*, **25**, 1329–1342.

Malik, H.S. and Henikoff, S. (2009) Major evolutionary transitions in centromere complexity. *Cell*, **138**, 1067–1082.

Nagaki, K., Cheng, Z., Ouyang, S. *et al.* (2004). Sequencing of a rice centromere uncovers active genes. *Nature Genetics*, **36**, 138–145.

Nagaki, K., Kashihara, K. and Murata, M. (2005) Visualization of diffuse centromeres with centromere-specific histone H3 in the holocentric plant *Luzula nivea*. *Plant Cell*, **17**, 1886–1893.

Ottaviani, D., Lever, E., Takousis, P. and Sheer, D. (2008). *Anchoring the genome. Genome Biology*, **9**, 201.

Paulson, J.R. and Laemmli, U.K. (1977) The structure of histone-depleted metaphase chromosomes. *Cell*, **12**, 817–828.

Ravi, M. and Chan, S.W.L. (2010) Haploid plants produced by centromere-mediated genome elimination. *Nature*, **464**, 615–619.

Ravi, M., Kwong, P.N., Menorca, R.M.G. *et al.* (2010). The rapidly evolving centromere-specific histone has stringent functional requirements in *Arabidopsis thaliana*. *Genetics*, **186**, 461–471.

Reece, J.B., Urry, L.A., Cain, M.L. *et al.* (2011). *Campbell Biology*, Benjamin Cummings.

Riha, K., Fajkus, J., Siroky, J. and Vyskot, B. (1998) Developmental control of telomere length and telomerase activity in plants. *Plant Cell*, **10**, 1691–1698.

Riha, K., McKnight, T.D., Griffing, L.R. and Shippen, D.E. (2001) Living with genome instability: plant responses to telomere dysfunction. *Science*, **291**, 1797–1800.

Shi, J., Wolf, S.E., Burke, J.M. *et al.* (2010) Widespread gene conversion in centromere cores. *PLoS Biology*, **8**, e1000327.

Singer, S.D., Liu, Z. and Cox, K.D. (2011) Minimizing the unpredictability of transgene expression in plants: the role of genetic insulators. *Plant Cell Reporter*, **31**, 13–25.

Singh, R.J. and Tsuchiya, T. (1981) Identification and designation of barley chromosomes by Giemsa banding technique: a reconsideration. *Zeitschrift für Pflanzenzüchtung*, **86**, 336–340.

Watson, J.M. and Riha, K. (2010) Comparative biology of telomeres: where plants stand. *FEBS Letters*, **584**, 3752–3759.

Wu, Y., Kikuchi, S., Yan, H. *et al.* (2012) Euchromatic subdomains in rice centromeres are associated with genes and transcription. *Plant Cell*, **23**, 4054–4064.

Yan, H., Jin, W., Nagaki, K. *et al.* (2005) Transcription and histone modifications in the recombination-free region spanning a rice centromere. *Plant Cell*, **17**, 3227–3238.

Chapter 5
Genomes of organelles

5.1 Plastids and mitochondria are descendants of free-living bacteria

Some of the most remarkable and astonishing components of plant cells are the chloroplasts and mitochondria. Together these are inextricably tied to cellular and organismal activities that make land plants such a dominant feature of the Earth and that provide the basis for sustaining the life of other terrestrial organisms such as animals.

Both plastids and mitochondria are the result of an ancient cell engulfing and establishing a symbiosis with a particular sort of bacteria. In the case of mitochondria, such an event seems to have happened only once and then spread to all organisms that are now called eukaryotes. In the case of chloroplasts, the number of events is less clear, but most evidence seems to be pointing to a single origin of the symbiosis as well. Together the mitochondria and chloroplasts allow even a single-celled organism to capture energy from the sun and then to use that energy to produce huge amounts of ATP.

The evidence for bacterial origin of the two organelles is now overwhelming and relies particularly on DNA sequences from both the organelles and the bacterial groups from which they are derived. In addition, many aspects of their physiology and cell structure are clearly bacterial rather than eukaryotic. For example, both have double membranes, their ribosomal function can be inhibited by chemicals (antibiotics) that affect bacteria but not eukaryotes, their DNA is concentrated in a defined region, or nucleoid, in the organelle, and it is arranged as a circular molecule, rather than the linear chromosomes of eukaryotes.

As described in Chapter 1, plastids and mitochondria retain their own (bacteria-like) genomes. In this chapter, we discuss the genomes of these organelles in more detail, and outline some aspects of their unique biology.

Each plant cell contains many plastids and mitochondria. In each cell of the shoot apical meristem there are about 20 undeveloped plastids (or proplastids); by the time the plastids differentiate into chloroplasts in mesophyll cells that number may increase to over 100. Mitochondria occur in even larger numbers, with hundreds to thousands in a single cell depending on the size of the cell and its energy requirements. Like their bacterial ancestors, plastids and mitochondria divide by fission, with timing that is not coupled to the division cycle of the cell in which they occur. Thus, when the cell divides, the number of organelles is reduced by approximately half, and then the organellar number gradually increases as the daughter cells enlarge and divide.

Replication of the organellar genome (Figure 1.6) is also regulated separately from replication of the nuclear genome. In meristematic cells, each plastid may have 50 to 150 copies of its genome and each mitochondrion may have 20 to 100 copies. As cell

Plant Genes, Genomes and Genetics, First Edition. Erich Grotewold, Joseph Chappell and Elizabeth A. Kellogg.
© 2015 John Wiley & Sons, Ltd. Published 2015 by John Wiley & Sons, Ltd.
Companion Website: www.wiley.com/go/grotewold/plantgenes.

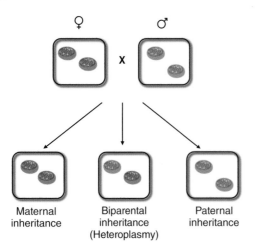

Figure 5.1 Organelles may be inherited maternally, paternally, or biparentally. A cell that has genetically distinct organelles is said to be heteroplasmic

division slows, however, DNA replication in the organelles also slows down and eventually stops while the organelles continue to divide. The combination of continued organellar division with lack of DNA replication means that organelles in differentiated tissues have only one or a few copies of their genomes.

In land plants, chloroplasts and mitochondria are almost always inherited via the female gametophyte, and are thus maternal (Figure 5.1). However, biparental inheritance of plastids has been reported in a few angiosperms such as *Pelargonium, Plumbago*, and *Oenothera*. The cells in these species are thus **heteroplasmic**, containing plastids that differ in genetic composition. In conifers, the mitochondria are transmitted through the female gametophyte, whereas chloroplasts are transmitted through the male gametophyte (pollen).

5.2 Organellar genes have been transferred to the nuclear genome

The ancient cells that captured the precursors of the mitochondria and chloroplasts underwent massive changes that prevented their newly acquired symbionts from ever again living as independent organisms. In one particularly striking response, genes were transferred from the original organellar genome to the nucleus of the host cell. While the average bacterial

genome encodes several thousand genes, chloroplast genomes encode about 130 genes and mitochondrial genomes generally only about 30–40. This means that several thousand former bacterial genes were transferred to the nucleus, where they are transcribed and translated along with nuclear genes; this estimate is supported by comparisons of plant nuclear and chloroplast genomes with those of cyanobacteria (Martin *et al.*, 2002) (Figure 5.2). For example, the small subunit of ribulose 1,5-bisphosphate carboxylase/oxygenase (Rubisco) is now a nuclear-encoded gene in land plants, and must be imported to the chloroplast to produce a functional holoenzyme. Some subunits of the chloroplast ATP synthase, and all the enzymes of the Calvin cycle are also encoded in the nucleus (Kleine *et al.*, 2009).

Genes were not transferred from the organelles to the nucleus all at once. Instead, the process happened one gene at a time over millennia and continues to occur at surprisingly high rates. One estimate suggests that one transfer event may occur for every 16 000 pollen grains produced (Huang *et al.*, 2003). When we consider that a single plant may produce thousands or millions of pollen grains in its lifetime, this represents a very high frequency.

Most of the time, a piece of organellar DNA that is transferred to the nucleus has no function. In the process of transfer, the organellar gene (if it is an entire gene at all) loses any bacterial regulatory sequences and does not easily pick up regulatory sequences from the nucleus. Thus the most common outcome of DNA transfer is that the transferred sequences simply exist in the nucleus as duplicate copies of sequences in the organelles. Eventually they begin to accumulate mutations and any similarity to the organellar genome is gradually lost.

Occasionally, an entire functional gene is transferred to the nucleus and acquires appropriate eukaryotic regulatory sequences, including a promoter and a polyadenylation signal (Figure 1.8). Because the proteins are still needed by the organelles, the transferred gene must also acquire a plastid targeting sequence, which will cause cellular transport machinery to move the protein back into the organelle (Figure 5.3). Once inside the organelle, the plastid targeting sequence is removed and the protein functions as it did in the bacterial ancestor. While most targeting sequences are specific to either plastids or mitochondria, approximately 100 proteins encoded by the nuclear genome have targeting sequences that take

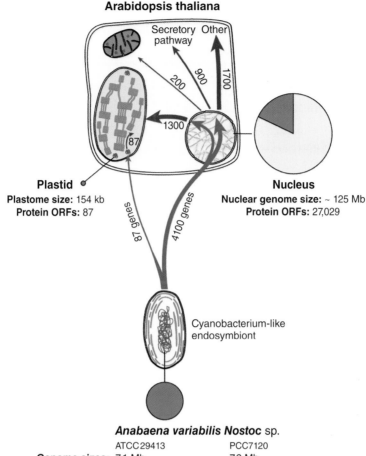

Arabidopsis thaliana

Plastid
Plastome size: 154 kb
Protein ORFs: 87

Nucleus
Nuclear genome size: ~ 125 Mb
Protein ORFs: 27,029

Cyanobacterium-like endosymbiont

Anabaena variabilis Nostoc sp.
ATCC 29413   PCC7120
Genome sizes: 7.1 Mb   7.2 Mb
Protein ORFs: 5661   6130

Figure 5.2 Comparison of the gene content of a cyanobacterium (*Anabaena variabilis*) with that of Arabidopsis thaliana. *A. variabilis* is thought to be similar to the bacterial ancestor of the chloroplast. About 4100 cyanobacterial genes now reside in the nucleus although they must have been in the genome of the original endosymbiont; they thus must have been transferred over evolutionary time. Red arrows indicate the numbers of genes now targeted to the chloroplast versus other cellular compartments. Green sector of pie diagram shows the fraction of nuclear genes thought to be of cyanobacterial origin. Kleine *et al.* (2009). Reproduced with permission from the Annual Review of Plant Biology, Volume 60 © 2009 by Annual Reviews, http://www.annualreviews.org

plastid signal peptide

envel	thyl	protein

mt signal peptide	protein

Figure 5.3 Two examples of proteins with targeting sequences. The upper diagram indicates a protein with two targeting sequences, one of which directs the protein to be transported to the chloroplast envelope and the other to the thylakoid membrane. The lower diagram indicates a protein with a mitochondrial targeting sequence. Once the protein reaches its destination the targeting sequences are removed

them to both types of organelles (or to mitochondria plus peroxisomes); most such dual-targeting sequences are shared among land plants (Xu *et al.*, 2013).

A well-documented example of the transfer process is provided by the chloroplast gene for a subunit of acetyl-CoA carboxylase (*accD*), which has been transferred to the nucleus independently in several different groups of flowering plants (Rousseau-Gueutin *et al.*, 2013). In the bluebell family (Campanulaceae) *accD* appears to have fortuitously been inserted just downstream of a pre-existing coding region that included a target protein and has been lost by the chloroplast (Figure 5.4). The new nuclear-encoded gene thus

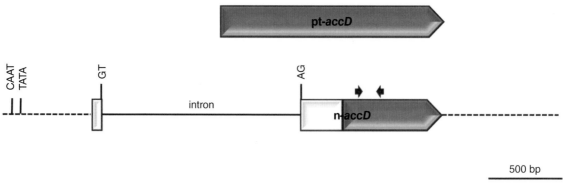

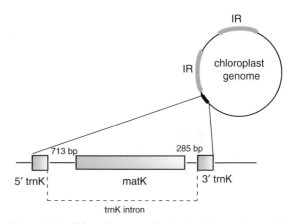

Figure 5.4 Comparison of the normal plastid accD gene (top) with the nuclear-encoded chimeric gene from *Trachelium caeruleum* (Campanulaceae). The plastid gene is a single open reading frame with no introns. In contrast, the nuclear gene includes a downstream region with high similarity to the plastid gene (gray) and upstream regions of nuclear origin (white). Presumed CAAT and TATA boxes are indicated as are splice sites for the single large intron. Although the portion of the nuclear gene that matches the chloroplast one is quite short, the protein appears to be functional and correctly targeted to the chloroplast. Rousseau-Gueutin *et al.* (2013). Copyright 2013 American Society of Plant Biologists. Used with permission

is a chimera created by joining parts of the coding sequences of nuclear and chloroplast genes.

5.3   Organellar genes sometimes include introns

Although most bacterial genes lack introns, introns do occur in both mitochondrial and chloroplast genomes. Introns and **splicing** will be discussed in detail in Chapter 13, but here it is important to mention the peculiar introns of the organellar genomes. Whereas introns of nuclear genes are removed by a protein complex known as the spliceosome, both mitochondrial and chloroplast genomes encode introns that can remove themselves from the mRNA; these are known as "**self-splicing**" or **group II introns**. For example, the chloroplast gene for the transfer RNA (tRNA) for lysine (*trnK*) is split by a large intron; within this intron is a protein-coding gene that encodes a maturase (Figure 5.5). This maturase encodes an enzyme that allows the intron to remove itself and to link together, or splice, the two parts of the *trnK* gene. In the mRNA, the intron forms a complex secondary structure with six domains that are involved in excision of the intron and splicing of the adjacent exons. However, even though the group II introns in plant organelles are capable of excising themselves, in fact their action is aided by several nuclear encoded genes. Thus, they are not as autonomous as might seem.

Figure 5.5 Chloroplast gene for trnK with a maturase in its intron. Introns are rare in bacteria but somewhat more common in plastids. Approximate sizes of genes and introns based on those for tobacco (Shaw *et al.*, 2005)

In addition, a few organellar genes contain group I introns, another sort of self-splicing intron.

5.4   Organellar mRNA is often edited

Another unique feature of organellar genes is RNA editing. In this process, individual base pairs of the mRNA are modified so that the message that is ultimately translated by the ribosome is different from the

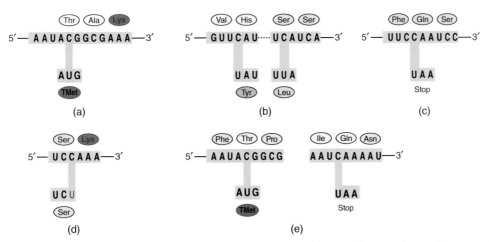

Figure 5.6 Substitution of U for C in RNA sequences can have several different effects on the resulting protein. These include: (a) creation of a start codon; (b) changes of amino acids; (c) creation of a stop codon; (d) creation of a silent change; or (e) creation of both start and stop codons. Buchanan *et al.* (2002). Reproduced with permission of John Wiley & Sons, Inc.

sequence that is encoded in the DNA. The most common change is from cytosine (C) to uracil (U), which is accomplished easily by removal of the amino group (Figure 5.6); in other words, the nucleotide at a particular position is not replaced but simply modified. Editing is particularly common in mitochondrial genes; it is estimated that 400–500 sites may be edited. In contrast, only about 30 sites in chloroplasts are edited (Gagliardi and Binder, 2007). The number of edited sites is much larger in *Selaginella*, and thus presumably in many other lycophytes, than in seed plants, but the significance of this observation is unclear (Banks *et al.*, 2011).

The function of RNA editing is unknown; it would seem much simpler to encode the correct sequence in the DNA. Editing appears to be characteristic of land plants, and not their green algal relatives. However, it is also quite common in plastids of non-plants such as trypanosomes (which are unicellular eukaryotes that infect animal cells), so it may simply have been lost in the algae that have been investigated. The resulting protein sequence from an edited gene is quite conserved, so that editing simply restores the message to what appears to be an ancestral sequence.

Much of the processing of RNA in mitochondria and plastids, including RNA editing, is controlled by proteins encoded in and imported from the nucleus. The **pentatricopeptide repeat** (**PPR**) proteins are one important group of these. The "repeat" referred to in the protein name consists of 35 amino acids, which occurs in 2–26 copies in Arabidopsis PPR proteins;

the repeats are degenerate, meaning that the sequence of amino acids is similar but not identical. In the mature protein, each repeat unit is folded in such a way that two amino acids make direct contact with the RNA (Barkan *et al.*, 2012). The particular amino acids specify the nucleotide to be bound. For example, a threonine (T) at one position in the repeat, combined with an asparagine (N) at a second position, will bind with adenine; the two amino acids are not right next to each other. Because each repeat can specify binding to a different nucleotide, each PPR protein can bind to a distinct RNA sequence. In addition, the PPR binding sites need not be contiguous base pairs; there may be a gap between binding sites for a single protein(Barkan *et al.*, 2012) (Figure 5.7).

The ability of PPR proteins to bind to a specific, relatively long, sequence of nucleotides is unusual among DNA binding proteins. Most DNA-binding proteins, including the transcription factors that control gene expression, recognize sequences of only a few base pairs (see Chapter 9), whereas PPR proteins can bind to much longer sequences depending on the number of repeats. In this respect, the PPR proteins are analogous to transcription activator-like effector (TALE) proteins (described in Chapter 9), which also contain repeats with an amino acid code for recognizing strings of nucleotides.

Sequenced plant genomes encode hundreds of PPR genes (450 in Arabidopsis, 650 in rice). The genome of *Selaginella* (a lycophyte described previously) includes

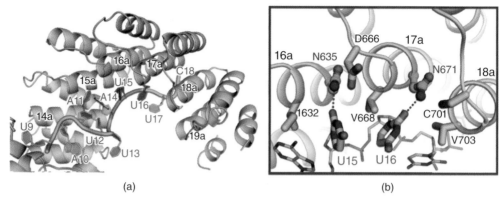

(a)   (b)

Figure 5.7 Binding of PPR proteins to RNA. Model of the interaction between repeats 14 through 19 of the protein (gray) and bases 9 through 18 of the RNA (yellow and orange). (a) Overview, showing the α-helices of the PPR repeats. (b) Hydrogen bonding of asparagine (N) 635 in repeat 16a to uracil 15, and asparagine 671 in repeat 17a to uracil 16. Yin *et al.* (2013). Reproduced with permission of Macmillan Publishers Ltd.

more than 800 of these, a number that may reflect the high number of edited sites in the RNA (Banks *et al.*, 2011). The large number of PPR proteins appears to be unique to vascular plants. Non-plants have at most a couple of dozen of PPR genes. In the moss *Physcomitrella patens* the number has expanded to 103; that number subsequently increased several fold in the present day angiosperms. Thus expansion of the PPR family occurred well after the colonization of land (Fujii and Small, 2011).

The PPR proteins fall into two classes that can be distinguished by their repeat patterns (Fujii and Small, 2011). In one group, nearly the entire protein is made up of copies of the standard 35-residue repeat; this group is known as the P class. In the other class, the P repeats alternate with long (L) and short (S) repeats, so the group is known as PLS. The PLS class is responsible for RNA editing, whereas the P class is involved in splicing, translation, and RNA turnover. Developmental processes regulated by PPR proteins include stress responses, and regulatory signaling from the organelles to the nucleus (known as **retrograde signaling**) (Laluk *et al.*, 2011).

5.5 Mitochondrial genomes contain fewer genes than chloroplasts

The ancestor of all mitochondria in eukaryotes must have shared many features with modern alpha Proteobacteria. Within the Proteobacteria, the species most similar to eukaryotic mitochondria are intracellular parasites, so it is not hard to imagine how one ancestral bacterium might have made the shift from being a parasite to being a symbiont (Gray *et al.*, 1999).

Mitochondria produce energy for the cell by passing electrons along an electron transport chain in the process of oxidative phosphorylation; plant mitochondria can also produce heat by use of the **alternative oxidase** (see Box, Alternative oxidase). Most of the proteins that make up this chain, including the ultimate powerhouse, ATP synthase, are encoded by the mitochondrial genome itself. In addition, the mitochondrial genome encodes ribosomal RNAs (rRNAs) and tRNAs so the mitochondria can produce their own proteins. However, eukaryotes differ in where the genes for the organellar ribosomal proteins reside. In animals and fungi, the ribosomal proteins are all encoded in the nucleus, whereas in other eukaryotes, including plants, some of the ribosomal proteins are encoded in the mitochondrion. Also plants have kept the genes for cytochrome c in the mitochondrion whereas they are nuclear in animals. All other proteins involved in the life cycle of the mitochondrion are imported from the nucleus. Plant mitochondrial genomes contain a variable number of genes but the number is usually less than 60 (Adams and Palmer, 2003) (Figure 5.8; cf. the chloroplast genome in Figure 1.6). This is more than vertebrate mitochondria, which contain only about 13, but is still a small number.

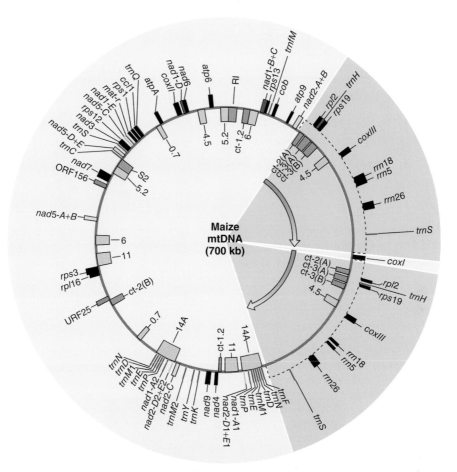

Figure 5.8 The mitochondrial genome of maize. This diagram shows the "master circle", which is the structure that would appear if all mitochondrial genomes in the cell were combined into a single circle. Genes are indicated on the outside of the circle, repeats on the inside. Blue, inverted repeats; green, direct repeats. Jones *et al.* (2012). Reproduced with permission of John Wiley & Sons, Ltd.

Alternative oxidase

The alternative oxidases are an intriguing feature of plant mitochondria. These enzymes are a plant-specific part of the mitochondrial electron transport chain, and are encoded by a small subfamily of nuclear genes. While they transfer electrons, they pump fewer protons across the membrane than the other parts of the electron transport chain, and thus contribute little to generation of the proton motive force that generates ATP (Figure 5.9). Instead, the energy produced is given off as heat (Vanlerberghe and McIntosh, 1997). The general function of the alternative oxidase is unclear, but it may serve to redirect electrons when the mitochondrial electron transport chain is overloaded.

In some plants the alternative oxidase is used to generate heat to help attract pollinators. The capacity of plants to generate heat was first recorded by the French botanist Lamarck (1815). Lamarck was an early leader in developing the idea of evolution and set the stage for the subsequent work of Darwin; he contributed a huge amount to our knowledge of plants and the way we think about the natural world. In plants such as skunk cabbage, which blooms in the early spring in North America, the alternative oxidase is able to raise the temperature of the inflorescence and cause it to emit a strong smell, which attracts pollinating insects. This warming mechanism is shared by other members of the skunk cabbage family (Araceae), in which the inflorescences become warm to the touch (Figure 5.10).

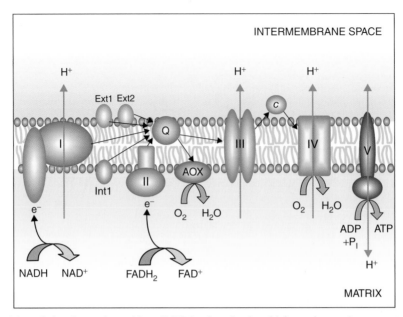

Figure 5.9 Position of the alternative oxidase (AOX) in the mitochondrial membrane. Components shown in pink are all involved in pumping protons across the membrane to create the proton gradient that drives ATP synthase (purple); these components are found in all eukaryotes. The AOX, shown in green, moves electrons but not protons; energy is given off as heat. The AOX is common in eukaryotes but does not occur in animals. iGem Team Georgia Tech 2010. http://2010.igem.org/Team:GeorgiaTech/Project. CC BY 3.0 (http://creativecommons.org/licenses/by/3.0/)

Figure 5.10 *Amorphophallus titanum* (Araceae). This giant plant produces huge inflorescences, each with a long appendage (large yellowish structures). The flowers are hidden down inside the ruffled purple spathe. Because of the action of the alternative oxidase in the mitochondria, the large appendage becomes warm to the touch and causes evaporation of putrid-smelling volatile compounds, which attract the flies that pollinate the plant. From https://bioscigreenhouse.osu.edu/titan-arum-woody

5.6 Plant mitochondrial genomes are large and undergo frequent recombination

Despite the small number of genes, plant mitochondrial genomes are appreciably larger than those in other eukaryotes. Whereas the mitochondrial genome of a human is about 16 kb and that of yeast is 80 kb, the smallest known plant mitochondrial genome is about 200 kb and the largest ones are 10 times that size. For example, the *Brassica hirta* mitochondrial genome is 208 kb whereas that of *Cucumis melo* (cucumber) is 2740 kb. Among plants in a single species, the mitochondrial genome can vary by nearly 40% in size (Allen *et al.*, 2007)

The entire mitochondrial genome can be diagrammed as a single circular molecule, but it rarely if ever exists in this form in the plant. Instead the genome exists as a complex set of linear and circular molecules that can be converted from one to the other by recombination involving extensive repeated sequences (Figure 5.11).

Repeated sequences are a major contributor to the large size of the mitochondrial genome. The repeats vary in length from a few base pairs to over 1 kb in size and are sometimes divided into classes according to their size. The repeated sequences provide the opportunity for recombination between different parts of the genome. Recombination between repeated sequences that are longer than 1 kb can create molecules in which the orientation of the two intervening sequences is flipped relative to each other, and can also lead to production of two circles of DNA instead of just one, with all forms co-existing in the same mitochondrion or cell (Figure 5.11). Recombination between shorter repeat sequences leads to small inversions of gene order. Because of this constant recombination, the order of genes is not fixed but is continually shuffled even among mitochondria within the same organism or cell. The extent of recombination is controlled by genes in the nuclear genome (Miller-Messmer *et al.*, 2012). In a few cases, the mitochondrial genome may also exist as a linear molecule (Allen *et al.*, 2007). The large mitochondrial genome of cucumber can be diagrammed not as one circle but three. The two smaller circles do not encode known genes and do

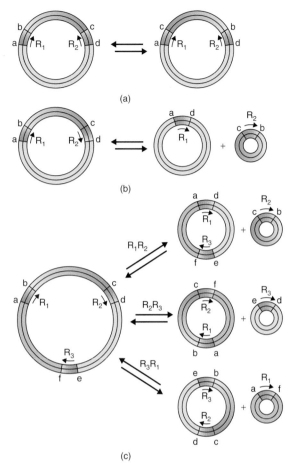

Figure 5.11 Recombination in the mitochondrial genome leads to rearrangements and small circular molecules. (a) Inverted repeats can recombine to produce two different arrangements of the circular molecule. (b) Recombination between direct repeats leads to two smaller circular molecules. (c) With more than one direct repeat, additional pairs of subgenomic circles are possible. Jones *et al.* (2012). Reproduced with permission of John Wiley & Sons, Ltd.

not recombine with the large one; their function is unknown (Alverson *et al.*, 2011).

The continuous recombination of the mitochondrial genome also leads to the production of chimeric genes, genes that have one portion from one previously existing gene and another from another one. These chimeric genes may lead to defects in mitochondrial development, which in turn affect any tissues that require a large amount of ATP. In plants, mitochondrial defects particularly affect pollen development, and can lead to malfunctions of either the cells that directly form the pollen (sporogenous cells) or the cells that surround and support them (tapetal cells) (Hanson and Bentolila, 2004; Bentolila and Stefanov, 2012). Because such mutations occur in the mitochondria, which are inherited via the mother along with most of the cytoplasmic contents of the cell, they are known as **cytoplasmic male sterility** or **CMS**. CMS occurs fairly frequently among wild plants and may become fixed (i.e., permanent) within populations of plants; such populations then produce a mixture of plants with bisexual flowers and plants with female-only (pistillate) flowers. CMS is also a valuable tool in plant breeding (see Box, Cytoplasmic male sterility).

Cytoplasmic male sterility

Modern agriculture depends heavily on the production of huge amounts of hybrid seed. When cultivars or races of crop plants are crossed, rather than allowing them to self-fertilize, hybrid offspring are produced; these have heterozygous genomes and greater yield than that of either parent. The increased yield of hybrids is known as hybrid vigor.

Cytoplasmic male sterility is a particularly valuable tool for production of hybrid seed. Normally, production of a hybrid requires that the anthers be removed from the plant used as the female (seed) parent to prevent self-pollination. In the case of maize (corn), the staminate and pistillate flowers are in separate inflorescences so the staminate inflorescence (the tassel) can be simply cut off. However, in most other crops the flowers are bisexual with both anthers and a pistil; to cross-pollinate these plants and to prevent self-pollination the anthers must be removed with tweezers. In either case, removal of anthers is difficult and time consuming. Thus a plant in which the anthers naturally fail to develop is useful as a seed parent.

Instances of CMS generally share a number of characteristics: (1) the defective protein is produced by a chimeric mitochondrial gene presumably created by recombination in the mitochondrial genome; (2) the protein is toxic to a component of normal mitochondrial metabolism and results in failure of pollen formation; and (3) fertility can be restored by one or more nuclear proteins imported into the mitochondrion.

For commercial seed production, it is ultimately necessary to restore fertility to the CMS line by creating a hybrid with a parent that carries a restorer gene (Figure 5.12). Depending on the nature of the chimeric mitochondrial gene that confers CMS, fertility may be restored by gene products encoded in either nucleus or the mitochondrion itself. In either case, restoration involves altering either transcription of the CMS gene or changing the accumulation of its protein product. For example, restorer genes may encode PPR proteins, described above as being important regulators of RNA processing in organelles (Hanson and Bentolila, 2004).

A recent study in rice has examined the mechanism of action of one CMS gene and its restoration (Luo *et al.*, 2013); while every CMS gene has a slightly different mechanism of action, this particular example shares many features in common with other instances of CMS. For the last 40 years, hybrid rice in China has been produced using a CMS allele known as WA for "wild abortive" pollen; these hybrids have contributed to substantial increases in yield. Like most other CMS mutants, CMS-WA pollen produces a chimeric mitochondrial protein. In this case the chimeric protein, which the authors call WA352, interacts with a protein, COX11, which is imported from the nucleus (Figure 5.13). COX11 normally is involved in metabolism of peroxide inside mitochondria. When WA352 binds to COX11, COX11 transcript is reduced, the anther tapetum fails to proceed through normal development, pollen does not form normally, and the plant is pollen-sterile. Two nuclear genes can restore pollen fertility, one by reducing the transcript level of WA352 and the other by reducing the protein level.

In most other cases of CMS the exact mechanism of action of CMS genes is unknown (Hanson and Bentolila, 2004). One widely studied example is the so-called Texas cytoplasm of maize, CMS-T or T-cytoplasm. This male sterile line was used widely to produce hybrid corn in the USA particularly in the 1950s to 1970s. Like CMS-WA, CMS-T is the result of a chimeric gene in the mitochondrion that produces

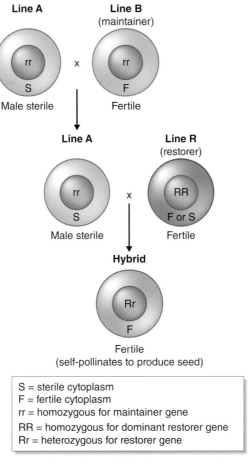

S = sterile cytoplasm
F = fertile cytoplasm
rr = homozygous for maintainer gene
RR = homozygous for dominant restorer gene
Rr = heterozygous for restorer gene

Figure 5.12 The use of CMS in crop breeding. Step 1, generating male sterile plants. A male sterile line is used as the female parent in the cross; the pollen comes from a plant that has a similar or identical nuclear genome but a non-mutant mitochondrion. All seed produced will be male sterile because the mitochondria are inherited only from the (male-sterile) female parent. Step 2, restoring fertility. The male sterile line is crossed with a line that is genetically distinct, and has restorer genes in its nucleus. The restorer genes allow pollen to develop normally. Step 3, commercial seed is now produced by the hybrid line with the restorer genes. Adapted from Canola Council of Canada (http://www.canolacouncil.org/crop-production/canola-grower%27s-manual-contents/chapter-2-canola-varieties/canola-varieties)

a toxic compound known as URF13. There are two restorer loci, one of which reduces the transcript of the *urf13* gene, and the other of which produces an aldehyde dehydrogenase.

An unexpected side effect of the T-cytoplasm was increased susceptibility to southern corn leaf blight. Because the T-cytoplasm was present in much of the hybrid corn, huge losses of yield occurred in the early 1970s as the leaf blight swept through populations. Since then the T-cytoplasm has been replaced by other male sterile cytoplasms, but the crop losses highlighted the peril of planting genetically identical plants over vast parts of the country.

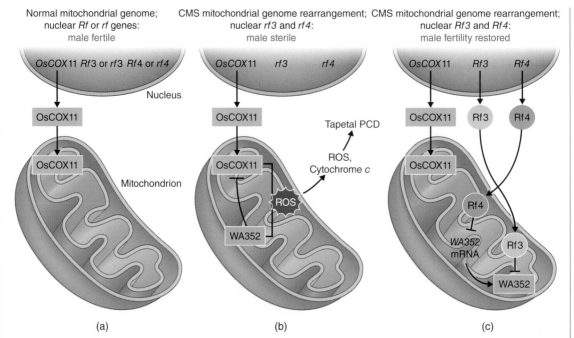

Figure 5.13 One mechanism of cytoplasmic male sterility in rice. (a) In normal plants, OsCOX11 is imported from the nucleus and leads to normal pollen. The restorer of fertility (*Rf*) genes are not involved. (b) In male sterile plants, the mitochondrial genome has been rearranged to produce a chimeric gene, which in turn produces a toxic protein, WA352. WA352 inhibits the proper function of OxCOX11; reactive oxygen species (ROS) are produced and lead to programmed cell death (PCD) in the cells surrounding the pollen-producing cells (tapetum). Male sterility occurs with recessive *rf* alleles. (c) With dominant *Rf3* or *Rf4* alleles, the amount of toxic WA352 protein is reduced. Male fertility is restored; the plant produces normal pollen. Ma (2013). Reproduced with permission of Macmillan Publishers Ltd.

Another cause of genome size variation is the ability of plant mitochondrial genomes to take up genes or gene fragments from many sources, most notably from the nucleus and chloroplast genomes in the same cell. In some plant mitochondria, the genes for tRNA have been derived from chloroplast-encoded genes that were presumably transferred. About 2–5% of the DNA in the mitochondrial genome of maize originated in the chloroplast (Allen *et al.*, 2007). In addition, mitochondrial genomes have acquired large pieces of genomes of other organisms. One early example was the discovery of an intron that has appeared multiple times in the *COX1* gene (which encodes part of the respiratory pathway); the source of the intron is likely fungal. It inserts into the same position in the gene each time it appears (Palmer *et al.*, 2000). The mechanism for this is unknown.

Mitochondrial genomes also lose genes to the nucleus. For some genes, this has clearly happened repeatedly over evolutionary time (Adams and Palmer, 2003). Movement of genes to the nucleus can be inferred by comparing the mitochondrial genome size and gene content of non-vascular plants, which are presumably similar in many ways to the earliest land plants, to those of the vascular plants. For example, the mitochondrial genome of *Marchantia* (a liverwort) is smaller than that of vascular plants (186 kb), consistent with the idea that the vascular plant mitochondrial genomes have acquired more repeat sequences and thus expanded. At the same time, the *Marchantia* mitochondrial genome contains more genes (94), as might be expected if the genomes of the vascular plants have lost many of them.

Despite the great variation in size and arrangement of mitochondrial genomes, the mutation rate – of both genes and intervening sequences – is lower than that in the chloroplast and the nucleus (Bentolila and Stefanov, 2012). Notable exceptions have been found in *Geranium* and plantain (*Plantago*) (Alverson *et al.*, 2011), in which the mutation rate for mitochondrial genes is 50–100 times greater than in other angiosperms.

Perhaps because of the extensive rearrangements of mitochondrial genomes, the exons of protein-coding genes are not always adjacent in the genome. Instead, parts of the coding sequence may be elsewhere, making splicing of the mRNA much more complex than in conventional intron-containing genes; the process of bringing together the disparate parts of genes is known as **trans-splicing**. For example, the gene *NAD5* is made up of 5 exons (A–E) and is spliced from three separate pieces. Exons A and B are separated by an intron, which is removed by conventional (*cis-*) splicing; a similar process occurs in exons D and E. Exon C is then spliced in between the a-b and d-e RNAs (Figure 5.14).

5.7   All plastid genomes in a cell are identical

A plant cell may contain different sorts of plastids, depending on the cell type and the stage of development. All plastids develop from proplastids in the meristematic cells; these proplastids may then differentiate into photosynthetic chloroplasts, but can also become starch-containing amyloplasts, or pigmented chromoplasts. Other types of plastids are leucoplasts, which synthesize monoterpenes, and etioplasts, which produce protochlorophyllide. Plastids can differentiate into one of these types, but can then dedifferentiate and develop into a different type (Figure 5.15). In addition, the plastids within a single cell can be interconnected by long extensions known as **stromules**. The role of these is unclear, but they appear to permit the exchange of substances including genetic material.

The most familiar of the plastids are the chloroplasts, which pack the mesophyll cells of leaves and also often of stems and parts of flowers and fruits – indeed any green tissue. Chloroplasts are generally infrequent in leaf epidermal cells, which are therefore translucent, except for the stomata. Chromoplasts are responsible for color in some plant tissues such as petals in some species. Amyloplasts are plastids that are packed full of starch. They occur most frequently in storage tissues such as tubers or endosperm.

Despite the morphological and physiological variability of plastids, all plastids within a cell and usually within a plant are generally identical genetically; they are clonal replicates and plant cells are usually

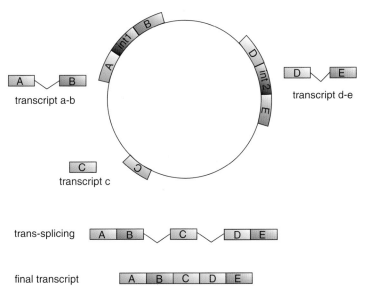

Figure 5.14 Trans-splicing. Exons A and B, D and E, and C are encoded by non-contiguous parts of the DNA, shown here as occurring at different places in a circular genome. Introns 1 and 2 (gray) are removed to produce an a-b transcript and a d-e transcript. The splicing machinery then brings all transcripts together

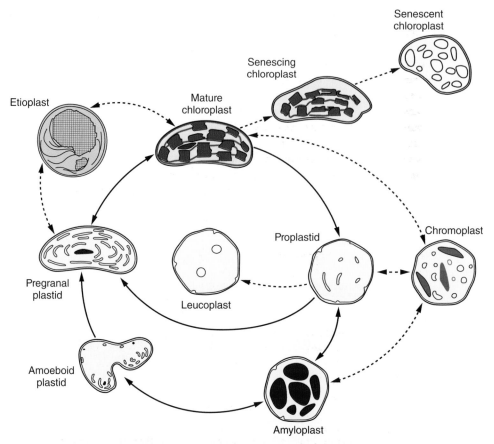

Figure 5.15 Plastid development and differentiation. Solid arrows, normal plastid development and differentiation. Dotted arrows, steps of differentiation that occur under certain specific conditions. Buchanan *et al.* (2002). Reproduced with permission of John Wiley & Sons, Inc

homoplasmic with respect to their plastids. Unlike the mitochondrial genome, the plastid genome is fairly stable; this makes it an attractive target for genetic engineering (see From the Experts, Manipulation of the chloroplast). In most land plants it ranges in size from ~120–160 kbp and contains about 100 protein-coding genes as well as genes for rRNA and tRNAs (Figure 1.6). A few plants have larger chloroplast genomes (e.g., *Geranium*, 217 kbp), and in some parasitic plants, for which chloroplasts are unnecessary for photosynthesis, the genome is much smaller than average. It is common for several plastid genes to be transcribed from a single promoter; such transcription units are polycistronic and are similar to bacterial operons. However, unlike operons, the gene products from a polycistronic transcription unit may not all be part of a single pathway or functional complex.

The most striking feature of the chloroplast genome is a pair of repeats that are arranged in inverse orientation; these are known as the inverted repeats (Figure 1.6). They divide the chloroplast genome into two segments, the small single copy (SSC) region and the large single copy (LSC) region. The inverted repeats contain the genes encoding the rRNAs, whereas the SSC and LSC include proteins involved in photosynthesis and in amino acid biosynthesis. The inverted repeat regions can pair and recombine such that the orientation of the SSC can be inverted relative to the LSC. This leads to two forms of the chloroplast genome, which are fully interconvertible. For reasons that are unknown, the chloroplast genome does not acquire gene sequences from other organisms or other parts of the cell as easily or frequently as the mitochondrial genome does.

5.8 Plastid genomes are similar among land plants but contain some structural rearrangements

While the chloroplast genome generally includes comparable sets of genes in most land plants, a few groups of plants show major changes in gene content. For example, some gymnosperms (pines and cedars) and some legumes have lost one copy of the inverted repeat; in these groups the chloroplast rRNA genes thus exist as single copy. Parasitic plants also generally have tiny plastid genomes because the genes for photosynthesis have been lost. The plastid is still needed for synthesis of some amino acids, however, so the plant cannot dispense with it entirely.

Chloroplast genomes have been widely used as a way to infer the history of plant species, and much of what we know about the details of plant phylogeny relies on data from plastid genome. One reason for this is simply ease of data acquisition: chloroplast DNA can make up as much as 30% of the DNA in the cell, so is easy to extract and the genes are easy to amplify by the polymerase chain reaction. In addition, the chloroplast genomes within a plant are generally clonally inherited and all identical, so allelic variation does not confuse sequence acquisition or interpretation. Finally, the rate of evolution of chloroplast genes is higher than that of the mitochondrion but slower than that of many nuclear genes. This has made sequences of the chloroplast genome a useful tool for estimating the evolutionary history of plants.

From the Experts

Manipulation of the chloroplast

The plastid genome (ptDNA) of higher plants is highly polyploid, and may contain thousands of identical genome copies localized in units of 10–100 plastids. Plastid engineering is a gradual process, in which transformation of one, or at most a few, plastid genomes is followed by the gradual diluting out of plastids with non-transformed ptDNA copies. Incorporating an antibiotic resistance gene in the transformation vector puts the native plastids at a disadvantage in tissue culture where a selective agent is present. When each of the plastid genomes is uniformly transformed, plants are regenerated from the cultured cells.

Plastid transformation is based on homologous recombination between the transformation vector and the ptDNA. In the transformation vector the marker gene and the gene of interest are flanked by ptDNA sequences, which target the insertion into the ptDNA. The commonly used plastid transformation vectors encode resistance to spectinomycin (*aadA* gene) or kanamycin (*neo* gene) antibiotics, encoded in genes engineered for expression in plastids. The antibiotics selectively inhibit translation on the plastid's prokaryotic-type ribosomes, but do not affect protein synthesis in the cytoplasm (Maliga, 2004, 2012). Manipulations of ptDNA include: (i) insertion of transgenes in intergenic regions (Figure 5.16a) (Lutz *et al.*, 2007); (ii) gene knockouts to probe plastid gene function (Figure 5.16b) (Allison *et al.*, 1996); (iii) cotransformation with multiple plasmids to introduce non-selected genes without physical linkage to marker genes (Lutz and Maliga, 2007); and (iv) post-transformation excision of marker genes to obtain marker-free plants (Figure 5.16c) (Lutz and Maliga, 2007).

Plastid transformation has been developed in *Chlamydomonas reinhardtii*, a unicellular alga (Boynton *et al.*, 1988) and tobacco (*Nicotiana tabacum*), a flowering plant (Svab *et al.*, 1990). Differences in the methodology in the two systems can be traced back to engineering the chloroplast genome of *Chlamydomonas* cells in photoautotrophic cultures and manipulating the plastid genome of higher plants in heterotrophically grown tissue culture cells (Maliga, 2012). Protocols for plastid transformation are available in tobacco, tomato, potato, eggplant, petunia, soybean, cabbage, lettuce, sugar beet, *Physcomitrella patens* (a moss), and *Marchantia polymorpha* (a liverwort) (Maliga, 2014). While initial success has been achieved in rice and maize, thus far no genetically stable transplastomic monocots have been obtained.

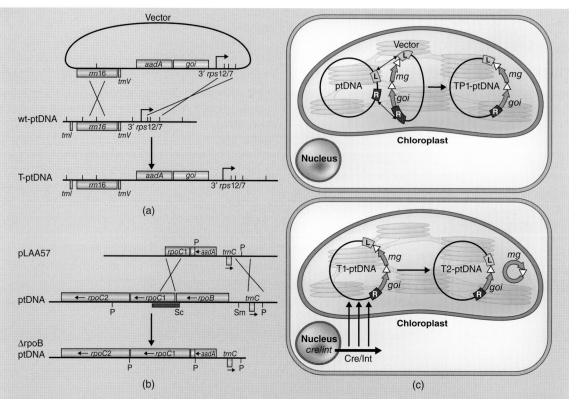

Figure 5.16 Manipulation of the plastid genome. (a) Vector targets insertion of a marker gene (*aadA*) and the gene of interest (*goi*) at a predetermined site in the plastid genome (wt-ptDNA) yielding a transplastomic ptDNA (T-ptDNA). Integration occurs by homologous recombination. Adapted from Lutz *et al.* (2007). (b) Gene knockout in the plastid genome. The *rpoB* gene is deleted by replacement with the *aadA* antibiotic resistance gene. Shown are the vector pLAA57, the wild type ptDNA and the deletion derivative ΔrpoB ptDNA. Deletion of the *rpoB* gene, encoding the catalytic subunit of the plastid-encoded RNA polymerase (PEP), revealed the existence of a second, nuclear-encoded RNA polymerase (NEP). Allison *et al.* (1996). Reproduced with permission of John Wiley & Sons, Inc. (c) Excision of plastid marker with phage site-specific recombinase enzymes. (Top) Marker genes in the plastid transformation vectors are flanked by *loxP* or *attP/attB* sequences (triangles) that are the targets for site-specific recombinases. (Bottom) The marker genes are efficiently removed when a gene encoding a plastid-targeted Cre or Int recombinase is expressed from a nuclear gene and the marker gene is excised by the plastid-targeted enzyme. The marker gene is subsequently segregated away in the seed progeny. Adapted from Maliga and Bock (2011). Copyright 2011 American Society of Plant Biologists. Used with permission

Expression of recombinant proteins in chloroplasts, instead of the plant nucleus, is advantageous because of the readily obtainable high protein levels, expression of pathways in polycistronic operons, and natural containment due to the lack of transmission of plastids via pollen in crops. Chloroplasts offer a unique system to express complex or cytotoxic proteins for therapeutic applications, including malaria antigens (Gregory *et al.*, 2012) and immunotoxin cancer therapeutics (Tran *et al.*, 2013).

Agronomic traits, such as herbicide and drought resistance, can be conferred by chloroplast expression (Maliga and Bock, 2011). Complex value-added traits can be expressed with particular efficiency from the chloroplast. For example, vitamin E content is enhanced by expressing three genes of the pathway in one operon (Lu *et al.*, 2013). The ultimate frontier of plant breeding is enhancing crop productivity by re-engineering photosynthesis (Whitney *et al.*, 2011; Hanson *et al.*, 2013).

By Pal Maliga

5.9   Summary

Chloroplasts and mitochondria are the descendants of symbiotic bacteria that have become permanent residents of plant cells; each plant cell has many of each. Their circular genomes are able to recombine and exist in different forms within each organelle. RNA editing, particular from C to U, is common in mitochondria and somewhat less common in chloroplasts. The genomes of both organelles contain self-splicing, or group II, introns. The mitochondrial genome contains only a few (30–40) genes separated by highly recombinogenic repeats, and usually exists as a large and complex set of interconvertible circular molecules. Recombination often leads to chimeric genes, some of which confer male sterility. Mitochondria often acquire genes from the chloroplast and from other plants, apparently by lateral gene transfer. The chloroplast genome has only one large repeated region, the inverted repeat, and is generally much more stable than the mitochondrial genome. The plastid genome contains about 100 protein-coding genes, most of which are well-conserved among angiosperms.

5.10   Problems

5.1 In what respects are organelles similar to bacteria?

5.2 Why can't the organelles revert to being free-living organisms?

5.3 What is RNA editing and which nucleotides are targeted most often? How does this mechanism help or hinder gene regulation?

5.4 What aspect of the mitochondrial genome makes it so prone to recombination?

5.5 What is a polycistronic transcription unit?

5.6 Why can't parasitic plants dispense with their chloroplasts entirely, even if they do not photosynthesize?

Further reading

Logan, D.C. (ed.) (2007) *Plant Mitochondria*. Annual Plant Reviews, Vol. 31, Blackwell.

References

Adams, K.L. and Palmer, J.D. (2003) Evolution of mitochondrial gene content: gene loss and transfer to the nucleus. *Molecular Phylogenetics and Evolution*, **29**, 380–395.

Allen, J.O., Fauron, C.M., Minx, P. *et al.* (2007) Comparisons among two fertile and three male-sterile mitochondrial genomes of maize. *Genetics*, **177**, 1173–1192.

Allison, L.A., Simon, L.D. and Maliga, P. (1996) Deletion of *Rpob* reveals a second distinct transcription system in plastids of higher plants. *EMBO Journal*, **15**, 2802–2809.

Alverson, A.J., Rice, D.W., Dickinson, S. *et al.* (2011) Origins and recombinatin of the bacterial-sized multichromosomal mitochondrial genome of cucumber. *Plant Cell*, **23**, 2499–2513.

Banks, J.A., Nishiyama, T., Hasebe, M. *et al.* (2011). The *Selaginella* genome identifies genetic changes associated with the evolution of vascular plants. *Science*, **332**, 960–963.

Barkan, A., Rojas, M., Fujii, S. *et al.* (2012) A combinatorial amino acid code for RNA recognition by pentatricopeptide repeat proteins. *PLoS Genetics*, **8**, e1002910.

Bentolila, S. and Stefanov, S. (2012) A reevaluation of rice mitochondrial evolution based on the complete sequence of male-fertile and male-sterile mitochondrial genomes. *Plant Physiology*, **158**, 996–1017.

Boynton, J.E., Gillham, N.W., Harris, E.H. *et al.* (1988) Chloroplast transformation in Chlamydomonas with high velocity microprojectiles. *Science*, **240**, 1534–1538.

Buchanan, B.B., Gruissem, W. and Jones, R.L. (2002) *Biochemistry and Molecular Biology of Plants*, American Society of Plant Biologists.

Fujii, S. and Small, I. (2011) The evolution of RNA editing and pentatricopeptide repeat genes. *New Phytologist*, **191**, 37–47.

Gagliardi, D. and Binder, S. (2007) Expression of the plant mitochondrial genome, in *Plant Mitochondria* (ed. D.C. Logan), Blackwell, pp. 50–96.

Gray, M.W., Burger, G. and Lang, B.F. (1999) Mitochondrial evolution. *Science*, **283**, 1476–1481.

Gregory, J.A., Li, F., Tomosada, L.M. *et al.* (2012). Algae-produced Pfs25 elicits antibodies that inhibit malaria transmission. *PLoS One*, **7**, e37179.

Hanson, M.R. and Bentolila, S. (2004) Interactions of mitochondrial and nuclear genes that affect male gametophyte development. *Plant Cell*, **16**, S154–S169.

Hanson, M.R., Gray, B.N. and Ahner, B.A. (2013) Chloroplast transformation for engineering of photosynthesis. *Journal of Experimental Botany*, **64**, 731–742.

Huang, C.Y., Ayliffe, M.A. and Timmis, J.N. (2003) Direct measurement of the transfer rate of chloroplast DNA into the nucleus. *Nature*, **422**, 72–76.

Jones, R., Ougham, H., Thomas, H. and Waaland, S. (2012) *The Molecular Life of Plants*, Wiley-Blackwell.

Kleine, T., Maier, U.G. and Leister, D. (2009) DNA transfer from organelles to the nucleus: the idiosyncratic genetics of endosymbiosis. *Annual Review of Plant Biology*, **60**, 115–138.

Laluk, K., AbuQamar, S. and Mengiste, T. (2011) The *Arabidopsis* mitochondria-localized pentatricopeptide repeat protein PGN functions in defense against necrotrophic funig and abiotic stress tolerance. *Plant Physiology*, **156**, 2053–2068.

Lamarck, J.B.d. (1815) *Flore Françoise*, L'Imprimerie Royale.

Lu, Y., Rijzaani, H., Karcher, D. *et al.* (2013) Efficient metabolic pathway engineering in transgenic tobacco and tomato plastids with synthetic multigene operons. *Proceedings of the National Academy of Sciences USA*, **110**, E623–E632.

Luo, D., Xu, H., Liu, Z. *et al.* (2013) A detrimental mitochondrial-nuclear interaction causes cytoplasmic male sterility in rice. *Nature Genetics*, **45**, 573–577.

Lutz, K.A., Azhagiri, A., Tungsuchat-Huang, T. and Maliga, P. (2007) A guide to choosing vectors for transformation of the plastid genome of higher plants. *Plant Physiology*, **145**, 1201–1210.

Lutz, K.A. and Maliga, P. (2007) Construction of marker-free transplastomic plants. *Current Opinion in Biotechnology*, **18**, 107–114.

Ma, H. (2013) A battle between genomes in plant male fertility. *Nature Genetics*, **45**, 472–473.

Maliga, P. (2004) Plastid transformation in higher plants. *Annual Review of Plant Biology*, **55**, 289–313.

Maliga, P. (2012) Plastid transformation in flowering plants, in *Genomics of Chloroplasts and Mitochondria* (eds R. Bock and V. Knoop), Springer, pp. 393–414.

Maliga, P. (2014) *Chloroplast Biotechnology: Methods and Protocols*, Springer Science+Business Media, LLC.

Maliga, P. and Bock, R. (2011) Plastid biotechnology: food, fuel and medicine for the 21st century. *Plant Physiology*, **155**, 1501–1510.

Martin, W., Rujan, T., Richly, E. *et al.* (2002) Evolutionary analysis of *Arabidopsis*, cyanobacterial, and chloroplast genomes reveals plastid phylogeny and thousands of cyanobacterial genes in the nucleus. *Proceedings of the National Academy of Sciences USA*, **99**, 12246–12251.

Miller-Messmer, M., Kühn, K., Bichara, M. *et al.* (2012) RecA-dependent DNA repair results in increased heteroplasmy of the Arabidopsis mitochondrial genome. *Plant Physiology*, **159**, 211–226.

Palmer, J.D., Adams, K.L., Cho, Y. *et al.* (2000) Dynamic evolution of plant mitochondrial genomes: mobile genes and introns and highly variable mutation rates. *Proceedings of the National Academy of Sciences USA*, **97**, 6960–6966.

Rousseau-Gueutin, M., Huang, X., Higginson, E. *et al.* (2013) Potential functional replacement of the plastidic acetyl-CoA carboxylase subunit (*accD*) gene by recent transfer to the nucleus in some angiosperm lineages. *Plant Physiology*, **161**, 1918–1929.

Shaw, J., Lickey, E.B., Beck, J.T. *et al.* (2005) The tortoise and the hare II: Relative utility of 21 noncoding chloroplast DNA sequences for phylogenetic analysis. *American Journal of Botany*, **92**, 142–166.

Svab, Z., Hajdukiewicz, P. and Maliga, P. (1990) Stable transformation of plastids in higher plants. *Proceedings of the National Academy of Sciences USA*, **87**, 8526–8530.

Tran, M., Van, C., Barrera, D.J. *et al.* (2013) Production of unique immunotoxin cancer therapeutics in algal chloroplasts. *Proceedings of the National Academy of Sciences USA*, **110**, E15–E22.

Vanlerberghe, G.C. and McIntosh, L. (1997) Alternative oxidase: from gene to function. *Annual Review of Plant Physiology and Plant Molecular Biology*, **48**, 703–734.

Whitney, S.M., Houtz, R.L. and Alonso, H. (2011) Advancing our understanding and capacity to engineer nature's CO_2 sequencing enzyme, Rubisco. *Plant Physiology*, **155**, 27–35.

Xu, L., Carrie, C., Law, S.R. *et al.* (2013). Acquisition, conservation, and loss of dual-targeted proteins in land plants. *Plant Physiology*, **161**, 644–662.

Yin, P., Li, Q., Yan, C. *et al.* (2013) Structural basis for the modular recognition of single-stranded RNA by PPR proteins. *Nature*, **504**, 168–171.

Part II
Transcribing Plant Genes

Chapter 6
RNA

6.1 RNA links components of the Central Dogma

RNA is an extraordinary, diverse molecule. It comes in many forms and shapes, each with a unique function. Some RNAs are able to transfer information, others can regulate cellular processes, and some catalyze chemical reactions by themselves or in combination with proteins. Perhaps more surprising, a single RNA molecule can perform more than one of these functions. RNA molecules can interact with DNA molecules and with proteins, frequently in a specific manner that is dictated by their structure, the sequence of nucleotides, or both. In this chapter, we will explore the many different kinds of RNAs and their roles in the plant cell. As in DNA molecules, some of the characteristics of RNA molecules are shared by all kingdoms of life, whereas others are specific to plants. Given the importance of the different types of RNA, how, when and what quantity is made is also of fundamental significance. Subsequent chapters in this section describe the enzymes responsible for making RNA from DNA templates, the DNA-dependent RNA polymerases.

In Part 1 of this book, we discussed DNA as the blueprint for all the instructions in a cell. Part 2 covers primarily how the DNA code is converted into meaningful directions that permit a cell to carry out all its functions. This flow of information is largely unidirectional, as eloquently pronounced by **The Central Dogma of Biology,** first enunciated by Francis Crick in 1958, that says "Once information has got into a protein, it can't get out again." Genetic information thus flows from DNA to RNA to proteins (Figure 6.1). Yet, as will be described below, it is now clear that information can also flow from RNA to DNA, and thus the Central Dogma is not as unidirectional as originally thought. RNA plays a central role in this flow of information. However, we will also see that RNA has a number of other fundamental cellular functions, including structural and catalytic roles. A typical eukaryotic cell, such as a plant cell, carries many distinct types of RNA molecules, which can be differentiated by some unique characteristics (Figure 6.2 and Table 6.1). **Messenger RNA** (mRNA), corresponding to just 2–4% of the total RNA in the cell, conveys the bulk of the genetic information from the DNA to the proteins. mRNAs can range in size from a few hundred to several thousand nucleotides. They typically carry a **cap** at the 5′ end, consisting of modified guanine nucleotide and a 3′ tail formed by up to several hundred adenosine nucleotides (**polyA tail**). RNA modifications are discussed at the beginning of Part 3 of this book, and mRNAs are a main focus of Chapter 8.

The particular structure of each kind of RNA can be exploited for applications in biotechnology. For example, the experimentalist can take advantage of the polyA tail present in mRNAs (Figure 6.2) to separate mRNAs from other cellular RNAs. Long strings of deoxy thymines (polydT's) can be covalently linked to a solid support, such as agarose or magnetic beads. By passing a suspension of RNAs through the beads, the polyA tails of the mRNAs will form Watson–Crick base pairs with the stationary polydT

Plant Genes, Genomes and Genetics, First Edition. Erich Grotewold, Joseph Chappell and Elizabeth A. Kellogg.
© 2015 John Wiley & Sons, Ltd. Published 2015 by John Wiley & Sons, Ltd.
Companion Website: www.wiley.com/go/grotewold/plantgenes.

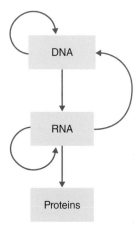

Figure 6.1 The Central Dogma of Biology. Arrows indicate the flow of information

tails. Thus, the mRNAs will bind to the polyT tract and other RNAs can be washed off. This method is known as affinity chromatography or magnetic bead capture (Figure 6.3a). In this particular application, the affinity chromatography is based on the unique pairing of polydT with the polyA present at the 3′ end of mRNAs. For many different applications, mRNAs need to be converted into complementary DNAs (**cDNAs**), by the action of **reverse transcriptase** (RT), corresponding to a RNA-dependent DNA polymerase, often from a viral origin.

As discussed earlier (see Chapter 1), determining the structure of a gene often implies establishing the DNA sequence corresponding to the 5′ end of the mRNA, which is known as the **transcription start site** (TSS) (do not confuse with the translation start site, which often consists of the AUG codon) and identified with position +1 (Figure 6.4), the intron/exon structure, and the 3′ end of the gene. This can be done by comparing the sequence of the mRNA with the corresponding genomic DNA sequence. This is conventionally done by producing a cDNA from the mRNA. If the cDNA is synthesized using oligo-dT as a primer for the reverse transcriptase, then the resulting cDNA will immediately show where the 3′ end of the transcript is (Figure 6.3). Like the mRNA from which it is copied, the cDNA will lack sequences corresponding to introns, and hence establish the intron/exon structure. In theory, the cDNA should also ascertain the 5′ end of the mRNA. However, this is often challenging because reverse transcriptase can be obstructed by RNA structures that prevent forward movement of the enzyme, so many cDNAs are truncated and fail

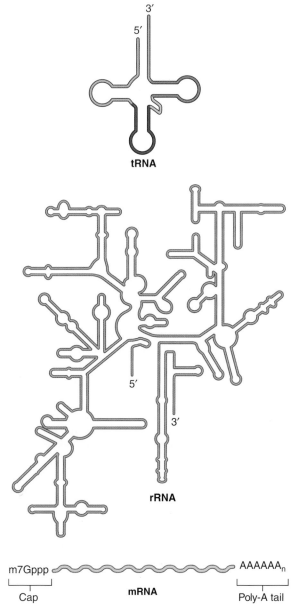

Figure 6.2 Diagrams showing main characteristics of various RNAs (tRNA, rRNA and mRNA). The different colors in the tRNA illustrate the various arms and loops. For example, the orange corresponds to the D arm and loop, the red corresponds to the anticodon arm and loop, the green corresponds to the T arm and loop, and the blue corresponds to the amino acid acceptor arm (the amino acid gets covalently linked to the 3′ end, which harbors the –OH group). For both the tRNA and rRNA, RNA structures are determined by base pairing along the stems of the RNAs

Table 6.1   Types and distribution of RNA types in a typical eukaryotic cell

RNA	Name	Function	RNA forms	Approximate size (nt)	Approximate cellular abundance (%)
rRNAs	Ribosomal RNAs	Part of ribosome (translation)	28S–25S 18S 5.8S 5S	3700 1900 160 120	80–85
mRNAs	Messenger RNAs	Protein coding		300–8000+	1–5
tRNAs	Transfer RNAs	Translation		73–93	15–20
snRNAs	Small nuclear RNAs	mRNA processing	e.g., U1, U2, U4, U5, U6	90–220	
snoRNAs	Small nucleolar RNAs	RNA processing		70–200	
miRNAs	Micro RNAs	RNA degradation, regulation of translation		22–23	
siRNAs	Small interfering RNAs	Chromatin modification, RNA degradation		21–27	

RNA and debates on the origin of life

Before we go deeper into the characteristics of the different types of RNAs inside a cell, let us go back 4.0 billion years, to the rise of life on Earth. As Richard Dawkins conjectured in *The Selfish Gene* (Dawkins, 1976): "At some point a particularly remarkable molecule was formed by accident. We will call it the Replicator. It may not necessarily have been the biggest or the most complex molecule around, but it had the extraordinary property of being able to create copies of itself." Was the *Replicator* an RNA molecule? Following the discovery of catalytic RNAs (Section 6.4), Nobel Laureate (1980) Walter Gilbert speculated that perhaps the early history of life occurred in an RNA world, where RNA served both as template and replication enzyme (Gilbert, 1986). While the idea of life beginning with RNA (the "RNA World") is a popular theory, other theories exist, including the possibility of alternative genetic systems (Robertson and Joyce, 2012) and in scenarios that consisted of metabolism-first. In the latter case, the spontaneous formation of compartments permits the occurrence of cycles of reactions which over time become more complicated, to the point at which polymers form that store the information (Shapiro, 2007). Proponents of the latter ideas point to the experiments of Stanley L. Miller and many others since the pioneering publication in 1953 (Lazcano and Bada, 2003), in which simple molecules were allowed to react in a way that might correspond to a prebiotic (before life) primordial soup. These experiments resulted in the formation of amino acids and other organic molecules, but not nucleotides or structures typically considered as having replicative potential. The debate in the area continues, and there is even a journal (*Origins of Life and Evolution of Biospheres*) that keeps the debate alive.

to reach the transcription start site. High-throughput DNA sequencing methods have made this less of a problem because the short reads can capture the 5′ ends of the cDNAs, even if they are underrepresented in the cDNA pool. In the not-so-distant past, however, other methods made use of the cap at the 5′ end of the mRNA to isolate full-length cDNAs. An example of how this is as follows: to select for cDNAs that have reached the 5′ end of the mRNA, first strand cDNA synthesis is followed by removal of any single-stranded mRNA, using for example the RNA nuclease RNase I (Figure 6.3c). Then, capped

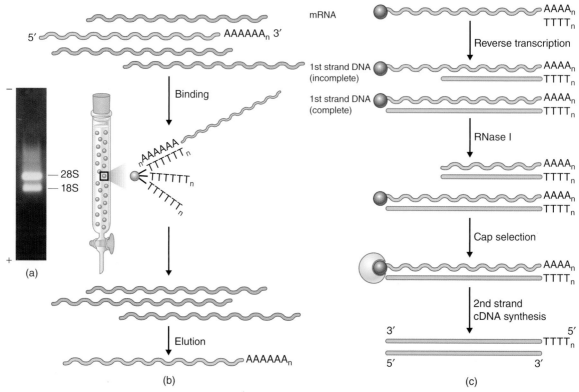

Figure 6.3 Purification of mRNA and full-length cDNA synthesis. (a) The total plant RNA separated by agarose electrophoresis ("−" and "+" electrodes indicated), stained with ethidium bromide and visualized under ultraviolet light. (b) Isolation of polyA RNA by capture on polydT beads. (c) Synthesis of complementary DNA (cDNA) starting from messenger RNA (mRNA)

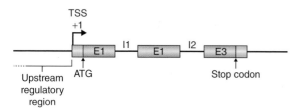

Figure 6.4 Structure of a typical plant gene. Exons (E1–3) are indicated as boxes and include the yellow boxes corresponding to the 5′UTR and 3′UTR, introns (I1 and I2) as lines, the transcription start site (TSS) corresponds to +1, and the translation start site is indicated by the ATG codon. A translation termination codon (TAG) is shown. Translation termination codons, or stop codons, include TAG, TAA and TGA. The 5′UTR and 3′UTR are shown in yellow. Many genes contain introns inside the 5′UTR; the ones illustrated here are in the protein coding region of the gene

mRNA-cDNA hybrids can be captured on beads linked to an antibody to the cap (Figure 6.3c). Truncated sequences are not capped and so are washed off, leaving only the desired full-length sequences. This method (or variants of it) is known as CAPture and has permitted the identification of the 5′ ends of a large number of mRNAs. Mapping mRNA 5′ ends to sequenced genomes is an important step in genome annotation.

6.2 Structure provides RNA with unique properties

Chemically, RNA has significantly different properties from DNA. The presence of the 2′-OH in the sugar that forms the nucleotide (Figure 6.5) makes RNA vulnerable to degradation at high (basic, >pH 7) pH. At basic pH, the 2′-OH of the ribose de-protonates to yield the oxide ion (O^-), which then attacks the adjacent 3′ phosphate (which has a partial positive charge due to the electrons being withdrawn by its resident oxygen atoms). The 2′,3′-cyclic phosphate group is unstable and readily hydrolyzes to an RNA

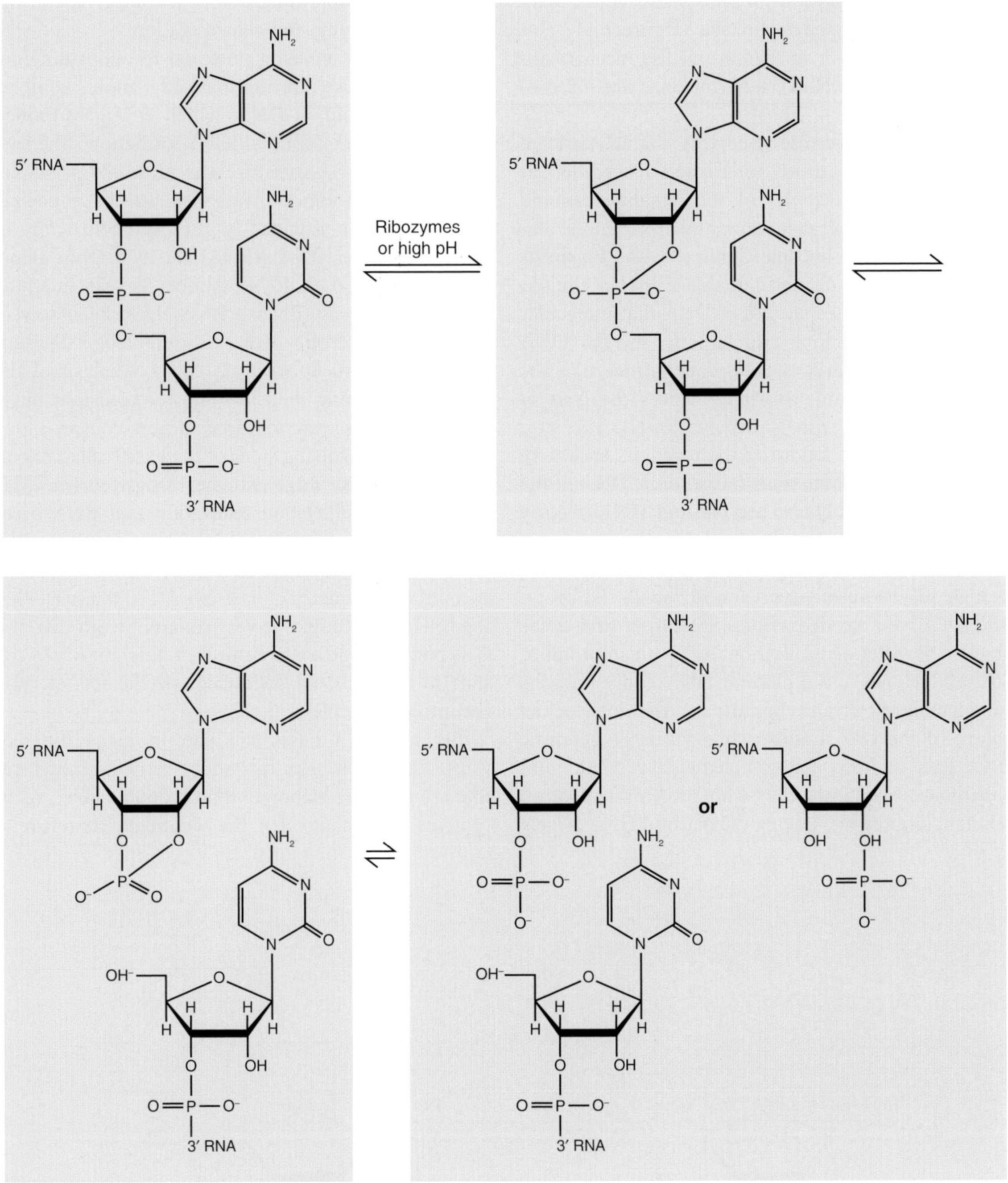

Figure 6.5 Hydrolysis of RNA in alkaline solutions

molecule with either 2′ or 3′ phosphates, thus break-ing the sugar–phosphate linkage (Figure 6.5). This acid–base reaction is at the core of the mechanisms used by **catalytic RNAs** (ribozymes, see Section 6-4) to process RNA.

Researchers have also used this characteristic of RNA extensively as a way to selectively degrade this nucleic acid. This is part of the "alkaline plasmid purification" method that many students in molecular biology labs have become familiar with. **Plasmids** are small (3–10 kb or more) double-stranded circular extrachromosomal molecules of DNA that molecular biologists use to clone and amplify specific DNA fragments in bacteria such as *Escherichia coli*. The first step in plasmid isolation involves the lysis of the bacterial cells with NaOH (or KOH). This also degrades the RNA following the reactions shown in Figure 6.5. At the same time, the genomic DNA of the bacteria and possibly also the plasmid DNA (we say possibly because it is **covalently closed circular**, or **ccc**) are denatured, but not degraded. DNA is resistant to high pH because it lacks the ribose 2′-OH group (the DNA base has deoxyribose instead of ribose, see Figure 6.5). After neutralization (e.g., with ammonium acetate, NH_4AcO), the plasmid DNA easily finds its complementary strand because of the interlocked nature of the two strands, while the large genomic DNA fails to finds the complementary strand and precipitates out of solution with proteins and other cellular components, leaving the plasmid DNA largely isolated from everything else in the supernatant, the solution remaining after centrifugation.

RNA has an amazing potential to adopt different structures. These are dramatically more complex than those found in DNA, which is largely limited by **heteroduplex** (double helix) formation. The first evidence that RNA structure was guided by principles beyond the canonical Watson–Crick base pairing (A-U, G-C) was provided by the crystal structure of the first transfer RNA (tRNA) in 1974 (Kim *et al.*, 1974; Robertus *et al.*, 1974). Almost 40 years later, the structures of many diverse RNAs have been solved, including the entire 70S ribosome from *Thermus thermophilus* (Yusupov *et al.*, 2001). Analysis of the structures forming the ribosome revealed that many of the RNA bases are involved in non-Watson–Crick base pairing, involving sides of the base (called edges) such as the sugar edge and the Hoogsteen (or C-H) edge (Figure 6.6). Upon realization that RNA bases can interact in ways beyond Watson–Crick pairing (allegedly with different strengths) and taking in to account the geometry of base pairs (i.e., the space that the U-U pair occupies is very different from that of a G-G pair), secondary structures of different RNAs can start to be predicted, following specific sets of rules (Leontis *et al.*, 2002).

Because RNA bases can pair in many different ways, RNA molecules can adopt intricate structures that are far more elaborate than a double helix. As in proteins (see Chapter 16), the **secondary structure** of

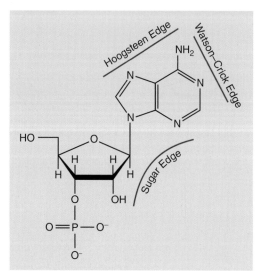
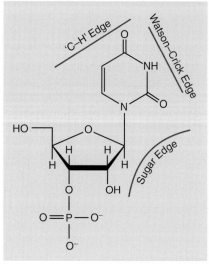

Figure 6.6 Watson and Crick and non-Watson and Crick pairing of RNA bases. The base on the left corresponds to adenosine and the one on the right to thymidine

RNA can exhibit double helices, hairpin (or terminal) loops, internal loops and multihelix loops. These secondary structures are further organized into tertiary structures, again stabilized by canonical (Watson and Crick) and non-canonical base pair interactions. The fold of the RNA provides surfaces for interaction with proteins and other nucleic acids, a case best exemplified again by the structure of the **ribosome** (Noller, 2005). An area of significant research today, helped by the increasing numbers of RNA structures being solved, is to identify recurrent RNA structural motifs that will permit prediction of the RNA structure from just the primary base sequence. This is a difficult computational problem that is not fully solved, because it requires assessing the minimum free energy structure of many possible configurations. The number of imaginable structures to be compared increases exponentially with the number of bases, and thus sophisticated algorithms must be developed to keep the amount of computational time required for predicting structures from spiraling out of control.

6.3 RNA has multiple regulatory activities

It was a great surprise when, in the 1970s, Tom Cech and Sidney Altman independently discovered that RNAs were able to catalyze chemical reactions (and were therefore awarded the Nobel Prize in Chemistry in 1989). Until that time, RNAs had been viewed as simple carriers of information, passive intermediaries in the activities of the cell. However, we now know that RNA molecules can fold into structures that can position specific functional groups, and thus can mimic the active sites of protein enzymes.

RNA catalysts are often divided into large and small. Besides the ribosome, which could be considered a large ribonucleic acid enzyme (or ribozyme) because the peptidyl transferase center is formed only by RNA (Steitz and Moore, 2003), other large RNA catalysts include RNase P and some types of introns. RNase P, required for the hydrolysis of the 5′ end of tRNA and other structural RNA precursors, has in most organisms both RNA and protein components that form what is known as a ribonucleoprotein complex. In 1983 Norman Pace and Sidney Altman demonstrated that the RNase P RNA by itself has catalytic activity, providing the first evidence for the existence of an RNA

enzyme (or ribozyme). However, in plant organelles (mitochondria and chloroplasts), RNase P activity is provided by a single protein without any RNA (Gobert *et al.*, 2010), calling into question the universality of RNase P as a ribonucleoprotein. All these ribozymes function through an acid–base mechanism very similar to that outlined in Figure 6.5, in which nucleotides serve as the general base and general acid for the reaction.

Several ribozymes that perform intramolecular cleavage reactions are known, and include the hammerhead, the hairpin, the hepatitis delta virus (HDV), and the Varkud satellite (VS) ribozymes (Ferre-D'Amare and Scott, 2010). Here, we describe the hammerhead ribozyme as an example of a catalytic RNA that was originally characterized from plants.

Hammerhead ribozymes were first identified as part of the satellite RNA of the tobacco ringspot virus (Prody *et al.*, 1986) where they participate in processing replication intermediates. Hammerhead ribozymes have since been associated with a number of viruses from plants and animals (Ferbeyre *et al.*, 2000). The 41 nucleotides that form the typical hammerhead ribozyme are organized in a tertiary structure composed of three A-form* helices (Stems I, II and III in Figure 6.7) around the catalytic core formed by 15 highly conserved nucleotides.

Structure-based approaches identified two complete hammerhead ribozymes naturally encoded in the genome of Arabidopsis (Przybilski *et al.*, 2005) (Figure 6.7). This observation was followed by a massive wave of hammerhead ribozyme discoveries in the repetitive sequences or introns of protein coding

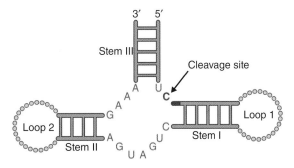

Figure 6.7 Structure of a typical hammerhead ribozyme

---

* *A-form helix.* A right-handed (clockwise) helix in which base pairs are significantly tilted with respect to the central axis of the helix (Rana, 2007). The more common structure of the Watson–Crick double helix is a DNA molecule in the B-form.

genes of a large number of plant, animal and fungal genomes (de la Pena and Garcia-Robles, 2010a, b; Seehafer *et al.*, 2011). Not all hammerheads need to be encoded in a contiguous sequence. For example, the CLEC2 ribozymes, so far identified in the 3′ UTRs of rodent C-type lectin type II genes, are split into two fragments separated by a long sequence unrelated to the ribozyme itself (Martick *et al.*, 2008).

Riboswitches are RNAs that function as sensors controlling gene expression in response to a physical (e.g., temperature) or chemical (e.g., a metabolite, tRNA or metal) signal, without the participation of a protein (Barrick and Breaker, 2007). Many riboswitches have been studied in plants (Bocobza and Aharoni, 2008). A riboswitch contains an RNA structure (aptamer) that recognizes a specific molecular target, such as a metabolite, with high affinity. It also can contain a region that modulates some aspect of gene expression such as transcription termination, translation initiation, RNA splicing, transcription interference by antisense action, or self-cleaving (Breaker, 2012).

A good example is the thiamine pyrophosphate (TPP) riboswitch, which is present in all kingdoms of life but has been particularly well characterized in plants; the crystal structure for the *Arabidopsis* molecule was solved in 2006 (Thore *et al.*, 2006). TPP, the active form of vitamin B_1, is an essential **cofactor** for many essential enzymes in all organisms. (Recall that a cofactor is a small molecule that is necessary for an enzyme to function.) The biosynthesis of thiamine involves several enzymatic steps, which are regulated by negative feedback. If TPP levels are high, then TPP itself reduces the accumulation of one of the enzymes, causing TPP levels to drop. In animals, fungi and plants, TPP controls primarily the processing (through alternative splicing, discussed in more detail in Chapter 13) and stability of the mRNA for one of the TPP biosynthetic enzymes. In plants, exemplified by the *Arabidopsis* TPP riboswitch structure shown in Figure 6.8 (which corresponds to a specific region of the *THIC* mRNA that encodes the thiamine C synthase enzyme), high TPP levels result in the inclusion of a 3′UTR intron, which decreases mRNA accumulation and protein production (Wachter *et al.*, 2007). Thus, in *Arabidopsis* and likely other plants, the TPP riboswitch functions by modulating the differential

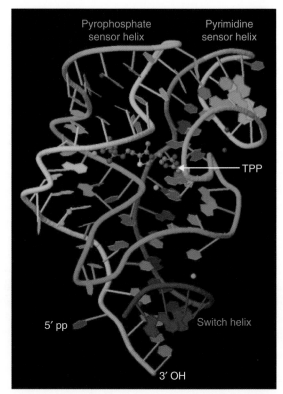

Figure 6.8 Structure of the *Arabidopsis thaliana* thiamine pyrophosphate (TPP) riboswitch. The TPP riboswitch recognizes both the pyrophosphate (left part of the structure) and pyrimidine sensor helices (right side of the structure). Note that even though the RNA is shown in different colors to highlight the various domains, it is a single RNA molecule (3′ OH and 5′ pp indicated). Image adapted from the RCSB PDB (www.rcsb.org) of PDB ID 3D2X (Thore, S., Frick, C. and Ban, N. (2008))

processing of the 3′ end of the *THIC* transcript. Although the position of the riboswitch is in the 5′ region of genes in fungi, mRNA processing mediated by the TPP-riboswitch has similar consequences (Bocobza *et al.*, 2007). Single base pair mutations that abolish the binding of TPP to the riboswitch render the structure insensitive to TPP levels. Riboswitches are becoming an important component in the toolbox of the molecular biologist, as they provide an opportunity for the generation of modules with predictable and specific regulatory behaviors. Such tunable regulatory components are becoming the basis for the emerging field of synthetic biology.

From the Experts

Viroids: multi-functional noncoding RNAs representing minimal replicating units

Viroids are single-stranded and covalently closed circular RNAs that infect plants. Their RNA genomes are 250–400 nt in size. These RNAs fold into secondary structures characterized by a series of loops flanked by helices and do not encode any proteins. A viroid RNA is thus both the genotype and phenotype. The approximately 30 viroid species currently known can be divided into those that replicate in the chloroplast or in the nucleus (Flores et al., 2005; Ding, 2009).

In nature, it is believed that a new infection starts when a viroid enters cells of a healthy plant that have been mechanically wounded by agricultural tools or other means that were contaminated with viroids by prior contact with infected plants. A viroid RNA must first resist cellular attacks from nucleases and RNA silencing. It then moves into the nucleus or the chloroplast to initiate replication. Some viroid progeny will remain in the organelles, but others will exit and move into neighboring cells via plasmodesmata to initiate new infections. Finally, viroids move into the phloem (vascular tissue) and spread throughout the whole plant. All or most of these biological processes involve interactions between viroid sequence/structural elements and cellular factors. Viroid infections may cause diseases in hosts, ranging from growth-stunting to plant death, with significant agronomical losses. Few other noncoding RNAs are known to express such an array of diverse biological functions. Viroids are therefore useful models to study many aspects of basic RNA biology including noncoding RNA regulation of gene expression, RNA structure–function relationships, RNA-mediated RNA replication, and intracellular and intercellular RNA movement, as well as the basic mechanisms of infection (Ding, 2010).

Viroid replication involves transcription of an RNA-template to produce molecules consisting of many repeated viroid sequences (concatemers); these are cleaved into unit-length molecules and subsequently ligated into circles. The viroid sequence/structural elements mediating each of these steps have been determined for several viroids in both families. The DNA-dependent RNA polymerase II (see Chapters 7 and 8) is used to transcribe nuclear viroids, whereas nuclear-encoded, chloroplast-targeted DNA-dependent RNA polymerase is used to transcribe the viroids found in that organelle. The cellular factors that cleave and ligate nuclear viroids are unknown. Chloroplastic viroids have intrinsic ribozyme activity, thus are able to cleave themselves. Whether ligation is catalyzed by the ribozyme or a cellular factor remains unclear (Flores et al., 2011).

A genome-wide mutational analysis has identified multiple loops in *Potato spindle tuber viroid* (PSTVd), a nuclear viroid, that are essential or important for single-cell replication and systemic infection (Zhong et al., 2008) (Figure 6.9). Several structural loops in the viroid RNA mediate movement of PSTVd between specific cells (Qi et al., 2004; Zhong et al., 2007; Takeda et al., 2011) but the cellular factors involved remain to be identified.

Specific viroid sequences or secondary structures play a role in host diseases. Although the specific mechanisms remain to be elucidated, viroid disease essentially results from the abnormal expression of host genes by these noncoding RNAs, which alter plant growth and development. Unlike many viruses, viroids usually have relatively narrow host ranges, with each species infecting one or a few plant species in the field. There is evidence for the continuing expansion of host ranges for some viroids. This represents continuous evolution of noncoding RNA sequences/structures to attain new biological functions (Ding, 2010).

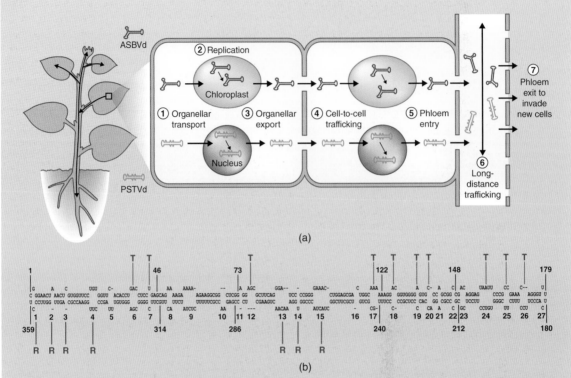

(a)

(b)

Figure 6.9 (a) Systemic infection of *Avocado sunblotch viroid* (ASBVd) and PSTVd that replicate in the chloroplast and nucleus, respectively. The infection includes intracellular and cell-to-cell trafficking, replication and long-distance trafficking. Adapted from Ding (2010). Reproduced with permission from John Wiley & Sons, Ltd. (b) Genomic map of PSTVd motifs for systemic trafficking (T) in a whole plant or for replication (R) in single cells of *Nicotiana benthamiana*. Adapted from Zhong *et al.* (2008). Copyright 2008 American Society of Plant Biologists. Used with permission

By Biao Ding

6.4   Summary

RNA molecules are much more than just messengers between the DNA cellular blueprint and the protein workhorses of the cell; they are key structural components with important catalytic activities. RNA structure is very important for their multiple functions, and often involves base pair interactions other than the canonical A-T and G-C. The origin of life probably involved some form of RNA molecule. Ribozymes are catalytic RNAs that form an important component of the cellular machinery. Viroids, which are plant specific, represent some of the smallest forms of replicating RNAs.

6.5   Problems

6.1 The following are terms that the student should be able to define: ribozyme, riboswitch, viroid, mRNA, tRNA, rRNA, cDNA, polyA tail.

6.2 Give an example of an RNA molecule that has both structural and informational functions.

6.3 List biological processes in which base pairing controls the specificity of a biological event.

6.4 Besides affinity purification using oligo-dT affinity chromatography, how could you utilize affinity chromatography in a different way to specifically enrich for mRNA from a total RNA preparation?

[Hint: Think about what is the large majority of RNA in a total RNA preparation.]

6.5 Biologists tend to speak about DNA or RNA solutions. Is this correct? Why?

6.6 Look for methods that would allow RNA to be denatured, for example to separate it by size on an agarose gel by electrophoresis.

6.7 Order the following by the strength of the double strand (consider same length molecules): RNA-RNA, DNA-DNA, RNA-DNA.

6.8 Indicate if the sequence of GGUCGACUUAGC is more likely to correspond to DNA or RNA.

6.9 What aspects of RNA biology are specific to plants?

References

Barrick, J.E. and Breaker, R.R. (2007) The power of riboswitches. *Scientific American*, **296**, 50–57.

Bocobza, S., Adato, A., Mandel, T. *et al.* (2007) Riboswitch-dependent gene regulation and its evolution in the plant kingdom. *Genes & Development*, **21**, 2874–2879.

Bocobza, S.E. and Aharoni, A. (2008) Switching the light on plant riboswitches. *Trends in Plant Science*, **13**, 526–533.

Breaker, R.R. (2012) Riboswitches and the RNA world. *Cold Spring Harbor Perspectives in Biology*, **4**. pii: a003566 (PMID: 21106649).

Dawkins, R. (1976) *The Selfish Gene*, Oxford University Press.

de la Pena, M. and Garcia-Robles, I. (2010a) Intronic hammerhead ribozymes are ultraconserved in the human genome. *EMBO Reports*, **11**, 711–716.

de la Pena, M. and Garcia-Robles, I. (2010b) Ubiquitous presence of the hammerhead ribozyme motif along the tree of life. *RNA*, **16**, 1943–1950.

Ding, B. (2009) The biology of viroid-host interactions. *Annual Review of Phytopathology*, **47**, 105–131.

Ding, B. (2010). Viroids: self-replicating, mobile, and fast-evolving noncoding regulatory RNAs. *Wiley Interdisciplinary Reviews: RNA*, **1**, 362–375.

Ferbeyre, G., Bourdeau, V., Pageau, M. *et al.* (2000) Distribution of hammerhead and hammerhead-like RNA motifs through the Genbank. *Genome Research*, **10**, 1011–1019.

Ferre-D'Amare, A.R. and Scott, W.G. (2010) Small self-cleaving ribozymes. *Cold Spring Harbor Perspectives in Biology*, **2**, a003574.

Flores, R., Grubb, D., Elleuch, A. *et al.* (2011) Rolling-circle replication of viroids, viroid-like satellite RNAs and hepatitis delta virus: variations on a theme. *RNA Biology*, **8**, 200–206.

Flores, R., Hernandez, C., Martinez de Alba, A.E. *et al.* (2005) Viroids and viroid-host interactions. *Annual Review of Phytopathology*, **43**, 117–139.

Gilbert, W. (1986). Origin of life: the RNA world. *Nature*, **319**, 618–618.

Gobert, A., Gutmann, B., Taschner, A. *et al.* (2010) A single *Arabidopsis* organellar protein has RNase P activity. *Nature Structural & Molecular Biology*, **17**, 740–744.

Kim, S.H., Suddath, F.L., Quigley, G.J. *et al.* (1974) Three-dimensional tertiary structure of yeast phenylalanine transfer RNA. *Science*, **185**, 435–440.

Lazcano, A., and Bada, J.L. (2003) The 1953 Stanley L. Miller experiment: fifty years of prebiotic organic chemistry. *Origins of Life and Evolution of Biospheres*, **33**, 235–242.

Leontis, N.B., Stombaugh, J. and Westhof, E. (2002) The non-Watson-Crick base pairs and their associated isostericity matrices. *Nucleic Acids Research*, **30**, 3497–3531.

Martick, M., Horan, L.H., Noller, H.F. and Scott, W.G. (2008). A discontinuous hammerhead ribozyme embedded in a mammalian messenger RNA. *Nature*, **454**, 899–902.

Noller, H.F. (2005) RNA structure: reading the ribosome. *Science*, **309**, 1508–1514.

Prody, G.A., Bakos, J.T., Buzayan, J.M. *et al.* (1986) Autolytic processing of dimeric plant virus satellite RNA. *Science*, **231**, 1577–1580.

Przybilski, R., Graf, S., Lescoute, A. *et al.* (2005). Functional hammerhead ribozymes naturally encoded in the genome of *Arabidopsis thaliana*. *Plant Cell*, **17**, 1877–1885.

Qi, Y., Pelissier, T., Itaya, A. *et al.* (2004) Direct role of a viroid RNA motif in mediating directional RNA trafficking across a specific cellular boundary. *Plant Cell*, **16**, 1741–1752.

Rana, T.M. (2007) Illuminating the silence: understanding the structure and function of small RNAs. *Nature Reviews Molecular Cell Biology*, **8**, 23–36.

Robertson, M.P. and Joyce, G.F. (2012) The origins of the RNA world. *Cold Spring Harbor Perspectives in Biology*, **4**. pii: a003608 (PMID: 20739415).

Robertus, J.D., Ladner, J.E., Finch, J.T. *et al.* (1974) Structure of yeast phenylalanine tRNA at 3 Å resolution. *Nature*, **250**, 546–551.

Seehafer, C., Kalweit, A., Steger, G. *et al.*(2011) From alpaca to zebrafish: hammerhead ribozymes wherever you look. *RNA*, **17**, 21–26.

Shapiro, R. (2007) A simpler origin for life. *Scientific American*, **296**, 46–53.

Steitz, T.A. and Moore, P.B. (2003) RNA, the first macro-molecular catalyst: the ribosome is a ribozyme. *Trends in Biochemical Sciences*, **28**, 411–418.

Takeda, R., Petrov, A.I., Leontis, N.B. and Ding, B. (2011) A three-dimensional RNA motif in potato spindle tuber viroid mediates trafficking from palisade mesophyll to spongy mesophyll in *Nicotiana benthamiana*. *Plant Cell*, **23**, 258–272.

Thore, S., Leibundgut, M. and Ban, N. (2006) Structure of the eukaryotic thiamine pyrophosphate riboswitch with its regulatory ligand. *Science*, **312**, 1208–1211.

Wachter, A., Tunc-Ozdemir, M., Grove, B.C. *et al.* (2007) Riboswitch control of gene expression in plants by splicing and alternative 3′ end processing of mRNAs. *Plant Cell*, **19**, 3437–3450.

Yusupov, M.M., Yusupova, G.Z., Baucom, A. *et al.* (2001) Crystal structure of the ribosome at 5.5 Å resolution. *Science*, **292**, 883–896.

Zhong, X., Archual, A.J., Amin, A.A. and Ding, B. (2008) A genomic map of viroid RNA motifs critical for replication and systemic trafficking. *Plant Cell*, **20**, 35–47.

Zhong, X., Tao, X., Stombaugh, J. *et al.* (2007) Tertiary structure and function of an RNA motif required for plant vascular entry to initiate systemic trafficking. *EMBO Journal*, **26**, 3836–3846.

Chapter 7
The plant RNA polymerases

7.1 Transcription makes RNA from DNA

In the previous chapter, we discussed the different types of RNAs present in a cell, their structures, chemical properties and regulatory activities. This chapter will summarize how RNA is made from DNA by a set of proteins known as DNA-dependent RNA polymerases, or more simply, **RNA polymerases**. The subject is one in which very significant discoveries have and continue to be made, and to which an entire textbook can be dedicated. Because there are outstanding reviews and books on this topic, this chapter will focus on providing the background necessary to understand the concepts in subsequent chapters, highlighting particular discoveries relevant to plants.

The process of copying a particular portion of the DNA into an RNA molecule is known as **transcription**, and the resulting RNA molecules are the transcripts. These transcripts have a nucleotide sequence complementary to the DNA strand from which they originate, the **template** strand, and as is the case for almost all nucleic acid biological synthesis, they are generated in a $5' \rightarrow 3'$ fashion. Thus, they are not only complementary in sequence to the DNA, but they are also antiparallel, meaning that the $5'$ end of the forming RNA molecule will correspond to the $3'$ region of the template DNA. Because the DNA is double-stranded, the transcribed RNA will be identical in sequence (with the exception of uracil in place of thymine bases) to the strand that is not being copied, which is called the **coding (or sense) strand** (Figure 7.1).

Transcription is the first step in the process of gene expression and it is perhaps the most highly regulated. Transcription involves four basic stages: (1) template recognition; (2) initiation; (3) elongation; and (4) termination. These four stages are characteristic of any of the enzymes that we will discuss in this chapter, and each one is precisely regulated to determine which templates must be transcribed, when, how many transcript copies need to be generated, and where those transcripts need to end. We describe later in the book the DNA characteristics that permit a gene to be transcribed with particular spatial and temporal patterns. Thus, for simplicity in this chapter we will refer to DNA sequences that direct and regulate transcription of a gene as a **promoter** (Figure 7.1). The promoter will correspond to gene regulatory DNA sequences located upstream ($5'$) from the transcription start site (TSS) (do not confuse the TSS with the first codon on which proteins will initiate). Gene transcription can also be controlled by DNA sequences (gene regulatory regions) located inside genes or in $3'$ regions of genes. Hence, the promoter is one of several possible gene regulatory sequences. By convention, we assign to the TSS the number +1, anything downstream ($3'$) is indicated with positive numbers, while sequences that are not transcribed and which are $5'$ to the TSS (or upstream) have negative numbers (Figure 1.8).

While the processes of DNA replication and transcription share some characteristics, there appears to be less selection for fidelity when making RNA, since its sequence usually does not get transmitted to the next generation. DNA replication is a very accurate process, with mutation rates <1 nucleotide change

Plant Genes, Genomes and Genetics, First Edition. Erich Grotewold, Joseph Chappell and Elizabeth A. Kellogg.
© 2015 John Wiley & Sons, Ltd. Published 2015 by John Wiley & Sons, Ltd.
Companion Website: www.wiley.com/go/grotewold/plantgenes.

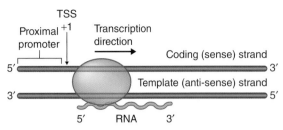

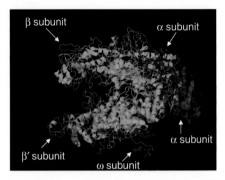

Figure 7.1 Diagram of RNA transcription. The two DNA strands are shown in blue, while the RNA is shown in red. The RNA generated by RNA polymerase II (RNP-II) is identical in sequence (with the exception that T nucleotides are replaced by U) to the coding strand, while the antisense strand serves as the template for RNP-II, which moves from left to right in this diagram. TSS corresponds to the transcription start site, which by convention is indicated by +1

Figure 7.2 Structure of the *Thermus aquaticus* RNA polymerase. The $\alpha_2\beta\beta'\omega$ core is shown. Protein Data Bank PDB ID 1HQM (http://www.wwpdb.org/) Minakin, L. *et al.* (2001)

every 10^9 nucleotides per cell division, but transcription introduces an average of one mutation every 10 000 nucleotides. However, while well studied in animal systems, very little is known about the fidelity of RNA polymerases in plants.

7.2 Varying numbers of RNA polymerases in the different kingdoms

Prokaryotes have a single RNA polymerase that is responsible for the synthesis of all ribosomal RNA (rRNA), transfer RNA (tRNA) and messenger RNA (mRNA). A typical bacterium such as *Escherichia coli* contains about 7000 RNA polymerase molecules, of which 2000–5000 can be engaged in transcription at any given time. The *E. coli* RNA polymerase has a molecular weight of 465 kDa, and is formed by five different subunits, each known by a different Greek letter ($\alpha_2\beta\beta'\omega\sigma$). The complex formed by the α_2, β, and β' subunits constitutes the **core polymerase** (Figure 7.2). While this core can synthesize RNA, it is unable to unwind the DNA on its own; therefore it can only use single-stranded or "relaxed" double-stranded DNA templates. For example, plasmids are normally in the covalently closed circular form (usually known as the ccc conformation) and are super-coiled; a nick in one of the two strands "relaxes" the super-coiling. Introducing nicks and unwinding the DNA is perhaps the function of some σ subunits, whose primary role

is in promoter recognition (see Chapter 8). The crystal structure of the $\alpha_2\beta\beta'$ core with one σ factor (σ^{70}) and DNA shows that σ^{70} interacts primarily with the surface provided by the $\beta\beta'$ subunits, and that σ^{70} makes extensive DNA sequence-specific contacts in the -10 to -35 region. In eukaryotes, the process of promoter recognition is largely performed by **transcription factors** (see Chapter 10). The β and β' subunits form the catalytic center (Figure 7.2), make extensive contact with each other, and have characteristics shared by eukaryotic catalytic subunits. The ω subunit functions in the assembly of the complex.

Metazoans have three distinct RNA polymerase activities, as demonstrated in seminal experiments by Roeder and Rutter in the 1970s involving the fractionation of sea urchin or rat liver extracts by ion exchange chromatography, a technique that permits the separation of macromolecules based on charge. These activities were subsequently shown to correspond to three distinct RNA polymerases, numbered I, II and III. In plants, fungi and animals RNA polymerase I (RNP-I) transcribes the genes for rRNA; RNA polymerase II (RNP-II) transcribes mRNA that will be translated to make proteins; and RNA polymerase III (RNP-III) transcribes tRNA and 5S rRNA (Table 7.1). RNP-I, RNP-II, and RNP-III share a number of subunits. In yeast, for example, all three RNA polymerases have 12 very similar subunits, with five of them being homologous to the $\alpha_2\beta\beta'\omega$ bacterial RNA polymerase core.

The analysis of the Arabidopsis genome (Arabidopsis Genome Initiative, 2000) revealed the presence of an unusual group of RNA polymerase subunits. These proved to encode two new RNA polymerases, RNP-IV and RNP-V, which are unique to the plant kingdom,

Table 7.1   Comparison of nuclear RNA polymerase functions in animals, yeast and plants

RNA polymerase	Yeast and animals	Plants
RNA polymerase I (RNP-I) (nucleolus)	28S rRNA 18S rRNA 5.8S rRNA	25S rRNA 18S rRNA 5.8S rRNA
RNA polymerase II (RNP-II)	mRNAs miRNAs snRNAs snoRNAs Other non-coding RNAs	mRNAs miRNAs snRNAs snoRNAs Other non-coding RNAs
RNA polymerase III (RNP-III)	5S rRNA tRNAs snRNAs (U3, U6) SINEs	5S rRNA tRNAs snRNAs (U3, U6) SINEs
RNA polymerase IV (RNP-IV)		siRNAs (24 nt)
RNA polymerase V (RNP-V)		Non-coding RNAs (involved in gene silencing)

rRNA, ribosomal RNA; tRNA, transfer RNA; miRNA, micro RNA; snRNA, small nuclear RNA; snoRNA, small nucleolar RNA; SINE, short interspersed nuclear element

yet evolved from the largest subunit of RNP-II (Luo and Hall, 2007). Plant genes encoding the subunits of RNP-I are named *NRPA* followed by a number that indicates the subunit. For example, *AtNRPA1* corresponds to the largest subunit of Arabidopsis RNP-I. Similarly, genes encoding RNP-II subunits are *NRPB*, for RNP-III are *NRPC*, for RNP-IV are *NRPD* and *NRPE* for RNP-V.

RNP-IV is primarily involved in the synthesis of small interfering RNAs (siRNAs), small (21–24 nucleotide long) RNA molecules involved in gene silencing (see Chapter 11). In contrast, RNP-V transcribes non-coding RNAs (ncRNAs), primarily derived from intergenic regions, which help guide components of the gene silencing machinery to specific DNA sequences (Wierzbicki *et al.*, 2008, 2009). These polymerases all have different locations inside the nucleus, reflecting the very different functions that they perform. However, as discussed later in this chapter, there is significant cross-talk between these polymerases, which share several subunits.

A major distinction between RNP-I, RNP-II and RNP-III is the sensitivity to the fungal toxin **α-amanitin**, a cyclic peptide formed by eight amino acids. RNP-II is inhibited by very low doses of

α-amanitin, while the activity of RNP-I is not affected at all. The sensitivity of RNP-III varies depending on the species of origin. For example, high-levels of α-amanitin inhibit animal RNP-III, but not the yeast enzyme. Plant RNP-III shows an intermediate sensitivity, and RNP-IV is insensitive to the cyclic peptide, highlighting that, despite sharing some subunits, the α-amanitin binding pockets of RNP-IV and RNP-V are different from that of RNP-II. Another drug commonly used to block RNA polymerases is **actinomycin D**, a polypeptide obtained from the bacteria *Streptomyces*. Unlike α-amanitin, actinomycin D blocks most effectively RNP-I, by binding to the double-stranded DNA template and intercalating itself between C-G and G-C base pairs. Actinomycin D will also block all other RNA polymerases at higher concentrations (Bensaude, 2011).

Actinomycin D belongs to a large group of compounds known as **antibiotics**, a term coined by Selman Waksman in 1942. We will mention antibiotics multiple times in this book, in the context of the main cellular process that they affect. Besides actinomycin D, another antibiotic that interferes with the process of transcription is rifamycin, which directly inhibits the bacterial RNA polymerase. As is often the case for antibiotics, synthetic derivatives (such as rifampicin) show increased activity and a broader action spectrum.

Study of the components necessary for DNA transcription has been greatly facilitated by *in vitro transcription systems*. In a typical *in vitro* transcription system, a template double-stranded DNA consisting of a minimal promoter region capable of driving transcription is fused to a small sequence that can be easily assayed. The minimal promoter must be empirically determined, but it usually comprises a region of 50–100 base pairs around the TSS. The template DNA is then supplied with the corresponding RNA polymerase and purified nuclear fractions (and of course nucleotides and other components that allow RNA synthesis to happen). These purified nuclear fractions have historically been named as TFIIX: a purified fraction X (obtained after some series of classical biochemistry purification steps) that allows RNP-II to activate transcription. Such fractions might be composed of one or many different proteins. While there are many advantages of using *in vitro* transcription systems as an approximation to identify necessary cellular components, they often fail to completely capture the exquisite specificity found *in vivo*.

In this chapter, we will describe basic features of RNP-I, RNP-III, RNP-IV and RNP-V, as well as the structure of the promoters controlled by RNP-I and RNP-III. Because of the complexity of RNP-II promoters and the huge literature addressing their functions, the following chapter (Chapter 8) specifically focuses on the regulation of transcription that result in mRNA formation by this polymerase.

7.3   RNA polymerase I transcribes rRNAs

The genes encoding the large **ribosomal RNAs** are organized in one or more tandem arrays consisting of several thousands of head-to-tail transcription units. The transcribed rDNA genes are spatially organized in the nucleolus, a non-membrane bound structure where the initial assembly of ribosomes occurs. In Arabidopsis, for example, there are two non-contiguous clusters of rDNA genes known as NORs (for nucleolus organizer regions), each containing ~375 rRNA genes (Doelling *et al.*, 1993). The regions corresponding to rDNA represent roughly 6% of the entire Arabidopsis genomic DNA. A single rDNA transcription unit results in a large RNA precursor that contains one copy each of the 18S, 5.8S and 25S rRNAs. Each rDNA gene is separated from the rest by intergenic spacers (IGSs, Figure 7.3). Within each rRNA gene, the regions encoding the 18S, 5.8S and 25S rRNAs are separated by two internal transcribed spacers (ITSs) (Figure 7.3) (see also Section 2.6). After transcription, the precursor RNA is cleaved, releasing the three large rRNAs. Neither the rRNA precursor nor the resulting rRNAs are capped, which is a 5′ RNA modification characteristic of RNP-II genes and discussed in more detail in Chapter 13. Each rDNA gene contains its own promoter.

Within one species, rDNA genes are almost identical to each other suggesting an active mechanism by which the organism keeps them highly similar, known as concerted evolution. However, when rDNA genes are compared between different species, even if closely related, very large differences are observed, and such variations have been extensively used to deduce evolutionary relationships between species (Alvarez and Wendel, 2003).

The unique organization of the rDNA genes contributes to the structure of the nucleolus, a non-membrane bound discrete subnuclear region in which the rDNA sequences are nucleated, rRNA transcription takes place, and the ribosomal subunits are assembled before being exported to the cytoplasm. In these NOR clusters, some of the rDNA genes are transcribed, while others are silent. At mitosis, the nucleolus is dissembled and transcription of rDNA genes is completely stopped.

The transcription of the large rRNAs is performed by nucleolar RNP-I. This polymerase is responsible for ~75% of the entire transcriptional output of a cell; it needs to transcribe massive quantities of rRNA to assemble the tens of thousands of ribosomes present in a eukaryotic cell. Transcription of rRNAs is tightly linked to the metabolic state of the cell, as well as the stage of the cell cycle.

Similar to all other RNPs, RNP-I is formed by many (14) subunits and has a molecular weight larger than 500 kDa. All the rDNA genes transcribed by RNP-I share a single type of promoter, which has been best characterized in animal cells. The RNP-I promoters are located in the IGS that separates adjacent rDNA genes (Figure 7.3). In plants and animals, but not in yeast, the IGS is rich in repetitive sequences, all of which contribute to enhancing rDNA transcription (Figure 7.3).

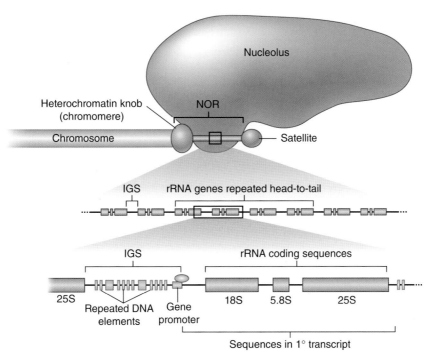

Figure 7.3 Organization of plant rDNA genes. Nucleolus organizer regions (NORs) are formed by long repeats containing the genes encoding the three largest rRNAs (18S, 5.8S and 25S). Within each NOR, the corresponding rRNA genes are almost identical in sequence. IGS, intergenic spacer. Adapted from Pikaard (2000). Reproduced with permission of Elsevier

The typical yeast or animal RNP-I promoter has a bipartite sequence formed by a core promoter that extends from −45 to +20, and it is sufficient for transcription to initiate. An upstream control element (UCE), located between −180 and −107, enhances the transcription furnished by the core promoter (Figure 7.4). The core promoter and the UCE are quite similar to each other (>85%) and are rich in G/C, which is a rare characteristic for promoters. However, the typical plant RNP-I promoter has a slightly different organization. In Arabidopsis, for example, sequences located between −55 and −33 and around +6 are sufficient for RNP-I transcription. At the transcription start site, there is a conserved sequence (TATAT$\underline{A}$^{A}/$_G$GGG, where $\underline{A}$ corresponds to +1) that has been well characterized in several plants, and is necessary for accurate transcriptional initiation (Doelling et al., 1993; Doelling and Pikaard, 1995). A similar sequence is also present around the transcription start site of animal and fungal rDNAs. So far, however, plant rDNA promoters appear to lack the UCE.

At least two protein complexes are required for RNP-I activity in animals. The upstream binding factor UBF1 recognizes both the core promoter and

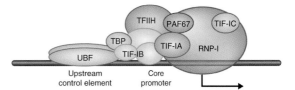

Figure 7.4 Cartoon of the pre-initiation complex of the vertebrate RNP-I promoter

the UCE, while selective factor 1 (SL1), formed by four distinct proteins, does not recognize any specific DNA sequences. Only when the UBF1-SL1-DNA complex is formed is RNP-I recruited to DNA and poised to initiate transcription (Figure 7.4). The SL1 complex contains a protein called the TATA-binding protein (TBP) that will be described in more detail later in the discussion about RNP-II, and other RNA polymerases. For activation of RNP-I promoters, the DNA-binding activity of TBP is not necessary, suggesting that it functions primarily by helping assemble the RNP-I complex. The structure of plant rRNA gene promoters is less well understood, and there is little evidence supporting the bipartite structure found in animals or yeast. The region surrounding the RNP-I TSS

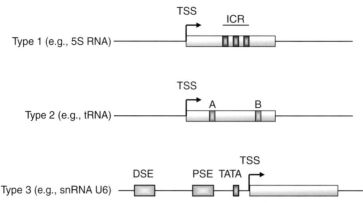

Figure 7.5 Different types of RNP-III promoters. The internal control region (ICR) of Type 1 promoters is formed by three distinct DNA sequence elements. DSE, distal sequence element; PSE, proximal sequence element; TATA, TATA box

appears to be very conserved between different plants and often includes the TATATA sequence, which can serve as a TBP-binding site, but functions even in its absence, suggesting another role as well (Doelling and Pikaard, 1996; Saez-Vasquez and Echeverria, 2006). In addition, many of the identified components of the yeast and animal complexes controlling RNP-I activity do not seem to be obviously present in sequenced plant genomes (Saez-Vasquez and Echeverria, 2006).

Because of the key role that ribosomes play during protein synthesis, RNP-I activity is modulated not only by the phase of the **cell cycle** in which the cells are, but also by stress, development and hormones. A key player that integrates hormonal and developmental signals with rRNA accumulation is the target of rapamycin (TOR) factor, a large serine/threonine protein kinase that is conserved in all eukaryotes. In Arabidopsis, TOR can enter the nucleus and through its HEAT domain [the name HEAT comes from the four proteins in which this domain was originally found: Huntingtin, elongation factor 3 (EF3), protein phosphatase 2A (PP2A), and the yeast kinase TOR1], directly bind to the 45S rRNA regulatory regions and hence control rRNA transcription.

7.4 RNA polymerase III recruitment to upstream and internal promoters

The promoters controlled by RNP-III are of two distinct classes (Figure 7.5). Promoters for 5S rRNA

and tRNA are downstream of the transcription start site (hence they are internal), while the promoters for snRNA are upstream from the start point. Transcription factors specific to RNP-III recognize these elements, which in turn recruit the polymerase to the respective promoters.

The complex architecture necessary for RNP-III to initiate transcription (also known as the **pre-initiation complex**, often denoted by PIC in the literature) depends on the particular promoter. Internal promoters can be divided in to two types, based on how RNP-III specific complexes assemble to recruit the polymerase. In Type 2 promoters, TFIIIC (transcription factor IIIC) recognizes two sequences within the transcribed region known as box A and box B (Figure 7.5). Binding of TFIIIC to these DNA sequences permits the recruitment of the TFIIIB large protein complex to the transcription start site, facilitating the recruitment of RNP-III. In Type 1 promoters, the TFIIIA protein complex binds to DNA box A, and this facilitates the recruitment of protein complex TFIIIC to DNA box C. As in Type 2 promoters, this then facilitates the recruitment of TFIIIB, followed by the recruitment of the polymerase. TFIIIB contains the TBP, which is also an integral part of the pre-initiation complexes for RNP-I and RNP-II. We discuss the TATA motif and TBP in significantly more detail in Chapter 8, as part of the control of RNP-II genes. However, in contrast to (some) RNP-II transcribed promoters, and similar to RNP-I promoters, there is no TATA box in the Type 1 or Type 2 RNP-III transcribed promoters. Type 3 promoters, which are external promoters, do contain a TATA motif (Figure 7.5).

7.5   Plant-specific RNP-IV and RNP-V participate in transcriptional gene silencing

RNP-IV was simultaneously discovered by the Baulcombe and Pikaard laboratories in 2005, using different approaches (Herr *et al.*, 2005; Onodera *et al.*, 2005). Upon examination of the Arabidopsis genome, the Pikaard group identified subunits for a distinct RNA polymerase that did not co-purify with RNP-I, RNP-II or RNP-III. Disruption of the *NRPD1* or *NRPD2* genes, corresponding to the RNP-IV largest subunit, showed significant decondensation and reduction in methylation of the 5S rDNA genes (Onodera *et al.*, 2005).

The identification of RNP-IV by the Baulcombe group originated from the identification of the *sde4* mutant, which exhibited reduced silencing of a transgene. The group subsequently demonstrated that SDE4 corresponded to NRPD1, the largest subunit of the RNP-IV (Herr *et al.*, 2005). As described more extensively in Chapter 11, plants have developed a sophisticated mechanism to protect their genome from DNA insertions, caused for example by transposable elements (see Chapter 3) or by the insertion of foreign DNA. This mechanism is known as **transcriptional gene silencing (TGS),** and involves the formation of small RNAs, particular of the siRNA class. **Gene silencing** significantly complicates maintaining expression of **transgenes** (genes artificially introduced into the plant genome).

The main identified function of RNP-V is to transcribe ncRNA, primarily derived from intergenic

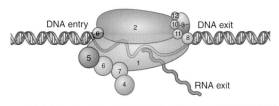

DNA entry   DNA exit

RNA exit

○ Subunits of RNP-II, RNP-IV and RNP-V that are encoded by the same gene
○ Subunits of RNP-II, RNP-IV and RNP-V that are unique to each enzyme
○ Subunits common to RNP-II and RNP-IV but unique in RNP-V
○ Subunits common to RNP-IV and RNP-V but unique in RNP-II
○ Subunits common to RNP-II and RNP-V but not observed in RNP-IV

Figure 7.6 RNP-II, RNP-IV and RNP-V share several subunits. Adapted from Haag and Pikaard (2011). Reproduced with permission of Macmillan Publishers Ltd

regions that help guide components of the gene silencing machinery to specific DNA sequences (Wierzbicki *et al.*, 2008, 2009). The mechanisms by which RNP-V recognizes intergenic regions to generate ncRNAs remain unknown. The functions of RNP-IV and RNP-V with regards to their participation in RNA-directed DNA methylation will be discussed in significantly more detail in Chapter 12. What is very interesting and worth stressing is the fact that many of the subunits are shared between RNP-II, RNP-IV and RNP-V (Figure 7.6).

7.6   Organelles have their own set of RNA polymerases

Plant organelles such as chloroplasts (plastids) and mitochondria were acquired in successive steps of **endosymbiosis** from prokaryotic organisms represented by cyanobacteria and bacteria, respectively (Gray, 1992) (see Chapters 1 and 5). Chloroplasts retained only ~100 genes from the thousands present in the **cyanobacterial** ancestor, and many of them conserved the **polycistronic** nature that characterizes bacterial **operons**. Chloroplasts also incorporated several unique elements that were not present in the original prokaryotic symbiont, such as the presence of introns and the modification of mRNAs by a process called **editing**, which consists of the change of cytidine to uridine at about 40 total positions in plastid RNAs (Barkan, 2011). Editing is a feature also shared by mitochondria. However, most of the proteins present in the chloroplast are encoded in the nucleus, synthesized in the cytoplasm and imported into the organelle.

Plastids require at least two types of RNA polymerases for plastidic gene transcription: a plastid encoded RNA polymerase (PEP) and one or two nuclear encoded polymerases (NEPs) (Hess and Borner, 1999; Barkan, 2011; Lerbs-Mache, 2011). Rather than a division of labor between NEP transcribing housekeeping genes and PEP transcribing photosynthesis genes, as originally thought, it is evident now that most chloroplast genes can be transcribed by either NEP or PEP, yet from different promoters (Barkan, 2011). However, mutation analyses indicate that both polymerase systems are necessary for making photosynthetically active chloroplasts.

PEPs have significant similarity to bacterial RNA polymerase subunits, and they have been found in all studied chloroplast genomes. As is the case in bacteria, promoters recognized by PEPs harbor DNA sequences that resemble σ factor-binding sites. These σ factors, which belong to the σ^{70} group, are encoded by the nucleus, and as many as six of them have been described in Arabidopsis, while only one is present in the unicellular alga *Chlamydomonas reinhardtii*. In Arabidopsis, each of the σ factors has a different function (Lerbs-Mache, 2011). NEPs are most similar to single-subunit T7 bacteriophage-like RNA polymerases important for mitochondrial gene transcription (Cermakian *et al.*, 1997) (see below). In dicotyledoneous plants, there are two NEPs, while none has so far been found in *C. reinhardtii*. This calls into question whether there is a unifying way in which all chloroplasts control gene expression (Lerbs-Mache, 2011).

Most mitochondrial genomes lack RNA polymerase genes, and nuclear genes encode the RNA polymerase activity of these organelles. However, the mitochondria of several fungi and plants (e.g., maize) also contain linear plasmids flanked by long terminal inverted repeats. These plasmids can contain open reading frames, some of them corresponding to single-subunit bacteriophage-like RNA polymerases.

7.7   Summary

Plants have three nuclear RNA polymerases, RNP-I, RNP-II and RNP-III, that are also found in other eukaryotes. In addition, they have RNP-IV and RNP-V, which are primarily involved in the control of gene expression through the recognition and synthesis of small RNAs. The mechanisms by which RNP-I, RNP-II and RNP-III transcribe DNA to generate rRNA, mRNA, and tRNA, respectively, are similar in many ways but how they are recruited to specific promoters and form pre-initiation complexes are distinct. Organelles have a distinct set of RNA polymerases, which in part reflect the origin of plastids and mitochondria from prokaryotic cells.

7.8   Problems

7.1 Using the information in this and previous chapters, define the following terms:
   (a) Gene
   (b) Promoter

   (c) Intron
   (d) Exon
   (e) mRNA
   (f) Transcription start site.

7.2 All polymerases function by synthesizing new DNA or RNA polymers in the $5' \rightarrow 3'$ polar fashion. Is there a similar polarity in the way nucleases function, for example, during proofreading?

7.3 Eukaryotic RNP-II can add 1000–2000 nucleotides in a minute. How long would the enzyme take to transcribe an average plant gene?
   [Hint: Remember that introns are also transcribed.]

7.4 Plastid transformation has become possible in the past decade. How would you design a promoter to be expressed in plastids, based on what is known about plastid regulation of gene expression?

7.5 What do you imagine to be the advantages of engineering genes into plastids, rather than in the nucleus of a plant?

References

Alvarez, I. and Wendel, J.F. (2003) Ribosomal ITS sequences and plant phylogenetic inference. *Molecular Phylogenetics and Evolution*, **29**, 417–434.

Arabidopsis Genome Initiative. (2000) Analysis of the genome sequence of the flowering plant *Arabidopsis thaliana*. *Nature*, **408**, 796–815.

Barkan, A. (2011) Expression of plastid genes: organelle-specific elaborations on a prokaryotic scaffold. *Plant Physiology*, **155**, 1520–1532.

Bensaude, O. (2011) Inhibiting eukaryotic transcription: which compound to choose? How to evaluate its activity? *Transcription*, **2**, 103–108.

Cermakian, N., Ikeda, T.M., Miramontes, P. *et al.* (1997) On the evolution of the single-subunit RNA polymerases. *Journal of Molecular Evolution*, **45**, 671–681.

Doelling, J.H., Gaudino, R.J. and Pikaard, C.S. (1993) Functional analysis of *Arabidopsis thaliana* rRNA gene and spacer promoters *in vivo* and by transient expression. *Proceedings of the National Academy of Sciences USA*, **90**, 7528–7532.

Doelling, J.H. and Pikaard, C.S. (1995) The minimal ribosomal RNA gene promoter of *Arabidopsis thaliana* includes a critical element at the transcription initiation site. *Plant Journal*, **8**, 683–692.

Doelling, J.H. and Pikaard, C.S. (1996) Species-specificity of rRNA gene transcription in plants manifested as a switch in RNA polymerase specificity. *Nucleic Acids Research*, **24**, 4725–4732.

Gray, M.W. (1992) The endosymbiont hypothesis revisited. *International Review of Cytology*, **141**, 233–357.

Haag, J.R. and Pikaard, C.S. (2011) Multisubunit RNA polymerases IV and V: purveyors of non-coding RNA for plant gene silencing. *Nature Reviews*, **12**, 483–492.

Herr, A.J., Jensen, M.B., Dalmay, T. and Baulcombe, D.C. (2005) RNA polymerase IV directs silencing of endogenous DNA. *Science*, **308**, 118–120.

Hess, W.R. and Borner, T. (1999) Organellar RNA polymerases of higher plants. *International Review of Cytology*, **190**, 1–59.

Lerbs-Mache, S. (2011). Function of plastid sigma factors in higher plants: regulation of gene expression or just preservation of constitutive transcription? *Plant Molecular Biology*, **76**, 235–249.

Luo, J. and Hall, B.D. (2007) A multistep process gave rise to RNA polymerase IV of land plants. *Journal of Molecular Evolution*, **64**, 101–112.

Onodera, Y., Haag, J.R., Ream, T. *et al.* (2005) Plant nuclear RNA polymerase IV mediates siRNA and DNA methylation-dependent heterochromatin formation. *Cell*, **120**, 613–622.

Pikaard, C. (2000) The epigenetics of nucleolar dominance. *Trends in Genetics*, **16**, 495–500.

Saez-Vasquez, J. and Echeverria, M. (2006) Polymerase I transcription, in *Regulation of Transcripion in Plants* (ed. K.D. Grasser), Blackwell Publishing Ltd, pp. 162–183.

Wierzbicki, A.T., Haag, J.R. and Pikaard, C.S. (2008) Noncoding transcription by RNA polymerase Pol IVb/Pol V mediates transcriptional silencing of overlapping and adjacent genes. *Cell*, **135**, 635–648.

Wierzbicki, A.T., Ream, T.S., Haag, J.R. and Pikaard, C.S. (2009) RNA polymerase V transcription guides ARGONAUTE4 to chromatin. *Nature Genetics*, **41**, 630–634.

Chapter 8

Making mRNAs – Control of transcription by RNA polymerase II

8.1 RNA polymerase II transcribes protein-coding genes

Chapter 7 provided an overall description of the five RNA polymerases that transcribe nuclear genes in plants. Among them, RNA polymerase II (RNP-II) has the primary responsibility for transcribing all messenger RNAs (mRNAs), many of which will be translated into proteins by the translational machinery described in detail in Chapter 15, while others will remain as non-coding RNAs. Transcripts generated by RNP-II can also be precursors to small RNAs (smRNAs), as will be described in more detail in Chapter 11. Because different sets of proteins and smRNAs are required in different cells at different times, the mechanisms by which RNP-II is recruited to the plethora of DNA regulatory sequences that characterize protein- and smRNA-coding genes are significantly more complex than those associated with RNP-I or RNP-III. Hence, the description of the DNA sequences that control RNP-II function, and the mechanisms by which RNP-II functions, merits a separate chapter and will be discussed here. We will start by briefly describing what is known about the structure of RNP-II, and then will focus on how RNP-II is recruited to DNA to form what is known as a **pre-initiation complex**. We will then describe **elongation** and **termination** of transcription by RNP-II. RNP-II transcribed genes are often referred to as Class II genes.

8.2 The structure of RNA polymerase II reveals how it functions

Much of our knowledge on the structure and mechanisms of action of RNP-II derives from studies carried out in yeast, pioneered by the work of Roger D. Kornberg who was honored with the Nobel Prize in Chemistry in 2006 for his studies on the molecular basis of eukaryotic transcription. RNP-II is the engine of the transcriptional machinery, capable on its own of unwinding the DNA, as well as polymerizing and proofreading the resulting RNA. RNP-II transcribes genes at the amazing speed of about 1500 nucleotides per minute. The enzyme consists of 12 subunits, a catalytic core plus a heterodimeric (hetero = distinct, dimeric = two) complex that often dissociates from the yeast enzyme during purification. In yeast, the

Plant Genes, Genomes and Genetics, First Edition. Erich Grotewold, Joseph Chappell and Elizabeth A. Kellogg.
© 2015 John Wiley & Sons, Ltd. Published 2015 by John Wiley & Sons, Ltd.
Companion Website: www.wiley.com/go/grotewold/plantgenes.

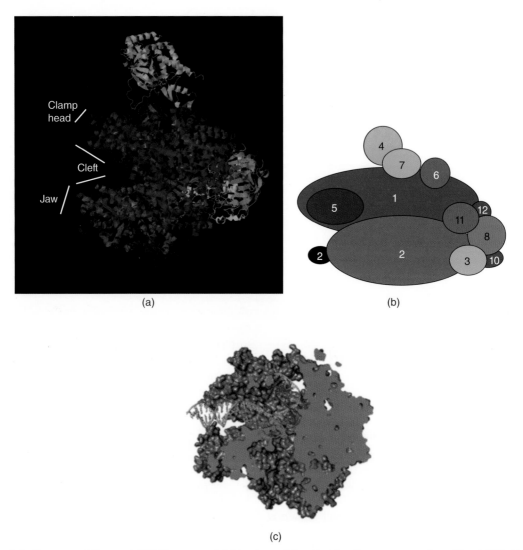

Figure 8.1 Structure of the yeast RNP-II enzyme. (a) Structure of the 12-subunit enzyme complex without DNA. Protein Data Bank 1WCM (http://www.wwpdb.org/). (b) Subunit representation in (a). Note that subunit 9 is beyond the complex and cannot be seen in this representation, and that subunits 12 and 10 are almost completely hidden as well. Colors in (b) correspond to those in (a). RNP-I, RNP-II, and RNP-III each have five core subunits (indicated by 1, 2, 3, 6, and 11) and five shared subunits (indicated by 5, 8, 9, 10 and 12). The two unique subunits of RNP-II (4 and 7) correspond to the Rpb4 and Rpb7 proteins. The "tail" (C-terminal domain, CTD) of subunit 1 is hidden in these representations. (c) Schematic view of the RNP-II elongation complex showing the entering DNA (left) and exiting DNA (top). Panel (c) Adapted from Wang *et al.* (2006). Reproduced with permission of Elsevier

genes encoding each of these subunits are named *RPB1-12*; and mutations in all but *RPB4* and *RPB9* are lethal. Rpb1 and Rpb2 correspond to the largest subunits (Figure 8.1). Strikingly, at least 10 of these yeast RNP-II subunits can substitute for the respective subunits in humans. This suggests a very high level of functional conservation, and provides good evidence that knowledge gained from studying the yeast proteins can be applied to plants or animals.

The structure of the 12-subunit yeast RNP-II was elucidated with and without DNA attached (Cramer *et al.*, 2000; Gnatt *et al.*, 2001). The structures are overall very similar, with the main difference being the position of the "clamp head" in the structures shown in Figure 8.1. The cleft is formed by the two largest sub-units, the clamp head and the jaw, and it is positively charged (remember that nucleic acids are negatively charged at cellular pH) to facilitate interactions with both the template DNA and the nascent RNA. The cleft contains the active site that the clamp swings over during transcription. As described in Chapter 7, nine of the 12 subunits are shared with RNP-I and RNP-III.

In addition to the 12 subunits (recall that subunits refer to proteins encoded by separate genes; these subunits assemble in a complex that constitutes the RNP), RNP-II contains a C-terminal domain (CTD) that is formed by a variable number of repeats. This is a feature that distinguishes the largest RNP-II subunit from those of RNP-I or RNP-III. The CTD is formed by a heptad repeat with the amino acid sequence $Tyr_1Ser_2Pro_3Thr_4Ser_5Pro_6Ser_7$, or $Y_1S_2P_3T_4S_5P_6S_7$ if using the single letter code for amino acids, where the numbers indicate the position within the repeat. Depending on the organism, the number of repeats varies. For example, yeast has 26 repeats, mammals have 52 and Arabidopsis has 34. Pretty much every amino acid in the CTD can undergo several types of post-translational modifications that modulate the activity of RNP-II. These modifications can also work in a combinatorial fashion, generating a complex regulatory code. We describe a similar regulatory code for the modification of histones in Chapter 12.

Here, we will be primarily focusing on the role of CTD phosphorylation. The status of CTD phosphorylation provides a good indication of whether transcription is initiating or elongating. In addition, CTD coordinates transcription with RNA processing mechanisms such as 5′-Cap addition, splicing, 3′-end formation, and mRNA nuclear export. These RNA processing activities are described in more detail in Chapter 13. The CTD is essential for the transcription of most genes, and whether it is required or not appears to depend on whether or not the core pro-moter, described in the following section, contains particular DNA motifs.

8.3   The core promoter

The promoter of a gene has been typically defined as the DNA sequences to which the RNA polymerase binds.

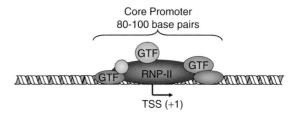

Figure 8.2 Diagram illustrating the core promoter of a RNP-II transcribed gene, indicating the RNP-II and some general transcription factors (GTFs). The transcription start site (TSS) is by convention indicated with the position +1

We shall expand this definition here by describing a promoter as all the DNA sequences flanking the transcription start site (TSS) that are required for the accurate expression of a gene. This includes the **core or basal promoter**, which is the minimum DNA sequence required for accurate transcription initiation by RNP-II (Figure 8.2). This definition of promoter also includes some of the DNA sequences to which a particular group of regulatory proteins, known as transcription factors, bind. Transcription factors are described in more detail in Chapter 9. It is worth noting that there are different definitions of gene promoter, and that the definition evolves as we learn more about how DNA sequences participate in the control of gene expression.

RNA polymerase enzymes on their own (the 10- or 12-subunit complex described in Section 8.2) can unwind DNA, and synthesize and proofread the nascent RNA. However, they lack any type of specificity in terms of what DNA to transcribe, or where to initiate transcription. In bacteria, RNA poly-merase binds to the DNA directly, and the specificity provided by the σ factor subunit of the polymerase (see Chapter 7) is sufficient to provide accurate tran-scription initiation. In contrast, eukaryotic RNP-II requires a number of other factors known as general transcription factors (GTFs) for promoter recognition and recruitment of RNP-II, since the purified RNP-II is catalytically active but cannot initiate transcrip-tion from a DNA template containing a promoter. However, if the purified RNP-II is provided with a cell extract, accurate transcriptional initiation occurs. Such an approach to assay RNA polymerase activity is called an *in vitro* transcription system. *In vitro* transcription systems provided an excellent assay for the identification of GTFs, which along with RNP-II and the appropriate DNA template, are necessary and sufficient for accurate RNP-II transcription. The RNP-II core promoter consists of 80–100 bp of DNA sequence centered around the TSS (Figure 8.2). The designation "GTF" is used to distinguish them from the gene-specific transcription factors described in Chapter 9. GTFs are shared among large numbers of promoters, and today it is thought that they contribute very little, if anything, to assisting RNP-II in establishing which genes it needs to transcribe at a specific time or in a particular cell; rather, they are simply part of the general transcription machinery. In contrast, the transcription factors described in Chapter 9 decode information contained in the DNA with regard to when, how much, and where a particular gene needs to be expressed. This decoding consists of transcription factors binding to specific promoter DNA sequences, and translating this code into a signal to the RNP-II through protein–protein interactions with components of the basal transcriptional machinery, which could be the GTFs, the Mediator complex described below, or RNP-II subunits.

The core promoter is also essential for the correct positioning and orientation of the RNP-II with respect to the TSS. Which sequences are characteristic of core promoters? To answer this question, we will first begin with a few definitions. DNA-sequence motifs that participate in the regulation of gene expression are often called *cis*-regulatory elements. They act in *cis* (from the Latin meaning on the same side) because they are located in the same DNA heteroduplex as the gene they control. This is different from the *trans*-acting factors (*trans* from the Latin across) encoded elsewhere in the genome. *Trans*-acting factors include transcription factors that have their site of action – which is to bind to specific DNA sequence elements – spread throughout the genome. One of the first *cis*-regulatory elements identified, and which is present in a significant fraction of RNP-II promoters, is the **TATA box**, originally identified by David Hogness in 1977. As its name indicates, the TATA box is a DNA motif rich in T and A nucleotides, and is usually positioned 25 to 30 base pairs upstream of the TSS (hence at positions −25 to −30). The TATA motif (with the consensus DNA sequence $TATA^T/_AA^T/_AA$ in Arabidopsis) (Molina and Grotewold, 2005) resembles the Pribnow box present in bacterial promoters, which is usually located at −10 and serves as a binding site for σ factors. As in animals, only about 30% of all plant Class II genes have a TATA motif. Core promoters or genes lacking a TATA motif are generally called **TATA-less**. The control of TATA-containing and TATA-less promoters happens by distinct mechanisms. For example, the RNP-II CTD appears to be essential for the transcription of TATA-less promoters, while it is thought to be dispensable in at least some TATA-containing promoters.

In Class II genes, the TATA motif is recognized by a GTF called TATA binding protein, or TBP. Remember that in Chapter 7 (Section 7.4) we discussed TBP in the context of transcription by RNP-I and RNP-III. Interestingly, TBP is essential for the assembly of the general transcription machinery complex, even in TATA-less promoters. We will learn more about TBP when we discuss the formation of the pre-initiation complex.

Plant core promoters generally lack many of the *cis*-regulatory elements that are common in animals, the main exception being the TATA motif. Thus, much remains to be learned about the structure of plant core promoters; it is even possible that structural components (e.g., chromatin) beyond just DNA sequence conservation play an important role.

We previously described the TSS as corresponding to the first nucleotide present in the mRNA. (You may want to look at Chapter 6 to remind yourself how TSSs are experimentally identified.) However, some genes have multiple weak TSS spread over hundreds of base pairs or more from one another. These two modes of transcription initiation are known as **focused** (one TSS) and **dispersed** (several TSS), respectively (Juven-Gershon and Kadonaga, 2010). Focused transcription appears to be characteristic of unicellular organisms and of many plant genes. In contrast, more than 70% of vertebrate genes are transcribed in the dispersed mode. Dispersed transcription is largely associated with the presence of CG islands, genomic regions rich in CG dinucleotides that can be 500–2000 bp long and within 150 base pairs immediately upstream of the TSS. CG islands are present in a very large number (>60%) of human genes and are recognized by the transcription factor Sp1, which then helps recruit the RNP-II pre-initiation complex to the corresponding genes. Importantly, CG islands, as defined in animals, are absent from plant genes, although there are clearly regions in plant genes with higher CG density.

8.4 Initiation of transcription

Before we describe how transcription initiates, we need to look at how the RNP-II complex assembles on the DNA in the genes it will control. In many ways, this resembles the launch pad for the (now extinct) space shuttle. The space shuttle, formed by the RNP-II complex, gets assembled and idles at the pad until the launch signal is given. The launch pad on which the RNP-II is assembled is the core or basal promoter. In the *in vitro* transcription system described in Section 8.3, this core promoter is sufficient to provide a very low level of transcript formation, and this is known as **basal transcription**. In our example, the idling of the shuttle could represent basal transcription. *In vitro* transcription systems resulted in the identification of several GTFs that are part of the complexes known as TFIIA, TFIIB, TFIID (which includes the TBP protein), TFIIE, TFIIF and TFIIH (Table 8.1). In this analogy, the launch signal is then provided by protein–protein interactions initiated by transcription factors that are recruited to other (regulatory) regions of the gene, in what we will generally call the promoter, which includes enhancers and other regulatory DNA elements to which other transcription factors bind.

The assembly of the pre-initiation complex, a prerequisite for transcription initiation, follows a defined sequence of events, as shown by studies carried out *in vitro* (Figure 8.3). The complex is nucleated by TFIID, which is recruited to DNA by the binding of TBP

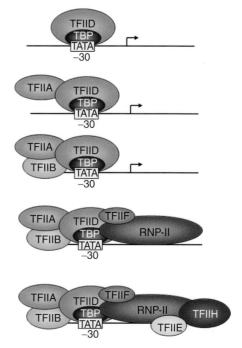

Figure 8.3 Assembly of the RNP-II pre-initiation complex on a TATA-containing promoter. TFIIA–TFIIH represent GTFs; TBP corresponds to the TATA binding protein. See the text for additional explanation

Table 8.1 Components of each general transcription factor complex. It is important to note that knowledge on these factors comes from studies in metazoans, with very little still known about their presence and functions in plants

TFII complex	Proteins forming the complex
TFIID	15 subunits: TBP and 14 TAFs (numbered TAF1 through TAF14)
TFIIA	2 subunits: TFIIα and TFIIγ
TFIIB	Single subunit
TFIIF	2 subunits: RAP74 and RAP30
TFIIE	2 subunits: TFIIα and TFIIβ
TFIIH	10 subunits: Cdk7 (CTD kinase), Cyclin H, XPD and XPB (helicases), p44 (E3 ubiquitin ligase), MAT1, p8, p52, p34, p62

to the TATA motif, in a TATA-containing promoter. This is followed by the recruitment of TFIIA, stabilizing the complex. In metazoans, TFIIB then binds, making contacts with the BRE element, if present. It remains unclear how this happens in plants. The TFIID-TFIIA-TFIIB complex is then sufficient to recruit the non-phosphorylated form of the RNP-II enzyme, which usually associates with TFIIF. The last steps in the assembly of the pre-initiation complex involve recruitment of TFIIH and TFIIE. It is largely unknown how assembly of the pre-initiation complex occurs in the absence of a TATA and other core promoter conserved elements. Based on a few studies in animals (George *et al.*, 2006), it is possible that nucleosome positioning and the three-dimensional structure of the chromatin helps distant *cis*-regulatory sequences (not necessarily restricted to the core promoter) come spatially together and help the pre-initiation complex form.

Let us look with a bit more detail at the components of each of the GTF complexes, as this will permit us to better understand the mechanisms by which RNP-II initiates and elongates transcription. Most of the TFII complexes are formed by multiple subunits, and some examples are provided in Table 8.1.

8.4.1 TFIID

TFIID is formed by TBP and 14 TBP-associated factors, or TAFs. The structure of the entire TFIID complex forms a horseshoe or clamp-like structure (Burley and Roeder, 1996). TBP has also been crystallized, originally from Arabidopsis (Nikolov et al., 1992), and subsequently from a number of other organisms. It has a saddle-like structure, and induces a sharp bend in the DNA. As a reminder, TBP is not just important for RNP-II transcription, but it is also an essential component of the transcriptional complexes of RNP-I and RNP-III (see Chapter 7). TATA-less genes often do not require TBP itself, but their expression can depend on TBP-like factors (TLFs). Nine of the 14 known TAFs have histone-fold motifs (see Chapter 12), which are involved in protein–protein interactions. TAFs with these motifs interact with each other forming five histone-like pairs, which are essential for the structural features of TFIID. Not all TAFs are universally required in all promoters, and which TAFs are essential in each case depends on particular components of the promoter. In fact, different TAFs can recognize different *cis*-regulatory elements. In metazoans, for example, TAF1 and TAF2 recognize the Initiator element, TAF6 and TAF9 recognize the DPE, and TAF12 makes contacts with specific transcription factors. TAFs can also function in the absence of TBP, in what are known as TBP-less TAF complexes.

8.4.2 TFIIA

TFIIA is formed by two subunits, and it is not essential for accurate transcriptional initiation. TFIIA functions in stabilizing the TBP-DNA complex and also blocks access to transcriptional inhibitors. Plants have a TFIIA complex that is more similar to that of animals than that of yeast (Li et al., 1999). Arabidopsis TFIIA mutants are unable to grow under conditions that require the oxidative stress response pathway (Kraemer et al., 2006). This pathway is important in the response of plants to environmental conditions (e.g.,

drought and cold) or developmental processes (e.g., seed maturation).

8.4.3 TFIIB

TFIIB is a single polypeptide that plays a key role in initiation by helping RNP-II identify the correct TSS (TSS selection); it also helps regulate promoter clearance, that is, the removal of GTFs and other DNA-binding proteins after RNP-II leaves the launch pad and begins productive elongation.

8.4.4 TFIIF

TFIIF is a heterodimer that associates with RNP-II before the RNP-II/TFIIF complex is recruited to the pre-initiation complex. TFIIF helps facilitate promoter opening, which involves the "melting" of DNA to allow the transcription bubble to form. Similar to TFIIB, TFIIF also participates in TSS selection.

8.4.5 TFIIE

Similar to TFIIF, TFIIE binds as a heterodimer of two subunits of 34 and 56 kDa in humans. It binds the promoter close to the TSS, creating a docking site for TFIIH. Based on sequence similarity, plants express proteins similar to the TFIIE subunits, but little is known about their function.

8.4.6 TFIIH

TFIIH is the most multifaceted of the six GTF complexes. It is formed by 10 subunits (Table 8.1) with a total molecular weight that exceeds 500 kDa. TFIIH participates in at least three distinct functions: promoter melting and RNP-II moving away from the TSS (escape) during initiation, transcription-associated DNA repair, and progression during the cell cycle. TFIIH is particularly involved in DNA excision repair, and is associated with at least three different genetic disorders in humans: Xeroderma pigmentosa, Trichothiodystrophy and Cockayne Syndrome. Each of these three disorders is caused by a mutation that impairs either the transcription or repair functions of TFIIH. Xeroderma pigmentosa results from mutations that only affect DNA excision repair. Trichothiodystrophy is a consequence of defects in transcription, while in Cockayne Syndrome, transcription-dependent repair is defective.

The TFIIH complex contains five catalytic activities. Cdk7 (Cdk stands for cyclin dependent kinase) is involved in phosphorylating the CTD of RNP-II, signaling the transition from initiation to elongation, and is described in more detail in Section 8.6. The DNA-dependent ATPase activity, and 5′ to 3′ and 3′ to 5′ DNA helicase activities participate in unwinding the DNA as RNP-II initiates and elongates transcription. The E3 ubiquitin ligase activity is associated with the transcription of genes associated with DNA repair. A better-known role of E3 ubiquitin ligases in targeted protein degradation is described in Chapter 17.

8.5   The mediator complex

Similar to the GTFs described in the previous section, Mediator is a large multi-protein complex that can bind directly to RNP-II and participates in pretty much every stage of transcription (initiation, elongation and termination). Although it can have a much more gene-specific function than other GTFs, it is nevertheless considered a GTF. The main function of Mediator (as its name suggests) is to provide a bridge between the gene-specific transcription factors (described in Chapter 9), the GTFs described in the previous section and RNP-II (Conaway and Conaway, 2011), primarily the CTD. The core Mediator complex is formed by a set of 20 or more conserved proteins (known as Med) arranged in three modules: head, middle, and tail (Figure 8.4). Proteins in these modules can interact with specific domains of transcription factors (known as transcriptional activation or repression domains, see Chapter 9), allowing the Mediator complex to adopt one of at least three distinct configurations (Figure 8.4): The "Mediator core" itself, the 'holoenzyme' in which Mediator is wrapped around RNP-II, and the 'core plus kinase' complex, which contains several kinases (including Cdk enzymes) and which block the interaction of Mediator with RNP-II. Experimental evidence indicates that the Mediator 'core' and the 'holoenzyme' configurations are primarily associated with transcriptional activation, while the 'core plus

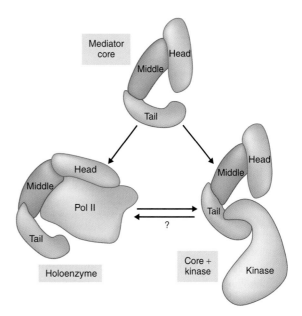

Figure 8.4 Mediator has been found to be present in yeast and animal cells in at least three different conformations. Conway and Conway (2011). Reproduced with permission of Elsevier

kinase' configuration participates in transcriptional repression.

Plant Mediator complex components have been characterized from Arabidopsis and rice (Backstrm *et al.*, 2007; Mathur *et al.*, 2011). Interestingly, mutants in several genes encoding Mediator components had been previously identified as affecting a number of developmental processes, plant metabolism or the response to environmental cues. One interesting example is provided by Arabidopsis Med25 corresponding to Phytochrome and Flowering Time 1 (PTF1), which was originally identified as a nuclear protein that functions in the phytochrome B pathway to induce flowering under suboptimal light conditions (Cerdan and Chory, 2003). PTF1 also controls the jasmonate signaling pathway (jasmonates are plant hormones, with jasmonoyl-isoleucine being the active form), thus participating in fungal resistance. Mutations in Arabidopsis Med25 render plants more resistant to drought, and this is a consequence of Med25 directly interacting with at least three transcription factors. One of them can either activate or repress transcription, with repression being mediated by Med25 (Elfving *et al.*, 2011).

8.6   Transcription elongation: the role of RNP-II phosphorylation

Now that we understand GTFs and the Mediator complex, we can come back to how transcription proceeds. Once the pre-initiation complex has formed, the helicases melt (separate the strands of) the DNA, allowing the one strand of the DNA to be positioned close to the catalytic site of the non-phosphorylated form of RNP-II. Keep in mind that RNP-II must not be phosphorylated to form the pre-initiation complex. The transition to transcriptional elongation is associated with phosphorylation of the serine residues at the second and fifth positions of the CTD (Ser2 and Ser5, respectively; Figure 8.5) of RNP-II (Buratowski, 2009). Phosphorylation of Ser5 is catalyzed by the kinase subunit of TFIIH (Cdk7; Table 8.1). This happens during the stage of promoter clearance (clearance refers to RNP-II breaking contact with the transcription factors that remain associated with the promoter region of the gene) at the initiation of transcription. In fact, the Ser5 phosphorylated form of RNP-II is primarily found bound to the 5′ end of genes. One way in which this has been determined is by conducting chromatin immunoprecipitation (ChIP, see Section 9.6) experiments using antibodies that recognize specific forms of RNP-II (such as RNP-II with Ser2 and Ser5 phosphorylated). Phosphorylation of Ser5 is also a prerequisite to the addition of the 5′-cap (see Chapter 13), providing one of several examples of how transcription and mRNA processing are intimately linked. Recent studies have shown that the TFIIH-associated kinase can also phosphorylate Ser7, which might also have a similar function during elongation as phosphorylated Ser5 (Figure 8.5).

As mentioned above, phosphorylated Ser5 is primarily associated with transcription of the first couple of hundred nucleotides. As RNP-II progresses through the DNA template, the levels of phosphorylated Ser5 decrease and phosphorylation of Ser2 increases (Figure 8.5), although RNP-II molecules with both marks can be found on the same template DNA molecule. Phosphorylation of Ser2 is carried out by another kinase (Bur1 in yeast), which appears to be recruited by phosphorylated Ser5, ensuring that Ser2 is phosphorylated after the phosphorylation of Ser5 has happened. Phosphorylation is a reversible mechanism,

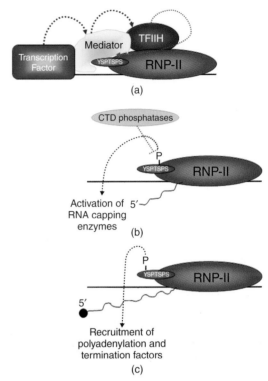

Figure 8.5   Simplified schema of the RNP-II C-terminal domain (CTD) cycle. (a) Transcription factors interact with the Mediator complex (protein–protein interactions represented by the blue lines), which in turn interacts with TFIIH kinase that phosphorylates Ser5 of the CTD of RNP-II [represented by the YSPTSPS sequence that is present in multiple repeats (34 in *Arabdiopsis*)]. (b) As RNP-II clears the promoter and the nascent RNA is formed (represented by the wavy line and the 5′), phosphorylated Ser5 participates in the recruitment of enzymes involved in RNA capping and is required for phosphorylation of Ser2. (c) Phosphorylated Ser2 participates in the recruitment of the transcriptional termination and RNA polyadenylation complexes. Not shown here is the effect of the different complexes on histone modifications

and CTD phosphatases that have distinct specificities for each of the phosphorylated residues in the CTD have been extensively described, primarily in yeast and animals. However, in the recent past, plant homologs of yeast/animal CTD phosphatases have been identified and are generally known as CTD phosphatase-like (CPL). Arabidopsis CPL1 and CPL2 are specific for Ser5 dephosphorylation (Koiwa *et al.*, 2004), which is important not just during transcriptional elongation, but is also essential during transcriptional termination to ensure that RNP-II is dephosphorylated before entering a new pre-initiation complex. CPL5 is specific for Ser2 dephosphorylation and positively regulates abscicic acid (ABA) dependent development (Jin *et al.*, 2011). Mutational analyses also revealed functions for Arabidopsis CPL3 in controlling ABA signaling, while CPL4 is required for normal Arabidopsis growth and development (Bang *et al.*, 2006).

8.7　RNP-II pausing and termination

RNP-II (with the phosphorylated Ser5 mark; Figure 8.5b) is often bound to the promoters of genes, even though transcription is not occurring. In this case, the RNP-II is described as either paused or arrested. The difference between pausing and arrest is that a paused RNP-II can continue transcription in the absence of other proteins, while arrested RNP-II needs other factors, such as transcription factor IIS (TFIIS). During pausing, RNP-II is engaged, but a collection of negative elongation factors (NELFs) make it stall. The effect of negative elongation factors is overcome by the action of the positive transcription elongation factor b (P-TEFb), which results in phosphorylation of Ser2. Stalling was first discovered in a heat shock gene (HSP70) from the fruit fly (*Drosophila melanogater*). Most heat shock proteins, including those from plants, accumulate very rapidly (within a few minutes) in response to a sublethal increase in temperature, known as a heat shock. Stalling was believed to be a mechanism by which, by having the RNP-II engaged in transcription, a very rapid response could be ensured. However, it appears that RNP-II is stalled at a very large fraction of all eukaryotic genes. In metazoans, RNP-II stalling is enriched in genes involved in signal transduction pathways (Adelman and Lis, 2012). It remains to be shown if this is also the case in plants.

RNP-II transcription does not progress along the template DNA strand in a continuous fashion, but rather the enzymes alternate between forward and backward movement. The reverse movement is known as **backtracking**, which leads to both pausing and arrest. Backtracking is less usual when RNP-II is in the paused state. Backtracking is also an important point in proofreading and correcting the RNA, if mistakes were introduced.

Transcription termination is significantly less well-understood than initiation or elongation. Termination has to be precisely controlled to ensure RNP-II recycling, and to prevent RNP-II from running into the next (downstream) gene, generating, for example, double-stranded RNAs that can result in gene silencing (see Chapter 11). Two pathways for transcription termination by RNP-II have been described (Kuehner *et al.*, 2011).

The first one is known as **polyadenylation (polyA)-dependent termination**, in which the end of transcription is intimately linked to the processing of the 3′ end of the forming transcript. This processing and degradation happens simultaneously with transcription and hence is called co-transcriptional. As RNP-II passes and transcribes the polyA site with the consensus sequence AAUAAA (incorporating the sequences required for polyadenylation into the nascent RNA), RNP-II pauses and RNA-binding proteins bind to the CTD of RNP-II and the polyA motif recruiting nucleases and other factors. Cleavage of the new transcript at the polyadenylation site facilitates access to an exoribonuclease (*exo* because it degrades from the end), which collides with the transcribing RNP-II, destabilizing the RNP-II from the template DNA. Transcription termination of full-length transcripts usually ends 500 bp–2 kb downstream (3′) of the polyadenylation site (see Chapter 13). Recently the Arabidopsis DICER-LIKE 4 (DCL4), an enzyme more prominently associated with gene silencing through microRNAs (miRNAs) (see Chapter 11), was shown to participate in the 3′ processing and transcriptional termination of the *FCA* gene (Liu *et al.*, 2012). More information on transcriptional termination can also be found in Chapter 13.

The second termination pathway is known as *Sen1*-dependent termination. This mechanism was first identified in yeast but has also been described in animals, where the *Sen1* homolog is called senataxin; whether this pathway also occurs in plants is unknown. *Sen1* processes non-coding RNAs such as small nuclear RNAs (snRNAs) and small nucleolar RNAs (snoRNAs) (see Chapter 6), which do not contain polyA tails. *Sen1* and other factors interact with the RNP-II CTD, providing conformational changes and unwinding the nascent RNA-DNA hybrid, which ultimately causes the polymerase to disengage. Genome transcript analyses in a number of organisms (but not yet in plants) have found short, promoter-associated transcripts. The early transcription termination that results in the formation of these cryptic unstable transcripts (or CUTs) is also dependent on the *Sen1* termination pathway, before the CUTs are degraded by the exosome complex. *Sen1* also participates in polyA-dependent termination, but its role in that termination mechanism remains unclear.

8.8   Transcription re-initiation

Transcription re-initiation happens much more rapidly than initial transcription initiation, and follows a different path. As mentioned earlier (Section 8.6), once the open complex forms after the assembly of the pre-initiation complex, RNP-II moves along the gene generating the mRNA, and TFIIB and TFIIF are cleared from the pre-initiation complex. What remains is known as the scaffold complex (Figure 8.6). Upon binding of TFIIB and TFIIF and recruitment of RNP-II, the complex is ready for a new round of transcription. One other component that remains in the scaffold complex is Mediator. As mentioned in Section 8.5, Mediator conveys signals from transcription factors to facilitate the assembly of the pre-initiation complex, and to activate transcription from genes with stalled RNP-II. Hence, Mediator only recognizes the unphosphorylated form of RNP-II. However, upon interaction with RNP-II, it stimulates the kinase activity of TFIIH, resulting in CTD phosphorylation, and hence decoupling from Mediator. It is unclear whether the same molecule of RNP-II that carried out the previous round of transcription is recruited in the following re-initiation round.

8.9   Summary

Transcription of Class II genes is an intricate process that requires the integration of signals by multiple macromolecular complexes that are recruited to the core promoter. Central to this is RNP-II and associated GTFs and Mediator complex. GTFs play a key role in recruiting RNP-II to pre-initiation complexes, and in modifying the phosphorylation status of the RNP-II CTD to ensure productive transcription. In the absence of other more specific transcription factors, RNP-II and GTF engage in what is known as basal transcription.

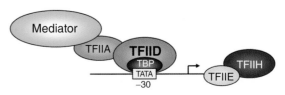

Figure 8.6 The scaffold complex facilitates re-initiation of transcription by RNP-II. The scaffold complex contains many of the protein components that remain on the promoter after RNP-II clears the promoter

8.10   Problems

Note, Problems 8.6–8.9 require reference to other chapters.

8.1 Define the following terms:
  (a) Pre-initiation complex (PIC)
  (b) GTF
  (c) TBP
  (d) TSS
  (e) CTD.

8.2 What is the importance of the TATA box?

8.3 How general is the TATA box?

8.4 How does CTD phosphorylation affect transcription? What is known in plants?

8.5 What is the role of the Mediator complex?

8.6 What is the difference between EMSA and ChIP?

8.7 What is the difference between a general transcription factor and a transcription factor?

8.8 How would you demonstrate that RNP-II is stalled at a particular gene?

8.9 How would you determine if a particular gene is transcribed by RNP-I or RNP-II?

References

Adelman, K. and Lis, J.T. (2012) Promoter-proximal pausing of RNA polymerase II: emerging roles in metazoans. *Nature Reviews Genetics*, **13**, 720–731.

Backstrom, S., Elfving, N., Nilsson, R. *et al.* (2007) Purification of a plant mediator from *Arabidopsis thaliana* identifies PFT1 as the Med25 subunit. *Molecular Cell*, **26**, 717–729.

Bang, W., Kim, S., Ueda, A. *et al.* (2006) Arabidopsis carboxyl-terminal domain phosphatase-like isoforms share common catalytic and interaction domains but have distinct in planta functions. *Plant Physiology*, **142**, 586–594.

Buratowski, S. (2009) Progression through the RNA polymerase II CTD cycle. *Molecular Cell*, **36**, 541–546.

Burley, S.K. and Roeder, R.G. (1996) Biochemistry and structural biology of transcription factor IID (TFIID). *Annual Review of Biochemistry*, **65**, 769–799.

Cerdan, P.D. and Chory, J. (2003). Regulation of flowering time by light quality. *Nature*, **423**, 881–885.

Conaway, R.C. and Conaway, J.W. (2011) Function and regulation of the Mediator complex. *Current Opinion in Genetics & Development*, **21**, 225–230.

Cramer, P., Bushnell, D.A., Fu, J. *et al.* (2000) Architecture of RNA polymerase II and implications for the transcription mechanism. *Science*, **288**, 640–649.

Elfving, N., Davoine, C., Benlloch, R. *et al.* (2011) The *Arabidopsis thaliana* Med25 Mediator subunit integrates environmental cues to control plant development. *Proceedings of the National Academy of Sciences USA*, **108**, 8245–8250.

George, A.A., Sharma, M., Singh, B.N. *et al.* (2006) Transcription regulation from a TATA and INR-less promoter: spatial segregation of promoter function. *EMBO Journal*, **25**, 811–821.

Gnatt, A.L., Cramer, P., Fu, J. *et al.* (2001) Structural basis of transcription: an RNA polymerase II elongation complex at 3.3 Å resolution. *Science*, **292**, 1876–1882.

Jin, Y.M., Jung, J., Jeon, H. *et al.* (2011) At CPL5, a novel Ser-2-specific RNA polymerase II C-terminal domain phosphatase, positively regulates ABA and drought responses in Arabidopsis. *New Phytologist*, **190**, 57–74.

Juven-Gershon, T. and Kadonaga, J.T. (2010) Regulation of gene expression via the core promoter and the basal transcriptional machinery. *Developmental Biology*, **339**, 225–229.

Koiwa, H., Hausmann, S., Bang, W.Y. *et al.* (2004) Arabidopsis C-terminal domain phosphatase-like 1 and 2 are essential Ser-5-specific C-terminal domain phosphatases. *Proceedings of the National Academy of Sciences USA*, **101**, 14539–14544.

Kraemer, S.M., Goldstrohm, D.A., Berger, A. *et al.* (2006) TFIIA plays a role in the response to oxidative stress. *Eukaryotic Cell*, **5**, 1081–1090.

Kuehner, J.N., Pearson, E.L. and Moore, C. (2011) Unravelling the means to an end: RNA polymerase II transcription termination. *Nature Reviews Molecular Cell Biology*, **12**, 283–294.

Li, Y.F., Le Gourierrec, J., Torki, M. *et al.* (1999) Characterization and functional analysis of Arabidopsis TFIIA reveal that the evolutionarily unconserved region of the large subunit has a transcription activation domain. *Plant Molecular Biology*, **39**, 515–525.

Liu, F., Bakht, S. and Dean, C. (2012) Cotranscriptional role for Arabidopsis DICER-LIKE 4 in transcription termination. *Science*, **335**, 1621–1623.

Mathur, S., Vyas, S., Kapoor, S. and Tyagi, A.K. (2011) The Mediator complex in plants: structure, phylogeny, and expression profiling of representative genes in a dicot (Arabidopsis) and a monocot (rice) during reproduction and abiotic stress. *Plant Physiology*, **157**, 1609–1627.

Molina, C. and Grotewold, E. (2005) Genome wide analysis of Arabidopsis core promoters. *BMC Genomics*, **6**, 25.

Nikolov, D.B., Hu, S.H., Lin, J. *et al.* (1992) Crystal structure of TFIID TATA-box binding protein. *Nature*, **360**, 40–46.

Wang, D., Bushnell, D.A., Westover, K.D. *et al.* (2006) Structural basis of transcription: role of the trigger loop in substrate specificity and catalysis. *Cell*, **127**, 941–954.

Chapter 9
Transcription factors interpret *cis*-regulatory information

9.1 Information on when, where and how much a gene is expressed is codified by the gene's regulatory regions

In previous chapters we described the plant RNA polymerases, RNP-I, RNP-II, RNP-III, RNP-IV and RNP-V. RNP-I and RNP-III are together responsible for transcribing specific genes to produce, after transcript processing, the greatest mass of cellular RNA [ribosomal RNA (rRNA) and transfer RNA (tRNA), respectively]. RNP-II transcribes genes that result in the greatest diversity of RNAs, namely the mRNAs, after processing of the corresponding pre-messenger RNA (mRNA) precursors (see Chapter 13). RNP-I recognizes only a single type of promoter shared by all large rRNA genes, and RNP-III recognizes just a handful of different promoter architectures. In sharp contrast, RNP-II needs to be recruited to each of the tens of thousands of promoters of the protein-coding genes. As described in Chapter 8, the core promoter of this class of genes provides only a very low (basal) level of transcription, at least when assayed in cell-free systems. In this chapter we discuss how RNP-II is instructed on which genes to transcribe, how much and when, by a large group of regulatory proteins known as **transcription factors**.

Every gene transcribed by RNP-II has a unique pattern of gene expression. Some genes are expressed at all developmental stages essentially in all cell types under all conditions. Such genes are said to be **constitutively** expressed. Examples include genes that encode histones (basic proteins necessary to assemble chromatin, see Chapter 12), genes encoding proteins involved in the cytoskeleton such as actin and tubulin, and genes that encode enzymes necessary for basic cellular metabolic functions, such as glycolysis. Other genes are expressed in only a few cells, or perhaps even just in one cell type in the plant. Examples include genes involved in the maintenance of shoot or root **meristematic** cells, those undifferentiated **pluripotent** "stem" cells that will ultimately divide and differentiate into every other cell type in the plant. In addition, a large majority of plant genes are not expressed at all under normal growth conditions, and instead are induced under particular stress conditions. Stress conditions can be **biotic**, caused by other living organisms like viruses, bacteria, fungi, insects, vertebrate herbivores, or even other plants. Other stress conditions can be **abiotic**, such as those caused by heat, cold or drought.

As is the case for most other cellular functions, the information of where, when, and how much of a gene is to be transcribed is provided by the sequence of the DNA. The region of a gene that controls its expression is called the gene **regulatory region**. Different from what happens for protein-coding regions, where the

Plant Genes, Genomes and Genetics, First Edition. Erich Grotewold, Joseph Chappell and Elizabeth A. Kellogg.
© 2015 John Wiley & Sons, Ltd. Published 2015 by John Wiley & Sons, Ltd.
Companion Website: www.wiley.com/go/grotewold/plantgenes.

genetic code allows us to immediately translate a DNA sequence into a protein sequence (see Chapter 15), the regulatory code in eukaryotes is not well understood and involves aspects of DNA structure and epigenetic modification, as well as the primary DNA sequence. Because of this complexity, the regulatory regions of a gene must be experimentally (empirically) determined. Note that we have defined the gene as "the complete set of instructions for building one RNA" (see Chapter 1); according to this definition, the regulatory sequences are a fundamental part of the gene.

There are multiple strategies for determining what region(s) of a gene are regulatory. In the past, it was assumed that all regulatory regions were immediately 5′ (upstream) of the transcription start site (Figure 1.8). While regulatory regions have also been identified 3′ of genes, inside introns, and in 5′- and 3′-untranslated regions (5′-UTR and 3′-UTR, respectively), it is simplest to begin the analysis with the region immediately upstream of the transcription start site (i.e., immediately upstream of the first exon). This region upstream of the gene is therefore referred to as the promoter (Figure 1.8), since it can *promote* gene transcription.

9.2 Identifying regulatory regions requires the use of reporter genes

As mentioned above, DNA regions important for gene regulatory functions cannot be easily identified by just looking at the DNA sequence and must be empirically determined. We illustrate this here with one example. In the case in point presented as part of Figure 9.1a, 300 base pairs (bp) of the promoter were initially used for analysis. Remember that sequences upstream of the transcription start site (+1) are usually indicated with negative numbers. We will refer to a region spanning 300 bp from the transcription start site (TSS) as the [−300; +1] region. In some publications, researchers use the "A" in the translational start codon (ATG) as +1. However, this provides no information on the length of the 5′-UTR (which cannot be determined just from the DNA sequence), making it impossible to predict where the promoter starts. Hence, in this book, we will use +1 to refer only to the transcription start site, not the translation start.

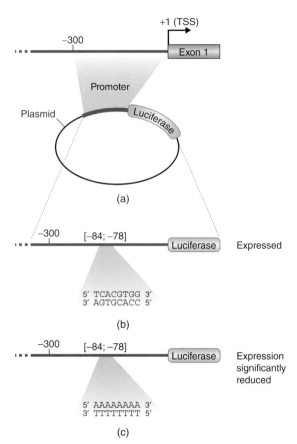

Figure 9.1 Dissecting promoter transcriptional activity. (a) A 300 bp DNA fragment upstream of the TSS is cloned in front of a reporter gene such as luciferase. (b) A DNA sequence is identified as potentially important for promoter function; and (c) when mutated, results in significant reduction of reporter gene expression

To identify cells in which the promoter is active and the gene is transcribed, the candidate promoter region is cloned (usually following polymerase chain reaction, or PCR) and placed in front (upstream) of a **reporter** gene. A reporter is a gene that provides a protein product that can be easily detected and visualized. One commonly used set of reporters is the fluorescent proteins, which were originally isolated from fluorescent jellyfish. When the tissue or cell expressing a fluorescent protein is exposed to ultraviolet light, the proteins fluoresce, showing clearly where the promoter is active. Different proteins fluoresce (literally) in different colors, such as green fluorescent protein (GFP), yellow fluorescent protein (YFP), cyan fluorescent proteins (CFP) and red fluorescent protein (RFP) (Figure 9.2a). The proteins can be used singly

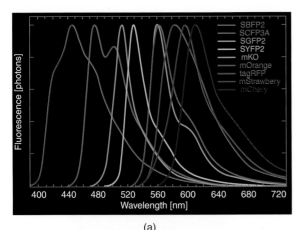

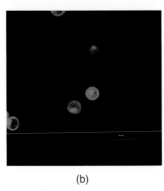

(b)

Figure 9.2 (a) Spectra of emission for frequently used fluorescent proteins. http://www.molecularcytology.nl/Joachim/Research-FP.html. Reproduced with permission of Joachim Goedhart. (b) Maize protoplasts expressing the green fluorescent protein (GFP). Photo courtesy of Isabel Casas. In this specific picture, GFP was expressed from the Cauliflower Mosaic Virus 35S (CaMV 35S) promoter

Once the promoter:reporter construct is generated using conventional molecular biology techniques, it is introduced into a plant or a plant cell by a number of methods, such as electroporation or bombardment. If the promoter region contains at least some of the necessary gene regulatory elements, the reporter gene will be transcribed and the resulting mRNA translated, and the cells will fluoresce or emit light. Following a similar strategy, the DNA fragment can be extended, truncated or mutated to determine which specific sequences are important for its regulatory properties (Figure 9.1). This approach is commonly known as **promoter bashing**. Ultimately, one ends with a map of the promoter in which the sequences important for gene regulation have been identified. For example, this type of study may establish that, for the imaginary example provided, a short sequence located between nucleotides −78 to −84 upstream of the transcription start site (TSS, see Figure 9.1b) is responsible for 50% of the expression of the gene in a particular cell type. The visual or computer-aided inspection of the sequence of this fragment of DNA may (or may not) provide some information on the molecular mechanisms by which it controls gene expression. For example, the sequence may reveal DNA motifs that have been previously associated with the expression of genes under different conditions or different cell types. Once identified, such small DNA regions are often referred to as *cis*-acting regulatory elements (or just *cis*-regulatory elements or *cis*-regulatory motifs) that might be recognized by transcription factors.

or together, depending on the intended application. Figure 9.2b shows maize protoplasts (cells lacking the cell wall) expressing GFP.

While the presence of fluorescence will indicate whether the gene is expressed or not, the level of expression cannot be determined because fluorescence quantification requires some sophisticated equipment. In many instances, quantitative data are needed to determine how efficient a promoter is in driving gene expression. For quantification, one commonly used type of reporter is a **luciferase**, one of a large group of oxidative enzymes that are responsible for many bioluminescence phenomena in nature. For example, firefly luciferase can convert the substrates luciferin and ATP into light (plus AMP and the oxyluciferin product). The amount of light emitted can then be measured with an instrument known as a **luminometer**.

9.3 Gene regulatory regions have a modular structure

The regulatory regions of most genes have a modular structure in which multiple *cis*-regulatory elements are organized in clusters known as *cis*-regulatory modules. In a modular regulatory region, each *cis*-regulatory module provides one component of the spatial (or temporal) output of the entire gene regulatory region. This concept is best illustrated by a constitutive promoter, which is expressed in (almost) every cell type of a plant. There are many well-characterized plant constitutive promoters. For example, promoters regulating expression of genes encoding for the structural proteins tubulin and actin have been generally described as constitutive. But some of the most commonly used

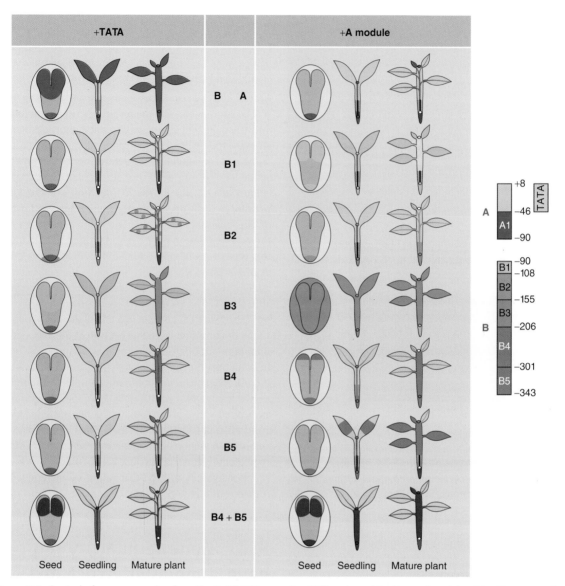

Figure 9.3 Control of gene expression from the CaMV 35S promoter. Different combinations of modules (the diagram on the right provides the position of the modules with respect to the TSS at +1) allow the CaMV 35S promoter to drive expression with different patterns. The left panel shows the expression pattern of combinations of the CaMV 35S minimal promoter (represented by the +TATA) with the various modules (B, B1, B2, B3, B4, B5, B4+B5) in Arabidopsis seeds, seedlings, or mature plants. The right panel shows the expression pattern of combinations of module A by itself (A) or with the various other modules (B, B1, B2, B3, B4, B5, B4+B5). Adapted from Benfey and Chua (1990)

constitutive promoters are from viruses, not from plants. Viruses have acquired elaborate regulatory mechanisms to ensure that most plant cells and tissues will effectively and efficiently express the viral genome, so the viral genome promoters are considered good candidates for strong, constitutive promoters.

One good example is the promoter of the Cauliflower Mosaic Virus (CaMV) that produces the viral 35S RNA (see Chapter 6 for the meaning of 35S as a measure of RNA size), and is therefore known as the *CaMV 35S promoter*, or just *35S*. Today, the 35S promoter is widely used to express genes in a constitutive fashion (ectopically, meaning that often not in the tissues/cell types where the gene is normally the expressed) in a variety of plants.

Seminal studies conducted by Philip Benfey, while working in Nam-Hai Chua's lab (Benfey and Chua, 1990; Benfey *et al.*, 1990a,b) revealed that different fragments (or modules) of the 35S promoter are responsible for directing gene expression in different plant parts (Figure 9.3). The constitutive activity of this promoter is thus provided not by a single universal promoter element, but rather by multiple *cis*-regulatory modules acting in concert. The 35S promoter continues to be one of the best-dissected constitutive promoters in plants, despite being of viral origin. From a practical perspective, 35S has proven not to be particularly active nor constitutive in all plant species. For example, it does not have activity in Arabidopsis pollen, nor does it drive high-level transcription in some monocotyledonous plants such as maize. Therefore, promoters such as the one from the maize *ubiquitin1* gene (Christenson and Quail, 1996) are more often used in maize and other grasses for constitutive expression.

The modular structure of promoters has important consequences. One practical consequence is that a specific module can be used to drive transcription in a particular tissue or cell type, along with a functional basal (or core) promoter (see Chapter 8). In Figure 9.3, this is represented by the fact that CaMV 35S minimal promoter (represented by the TATA region) can be combined with multiple other domains to provide transcription. The individual domains in the absence of a minimal promoter would not be functional. A corollary of this is that mutations in a particular module may affect only expression in a specific tissue or cell type. Thus, when conducting the promoter bashing experiments described earlier, a particular mutation or deletion may appear not to have any effect, but this could mean simply that the correct cell type was not investigated. A second consequence of the modular structure of promoters is that gene regulation in one part of the plant can be decoupled from that in other plant parts. This can result in the same gene affecting different tissues or cell types in different species of plants. The modular structure of promoters also has implications for the evolution of plant gene expression, in that selection can modify one component and leave the others alone. This permits diversification of expression in some parts of the plant without disrupting others.

9.4   Enhancers: *Cis*-regulatory elements or modules that function at a distance

Enhancers have been traditionally defined as DNA regions that can enhance or augment a gene expression pattern in a quantitative manner, but do not initiate gene expression or regulate the qualitative nature of gene expression. Enhancers may operate at large distances away from the transcription start site (hundreds to thousands of base pairs away). Enhancers can be located both in the 5′ or 3′ regions of genes and can function even if inverted. In many textbooks, a distinction is made between promoters and enhancers, leaving readers with a feeling that the mechanisms involved in promoter and enhancer function are different. However, it is often very difficult to classify a *cis*-regulatory element as one or the other. For example, is a *cis*-regulatory element that functions 10 kbp upstream of the TSS part of the promoter or an enhancer?

From this book's perspective, promoter *cis*-regulatory elements (excluding of course the core promoter, see Chapter 8) and enhancer *cis*-regulatory elements are one and the same. As we will discuss in Section 9.6, they are both recognized by a specialized type of proteins, the transcription factors, which in both cases need to interact with components of the basal transcriptional machinery to modulate gene expression. By realizing that chromatin is not just a linear string of DNA, but rather is bent and folded in complex ways, one can easily see how a transcription factor recognizing a *cis*-regulatory element 250 kbp away from the TSS could be positioned very close to the core transcriptional machinery, allowing the same protein–protein interactions as would direct transcription from a more proximal *cis*-regulatory element.

Superimposed on the code furnished by the DNA sequence is what has come to be known as the histone code (see Chapter 12). Histones are arranged into nucleosomes, and chromatin structure plays a key role in modulating every aspect of gene expression. For example, there is evidence that there are fewer nucleosomes in the regulatory regions proximal to the TSS in genes that are primed for transcription.

9.5   Transcription factors interpret the gene regulatory code

So far, we have looked at gene regulation just from the perspective of regulatory DNA. But the regulatory code in the DNA must be somehow recognized, interpreted and converted into a signal to the RNA polymerase to transcribe a gene. This interpretation of the DNA blueprint is carried out by a large and diverse group of proteins collectively known as **transcription factors**. Here we define transcription factors as proteins that recognize and bind to a particular DNA sequence. We have already briefly mentioned transcription factors in the context of RNP-II in Chapter 8. Many other proteins participate in the control of gene expression, and many studies include them under the definition of transcription factors. However, for clarity, we will refer to those other proteins as co-activators or co-repressors, chromatin readers or writers (see Chapter 12), depending on the specific function attributed to them.

Transcription factors are unique to eukaryotic organisms. While the seven σ factors in the bacterium *Escherichia coli* are essential for promoter recognition, they cannot bind promoter DNA on their own. Instead, the main σ factor (called σ^{70} because of its molecular weight of 70 kDa) helps the core RNA polymerase bind to promoters of housekeeping genes. On account of the chromatin structure that characterizes eukaryotic genes, transcription in eukaryotes tends to be turned off in the absence of a signal (or transcription factor), while bacterial genes tend to be turned on as a default. Thus, the prevalent function of transcription factors in eukaryotes has often been assumed to be the activation of transcription, overcoming repressive mechanisms that otherwise would keep all genes off. Archaea are single-celled microorganisms that are distinct from bacteria and eukaryotes. For many aspects of gene expression, Archaea are more similar to eukaryotes. Archaea have just one RNP with a domain structure that resembles very much that of eukaryotic RNP-II. Transcription by archaeal RNP requires only a few transcription factors, suggesting a system similar to, but significantly less complicated than, that in eukaryotes (Korkhin *et al.*, 2009; Grohman and Werner, 2011).

Although a main function of transcription factors is to activate transcription, this does not imply that all transcription factors are just activators. Indeed, they can equally well act as activators or repressors of transcription, and a theme that is emerging in both animals and plants is that many, if not all, transcription factors can function both as activators and as repressors, depending on the proteins with which they interact. As will become evident throughout the remainder of this chapter, protein–protein interactions are central in any gene regulatory mechanism.

9.6   Transcription factors can be classified in families

Many transcription factors bind DNA as homo- or heterodimers (homo = same, hetero = different). Such dimerization results from specific protein–protein contacts and is necessary to stabilize the interaction with the DNA, and to expand the specificity of DNA recognition, as described in the next section. Based on amino acid sequence conservation of the DNA-binding domain or of the protein–protein interaction region, transcription factors are classified into families. The principles for classification vary slightly from author to author, but in general, 50–60 families of transcription factors can be recognized in multi-cellular organisms.

The names of transcription factor families are based on criteria such as the structure of the DNA-binding or dimerization domain [*e.g.*, basic helix-loop-helix (bHLH); basic leucine zipper (LZip)], the protein or proteins from which they were originally discovered (*e.g.*, the name MADS derives from four transcription factors in which a highly conserved amino acid sequence or domain was first identified: yeast M̲CM1, Arabidopsis A̲GAMOUS, snapdragon D̲EFICIENS and human S̲RF), or the function of the first member of the family, such as homeodomain (HD) transcription factors (from the word homeosis, which refers to form). Indeed, HD transcription factors were first identified in genes that control the pattern of body formation in the fruit fly *Drosophila melanogaster*.

A slightly larger percentage of genes encode transcription factors in plants than in animals, and the particular families and the number of members of each family are often quite different. For example, certain families are unique to plants (*e.g.*, NAM or AP2)

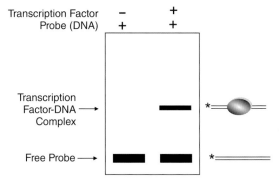

Figure 9.4 Electrophoretic mobility shift assay (EMSA) provides one method by which the binding of a transcription factor to DNA can be studied. A DNA fragment that is labeled (for example with ^{32}P, which is radioactive) is run by itself (free probe) or incubated with the DNA-binding transcription factor, forming the transcription factor-DNA complex that runs slower in the polyacrylamide electrophoresis than the free probe

while others have only been identified in animals (*e.g.*, NF-κB). When a transcription factor family is present in both plants and animals, we infer that their common unicellular ancestor had at least one member of the family. Often, transcription factor genes accumulate in a genome by gene duplications (see Chapter 2), but genes that have duplicated extensively in the history of plants may not be duplicated in animals and vice versa. For example, the MYB family is very large in plants (160 members in Arabidopsis, 220+ in rice), while there are only a handful of MYB transcription factors in humans. This has led to the speculation that amplification of specific families of transcription factors in the plant lineage could have been associated with plant-specific functions. As an example, amplification of MYB transcription factors was hypothesized as being important in the control of plant form and metabolic diversity (Martin and Paz-Ares, 1997).

9.7 How transcription factors bind DNA

How transcription factors bind DNA is central to their function. Methods to investigate the binding of transcription factors to DNA are numerous, and can be divided into *in vitro* and *in vivo*. *In vitro* methods are those in which the protein and the DNA are isolated from the plant cells and protein–DNA

interactions are examined. An example of an *in vitro* method is the electrophoretic mobility shift assay (EMSA), also known as gel shift, in which a labeled (e.g., radioactively) double-stranded fragment of DNA is incubated with the purified transcription factor, or with a cellular extract containing it. Then, the complex is separated by polyacrylamide gel electrophoresis, and the radioactivity is visualized by X-ray autoradiography. If the transcription factor binds to the radioactive DNA probe, then the protein-DNA complex will migrate more slowly than DNA (probe) without the protein (Figure 9.4). DNA fragments in which the transcription factor-binding site is mutated can be used either as probes or as competitors to examine the DNA-binding preference of the transcription factor. While such experiments can be informative, they only show that the protein *can* bind to the DNA, but not whether it actually *does* so in the plant. To determine whether binding occurs inside the cell, an *in vivo* approach is needed.

An increasingly popular *in vivo* method is based on chromatin immunoprecipitation (ChIP). The basic principle here is to capture the transcription factor–DNA interactions as they are happening inside the nucleus of a plant. To achieve this, researchers chemically "freeze" the protein–DNA interactions using formaldehyde, which cross-links proteins with other proteins as well as with DNA. Following cross-linking, the chromatin (DNA with proteins) is fragmented, and then the transcription factor of interest is immunoprecipitated using antibodies against it (Figure 9.5). The beauty of formaldehyde cross-linking is that it can be reversed by heat; hence once the immunoprecipitate is collected, heating at 65°C for a few minutes releases the proteins from the DNA. The DNA is then purified by conventional methods and analyzed for the presence of a particular regulatory sequence using a PCR assay, in what is commonly known as ChIP-PCR (Figure 9.5). In this process, the researcher can query the immunoprecipitate using PCR with primers that correspond to the regulatory region of a gene of interest. If a PCR product is obtained, it indicates that the promoter is recognized by the specific transcription factor. Some high throughput methods involve hybridization to oligonucleotide arrays that comprise sequences for the entire genome, a technique called ChIP-chip (Figure 9.5). Alternatively, all the immunoprecipitated DNA can be sequenced by one of several high throughput methods; these results simultaneously evaluate all the

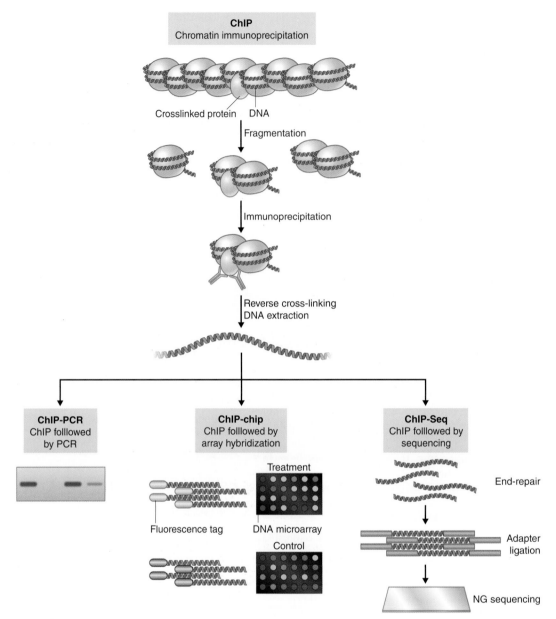

Figure 9.5 Chromatin immunoprecipitation (ChIP)-based technologies to investigate transcription factor-DNA interactions *in vivo*. See the text for additional information. Next generation (NG) includes high-thoughput DNA sequencing methods such as provided by Illumina

different places in the genome where the transcription factor was located in that particular tissue and time. The method is called ChIP-Seq (Figure 9.5), and is becoming increasingly popular.

Nature has developed a limited number of structures that participate in sequence-specific binding to DNA. In many instances, these structures involve a short basic (positively charged at pH 7) α-helix that makes direct contact with 3–4 nucleotides in both strands of the major grove (Figure 1.1b) of the DNA. The basic helix allows ionic interactions with the negatively charged phosphate groups of the DNA to stabilize the protein-DNA complex. Other components of the DNA-binding domain often function

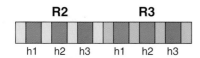

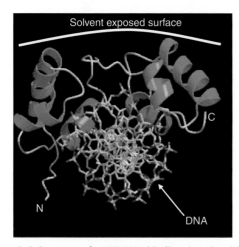

Figure 9.6 Structure of a MYB DNA-binding domain with two MYB repeats (R2 and R3). The three α-helices that form each MYB repeat are indicated as h1–3; the α-helix that makes the DNA contacts is indicated in purple, while the structural α-helices are in green. Both MYB repeats are joined together by a linker. The DNA is coming out of the paper (i.e., it is perpendicular to the plane of the page)

solely to properly position the DNA-binding α-helix. One example is shown in Figure 9.6, depicting a typical MYB domain formed by two MYB repeats. Each MYB repeat is formed by three α-helices, the second and third forming what is known as a helix-turn-helix structure, with the third helix making contact with the DNA, as is evident in Figure 9.6. The other two helices in each MYB repeat keep the DNA-recognition helix properly placed. Thus, a DNA-binding domain like the one in Figure 9.6 has two clear surfaces: one is involved in binding DNA, while the other (often described as solvent-exposed) is available for protein–protein interactions.

Given the very short DNA sequence that a single α-helix recognizes, interactions with other transcription factors are essential to provide the exquisite regulatory specificity that they have *in vivo*. In the case of R2R3-MYB factors, since two α-helices (one from each MYB repeat, pink in Figure 9.6) are involved, the DNA sequence that they recognize is 6–8 bp. Basic α-helices are also a hallmark of bHLH and bZIP transcription factors (Figure 9.7a, b). In those cases, the HLH and ZIP dimerization domains allow these proteins to bind DNA as dimers. Zinc-finger domains are also a large class of transcription factors with ample representation in the plant kingdom. In zinc-finger domains, a zinc

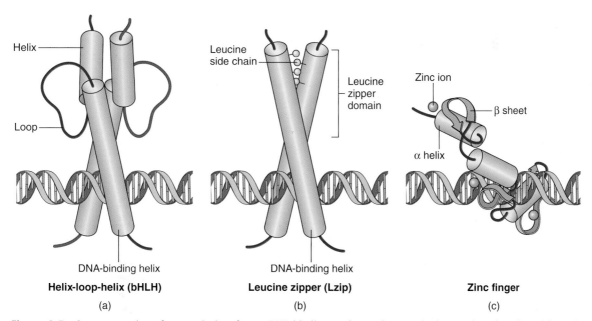

Figure 9.7   Some examples of transcription factor DNA-binding and protein–protein interaction domains. (a) Basic helix-loop-helic (bHLH), (b) leucine zipper (Lzip), and (c) zinc finger. (a–c) Adapted from Buchanan *et al.* (2002). Reproduced with permission of John Wiley & Sons, Inc

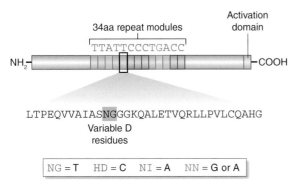

Figure 9.8 Structure of a canonical TAL effector protein. Each of the 13 repeats in the TAL protein is illustrated by colors, and the preferred nucleotide that they recognize is indicated. Adapted from http://taleffectors.com/

ion stabilizes the structure (Figure 9.7c), and similar to the role of the bZIP, it serves to orient the α-helix for DNA binding. Zinc ions can be coordinated (coordination is a type of non-covalent chemical interaction) by histidines, cysteines, or combinations of the two. For example, a zinc finger coordinated by two cysteine and two histidine residues is known as Cys_2His_2 finger, and a four-cysteine finger is referred to as Cys_4.

An interesting consequence of how transcription factors operate is provided by an example of a microbial pathogen usurping plant gene regulation. *Xanthomonas campestris*, a serious bacterial pathogen of peppers and other Solanaceous species, expresses a protein product, the transcription activator-like protein known as TAL effector (or TALE), which is inserted by the bacterium into the cells of its plant hosts and modifies transcription of defense genes (Boch *et al.*, 2009). TAL effectors move to the nucleus and serve as transcription factors to mitigate the host's defense response. TAL effectors are formed by a central DNA-binding domain of tandem repeats, and an acidic activation domain (see Section 9.7). The TAL effector DNA-binding domain is unique in having a conserved region consisting of repeats of 33–34 amino acids. Each repeat provides specificity to one nucleotide in the target DNA sequence. Hence, in the example shown in Figure 9.8, 13 repeats (indicated in colors) permit this TAL effector to bind to the DNA sequence TTATTCCCTGACC. Researchers have taken advantage of this to artificially generate custom TALEs that can be tethered to other transcriptional regulator domains or other activities (e.g., nucleases) to effect genome remodeling.

From the Experts

Manipulating genes and genomes with TALEs

The TAL effector domain interacts with DNA by a simple, elegant mechanism. Each of the 13–28 tandem TAL effector domains present in naturally occurring TALEs is nearly identical in amino acid sequence. The exceptions are the so-called repeat variable di-residues (RVDs) located at amino acid positions 12 and 13 in the repeat. The TAL effector domain forms two α-helices that create a hairpin in which the RVDs are positioned at the hairpin's apex. Residue 12 reaches back to help stabilize the hairpin, whereas residue 13 projects from the tip and makes a base specific contact in the major groove of DNA. The most common RVDs – amino acid residues NI, HD, NG, and NN – bind to adenosine, cytosine, thymine, and either guanine or adenosine, respectively. Because there is a one-to-one correspondence between the repeats in the protein and the bases in the DNA they contact, TALEs bound to DNA make a right-handed superhelix that embraces the major groove, each TAL effector domain contacting one consecutive base in the DNA helix.

The TAL effector domain has emerged as a powerful reagent for manipulating DNA *in vivo*, because custom TAL effector arrays can be easily assembled that bind to novel DNA sequences. Consequently, artificial transcriptional regulators can be created, by simply replacing the repeat arrays of native TALEs with custom arrays that activate transcription of target genes of interest. The transcriptional activation domain of custom TALEs can be replaced with other effector domains, including those that mediate transcriptional repression (Figure 9.9). One widely used reagent for genome manipulation is fusions between custom TAL effector repeat arrays and nucleases (TAL-ENs). When expressed in cells, TALENs create chromosomal breaks at specific target loci. If the broken chromosomes are rejoined imprecisely, then mutations are introduced at the break site that can knock out gene function. If a DNA template is provided with the TALEN, then the break

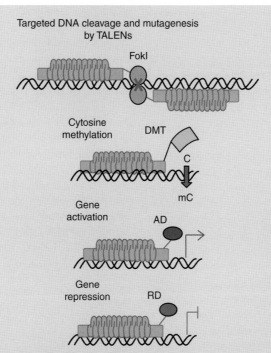

Figure 9.9 Fusions between custom TAL effector DNA binding domains and various proteins make it possible to manipulate genes and genomes *in vivo*. TAL effector nucleases (TALENs) are fusions to the catalytic domain of *Fok*I that create targeted double-strand breaks, enabling gene knockouts and gene editing. TAL effector-based epigenetic modifiers are created, for example, by fusing custom DNA binding domains to DNA methyltransferases (DMT) that methylate cytosine residues. TAL effector proteins fused to transcriptional activation (AD) and repression domains (RD) provide artificial regulators of gene expression

can be repaired through homologous recombination. Consequently, specific DNA sequence alterations present in the repair template can be incorporated at or near the TALEN cleavage site. Many other TAL effector-based reagents are being developed, including those that create targeted epigenetic modifications such as the methylation of cytosines or the modification of histones. It is clear that this simple DNA binding motif derived from a plant pathogen is rapidly providing new-found control over the genetic material, enabling both directed manipulation of DNA sequences and their expression.

By Dan Voytas

The newest tool for engineering genomes is provided by a system in which the DNA-binding specificity is provided by an RNA that efficiently directs a DNA nuclease to its target. This system, first identified in bacteria and Archaea and employed by researchers to alter complex genomes (e.g., humans), is known as CRISPR (for clustered regularly interspaced short palindromic repeats). Since this chapter is about protein-DNA recognition and CRISPRs are based on DNA-RNA complementary, we will not go any deeper here in the discussion of this promising new method

for genome engineering (Le Cong *et al.*, 2013; Mali *et al.*, 2013; van der Oost, 2013).

9.8 Modular structure of transcription factors

Transcription factors bind to DNA, but also must interact with other proteins, either corresponding to

Figure 9.10 Structure of a hypothetical transcription factor containing a DNA-binding domain (DBD) and a transcriptional activation domain (TAD). The DNA-binding domain and transcriptional activation domains can be found in any part of the protein, and the latter can be in any position with respect to the DBD

GTFs, the Mediator complex (see Chapter 8), components of the basal transcription machinery, or other transcriptional regulators. The combined effect of the transcription factor and its interacting proteins then either activates or represses transcription. The protein–protein interactions are determined by specific and often evolutionarily conserved sequences of amino acids. If the sequence leads to an interaction that then triggers transcription, it is known as a **transcriptional activation domain** (Figure 9.10). Conversely, if the interaction blocks transcription, the string of amino acids is known as a **repressor domain**. In most transcription factors, the DNA-binding domain and transcription activation domains are clearly separated from each other, hinting that the protein may be modular in structure.

Experiments in which a particular domain of one protein was exchanged with the respective domains of others (known as domain-swap experiments) in the mid-1980s established that eukaryotic transcription factors have a modular structure (Brent and Ptashne, 1985). The modular structure of transcription factors allows the DNA-binding domain from one transcription factor to be combined with the transcriptional activation domain(s) from another. This is important from both experimental and evolutionary perspectives in shaping the way in which regulatory proteins assemble on different promoters.

In contrast to DNA-binding domains, which commonly have well defined folds to conform to the very uniform structure of the double helix, transcriptional activation domains are usually much more disorganized in structure, adopting defined conformations only upon interaction with components of the transcriptional machinery, such as for example subunits of the Mediator complex (see Chapter 8). Because the conformation of transcriptional activation domains depends on their particular interactions with other transcriptional proteins, and because those interacting proteins are themselves not well known,

it is difficult to determine from the primary amino acid sequence alone what region of a transcription factor corresponds to the transcriptional activation domain. One approach to identify a transcriptional activation domain involves generating a chimeric protein containing a putative activation domain and a well-characterized DNA-binding domain such as the *Saccharomyces cerevisiae* (yeast) GAL4 DNA-binding domain (GAL4 DB, Figure 9.11). This chimeric protein is then assayed in plant cells using the methods described in Section 9.1 to investigate whether it can activate transcription of a reporter construct that harbors DNA sequences recognized by GAL4 (Figure 9.11). If activation is obtained (reflected for example in increased luciferase activity driven by a promoter containing GAL4-binding sites), then one can conclude that the region of the transcription

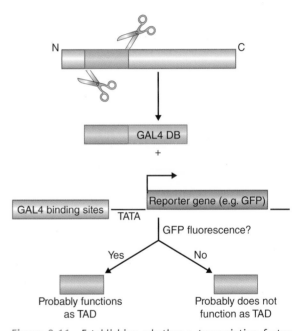

Figure 9.11 Establishing whether a transcription factor contains a region capable of activating transcription. The region in a protein suspected to contain a transcriptional activation domain is cloned as a translational fusion with the GAL4 DNA-binding domain. The transcriptional activity of the fusion protein is then assayed (ideally in plant cells) on a synthetic promoter that contains GAL4 binding sites controlling the expression of a reporter gene. TATA indicates that this promoter should also contain a functional minimal promoter. If activation is observed, then the researcher can conclude that the region could correspond to a transcriptional activation domain

factor assayed might contain a transcriptional activation domain.

Transcriptional activation regions are classified as acidic, proline- or glutamine-rich domains (Roberts, 2000). Only a few transcriptional activation domains, primarily of the acidic type, have been dissected in detail in plant transcription factors. Acidic transcriptional activation domains are usually characterized by the presence of an amphipathic (from the Greek: *amphi* = both, *pathos* = suffering; a molecule that has two sides with very different properties) α-helix, in which one side is acidic and the other hydrophobic.

The yeast two-hybrid system to investigate protein–protein interactions is based on the idea that the DNA-binding and transcriptional activation domains of transcription factors need to be in close proximity, but not necessarily on the same protein (Fields and Song, 1989). Based on this notion, the protein of interest (usually called the bait, indicated by X in Figure 9.12) is fused to a known DNA-binding domain such as the GAL4 DNA-binding domain (DB in Figure 9.12) and expressed in yeast. The yeast cells are then transformed with a library of all, or a subset, of proteins fused to a transcriptional activation domain which functions in yeast (AD in Figure 9.12). If protein X can interact with a protein Y, the DNA-binding and transcriptional activation domains are brought in proximity, and activation of transcription occurs. This system has been extensively used in both plants and animals to determine what is known as the **interactome**, the space of all possible protein–protein interactions in a cell.

Transcription factors can also have transcriptional repressor regions, by themselves or in combination with a transcriptional activation domain. Among the best-characterized plant transcriptional repression domains is the EAR (ERF-associated amphiphilic repression) domain. This domain is present in a large number of plant transcription factors associated with stress and defense functions (Kazan, 2006). The EAR domain is sufficient to convert activators into repressors (Hiratsu *et al.*, 2003), providing a powerful biotechnological tool. Indeed, researchers often use the minimal functional region of the EAR domain of the SUPERMAN zinc-finger transcription factor, which has been optimized for specificity. This has come to be known as the SRDX motif, a short 12 amino acid region with the sequence LDLDLELRLGFA (Hiratsu *et al.*, 2003). Another repression domain consists

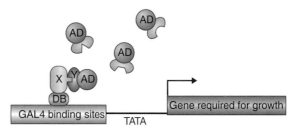

Figure 9.12   The yeast two-hybrid system to explore protein–protein interactions. The protein of interest (bait, indicated as X here) is fused to the GAL4 DNA-binding domain (DB) and expressed in yeast. The yeast cells are transformed with a library of all, or a subset, of proteins fused to a transcriptional activation domain that functions in yeast (AD). Only if X can interact with a protein Y, the DNA-binding domain and transcriptional activation domain are brought in proximity, and activation of transcription occurs

of the BRD (B3 repression domain) domain, which contains the R/KLFGV core motif. BRD repression domains are present in a large number of plant transcription factors (Ikeda and Ohme-Takagi, 2009). Similar to the way by which transcriptional activation domains interact with co-activators such as histone acetyl transferases (HATs), transcriptional repression domains interact with co-repressors such as histone deacetylases (HDACs). HATs and HDACs function by modifying histones, making the chromatin more or less permissive for gene expression (see Chapter 12 for a description of chromatin and effects on gene expression).

Traditionally, transcription factors have been characterized as activators or repressors. However, the ability of a transcription factor to activate or repress transcription is provided not only by the presence of activation or repression domains, but also, and perhaps even more importantly, by the way the transcription factor interacts with other proteins (e.g., other transcription factors) or co-regulators. Thus, the distinction between an activator and a repressor may not be a property of the transcription factor itself but rather of its cellular environment. One good example is provided by the WUSCHEL (WUS) transcription factor (Ikeda *et al.*, 2009), which in Arabidopsis controls the maintenance of a particular group of pluripotent (stem) cells in the shoot meristem. Depending on the particular target, WUS can function as either an activator or a repressor.

9.9 Organization of transcription factors into gene regulatory grids and networks

One transcription factor often controls the expression of other transcription factors (including itself), resulting in a complex grid of interactions. The grid can be imagined as a static representation of all the possible interactions between transcription factors and the corresponding target genes. Under certain circumstances, a part of that grid would be executed, resulting in a network. Thus, a gene regulatory network can be conceptualized as a temporal or spatial execution of a portion of the gene regulatory grid. Such networks can be hierarchical in nature, meaning that some transcription factors act high in the network, controlling other transcription factors, while others are closer to the bottom, controlling directly the genes encoding for enzymes or structural proteins. In such a hierarchical network it is intuitively easy to understand how the regulators higher in the network (often referred as master regulators) control a few other transcription factors, which in turn control many more, resulting in amplification of the message to be conveyed by the upper echelon factors. One can, for example, compare such a hierarchical network with the structure of a USA university. The university president is close to the top and instructs one or a few close advisers (e.g., a provost), who in turn communicate to vice provosts and deans, who then communicate with departmental chairs, and then to the faculty. The faculty would be the last layer of transcription factors in this network, and they would talk directly with the students. Of course, such a hierarchical network does not preclude the deans or provosts talking to the students directly (which happens sometimes), but the faculty carries out most of the interactions with the students.

How does thinking in terms of grids and networks help us understand the process of gene regulation in a cell? Networks can be imagined as directional graphs (digraphs in mathematical terms) in which proteins/genes represent nodes and the communication between the nodes are the edges (or connecting line, Figure 9.13). For gene regulatory networks, the edges are directional, because the transcription of a

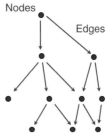

Figure 9.13 Schematic of a particular motif in a gene regulatory network. Nodes (circles) represent proteins, and edges (arrows that join nodes, this is a directional network, otherwise they would be lines instead of arrows) represent the connections between the protein encoded by one gene and the gene it controls (in a gene regulatory network)

gene is controlled by a transcription factor, and not vice versa. Mathematics has developed an entire field called **Graph Theory** to explain the behavior of graphs or networks, and it is worthwhile exploring further how graphs are pervasive in many aspects of biology (Barabasi, 2003).

9.10 Summary

Transcription factors are sequence-specific DNA-binding proteins that determine when, where, and how often a gene is expressed. Transcription factors are recognized by the presence of conserved domains, which also contribute to their classification into families. Transcription factors often contain, in addition to the DNA-binding domain, a region that interacts with protein components of the basal transcriptional machinery. Methods are available to determine which DNA sequences and where in the genome transcription factors bind. This permits determination of the structure of gene regulatory networks.

9.11 Problems

9.1 A transcription factor recognizes the DNA consensus sequence $CC^T/_AACCA$. Assuming a random distribution of nucleotides in the Arabidopsis genome, how many times in the genome would you expect this transcription factor to bind?

9.2 Now consider the maize genome. Given the difference in the number of potential binding sites, would you expect that maize and Arabidopsis transcription factors function in very different ways? Explain your answer.

9.3 In the yeast two-hybrid system, what is the problem if the bait is a transcription factor that has its own transcriptional activation domain? Explain how you would still be able to use the yeast two-hybrid system in that case.

9.4 A transcription factor is:
(a) A protein
(b) An RNA
(c) A lipid
(d) None of the above.

9.5 Binding of a transcription factor to DNA requires:
(a) ATP
(b) A specific DNA sequence
(c) A favorable transcription factor concentration
(d) The presence of the core promoter
(e) The RNA polymerase
(f) None of the above.

9.6 The transcription start site corresponds to:
(a) The CAP
(b) ATG
(c) A random set of nucleotides upstream of ATG.
(d) None of the above.

More challenging problems

9.7 The *in vitro* DNA-binding specificity of a rice WRKY transcription factor, expressed as a recombinant protein in *E. coli*, was determined to be TTTGACC/T. ChIP-chip experiments performed with antibodies against the non-DNA binding region of this WRKY factor resulted in the identification of 524 putative targets. However, most of the regions in the promoters displaying the strongest signals in the ChIP-chip experiments failed to show DNA-sequence motifs that fit the TTTGACC/T consensus.
(a) Provide one reason why antibodies against the non-DNA binding region of the WRKY transcription factor may have been used.
(b) Provide one reason why the promoters identified by ChIP-chip may not have motifs fitting the *in vitro* DNA-binding consensus for this WRKY transcription factor.

(c) Is the sequence of the rice genome necessary for performing the ChIP-chip experiments? Why?

9.8 A graduate student identified a rice gene (*RGA*) that, based on loss-of-function mutations, was implicated in the control of leaf development. Computer-assisted sequence analyses (such as those provided by the tool BLAST, http://blast.ncbi.nlm.nih.gov/Blast.cgi) revealed similar proteins expressed in other plants, but no recognizable motifs that would indicate how the RGA protein participates in leaf development. The student decided to test whether RGA may encode a DNA-binding protein.
(a) Briefly explain how the student would do this.
    After establishing that RGA indeed has the potential to bind to DNA, the student decided to look for possible target genes for this protein.
(a) What method(s) would the student use for this?
(b) Is DNA-binding sufficient to implicate RGA in the regulation of gene expression? Explain briefly.
    To further establish the function of RGA, the student fused RGA to a transcriptional repressor motif (REP) and generated transgenic rice plants. To the student's disappointment, wild-type plants expressing p35S::RGA-REP look like wild-type. However, p35S::RGA-REP complemented the phenotype of *rga* mutant plants (i.e., *rga* p35S::RGA-REP looked like wild type).
(d) What information may these results be suggesting with regards to the function of RGA? How would you test your hypothesis?

9.9 How would you adapt the system shown in Figure 9.11 to investigate whether a protein domain has transcriptional repression activity?

References

Barabasi, A.L. (2003) *Linked*, Penguin Group.
Benfey, P.N. and Chua, N.H. (1990) The Cauliflower Mosaic Virus 35S promoter: combinatorial regulation of transcription in plants. *Science*, **250**, 959–966.
Benfey, P.N., Ren, L. and Chua, N.H. (1990a) Tissue-specific expression from CaMV 35S enhancer subdomains in early stages of plant development. *EMBO Journal*, **9**, 1677–1684.

Benfey, P.N., Ren, L. and Chua, N.H. (1990b) Combinatorial and synergistic properties of CAMV 35S enhancer subdomains. *EMBO Journal*, **9**, 1685–1696.

Boch, J., Scholze, H., Schornack, S. *et al.* (2009) Breaking the code of DNA binding specificity of TAL-type III effectors. *Science*, **326**, 1509–1512.

Brent, R. and Ptashne, M. (1985) A eukaryotic transcriptional activator bearing the DNA specificity of a prokaryotic repressor. *Cell*, **43**, 729–736.

Buchanan, B., Gruissem, W. and Jones, R. (eds) (2002) *Biochemistry & Molecular Biology of Plants*, John Wiley & Sons, Ltd

Christensen, A.H. and Quail, P.H. (1996) Ubiquitin promoter-based vectors for high-level expression of selectable and/or screenable marker genes in monocotyledonous plants. *Transgenic Research*, **5**, 213–218.

Fields, S. and Song, O.-K. (1989) A novel genetic system to detect protein-protein interactions. *Nature*, **340**, 245–246.

Grohmann, D. and Werner, F. (2011) Recent advances in the understanding of archaeal transcription. *Current Opinions in Microbiology*, **14**, 328–334.

Hiratsu, K., Matsui, K., Koyama, T. and Ohme-Takagi, M. (2003) Dominant repression of target genes by chimeric repressors that include the EAR motif, a repression domain, in *Arabidopsis*. *Plant Journal*, **34**, 733–739.

Ikeda, M., Mitsuda, N. and Ohme-Takagi, M. (2009) Arabidopsis WUSCHEL is a bifunctional transcription factor that acts as a repressor in stem cell regulation and as an activator in floral patterning. *Plant Cell*, **21**, 3493–3505.

Ikeda, M. and Ohme-Takagi, M. (2009) A novel group of transcriptional repressors in Arabidopsis. *Plant Cell Physiology*, **50**, 970–975.

Kazan, K. (2006) Negative regulation of defence and stress genes by EAR-motif-containing repressors. *Trends in Plant Science*, **11**, 109–112.

Korkhin, Y., Unligil, U.M., Littlefield, O. *et al.* (2009) Evolution of complex RNA polymerases: the complete archaeal RNA polymerase structure. *PLoS Biology*, **7**, e102.

Le Cong, F., Ran, A., Cox, D. *et al.* (2013) Multiplex genome engineering using CRISPR/Cas systems. *Science*, **339**, 819–823.

Mali, P., Yang, L., Esvelt, K.M. *et al.* (2013) RNA-guided human genome engineering via Cas9. *Science*, **339**, 823–826.

Martin, C. and Paz-Ares, J. (1997) MYB transcription factors in plants. *Trends in Genetics*, **13**, 67–73.

Roberts, S.G. (2000) Mechanisms of action of transcription activation and repression domains. *Cell and Molecular Life Sciences*, **57**, 1149–1160.

van der Oost, J. (2013) Molecular biology. New tool for genome surgery. *Science*, **339**, 768–770.

Chapter 10

Control of transcription factor activity

10.1 Transcription factor phosphorylation

In the previous chapter, we described transcription factors and other proteins that participate in the regulation of transcription. Once the mRNA for a transcription factor is translated, multiple regulatory mechanisms can determine when, where and how the transcription factor becomes active to control gene expression. In this chapter, we describe some of the most common and best-described mechanisms that modulate transcription factor activity.

Stress conditions induced by the environment (i.e., abiotic) or by pathogens (i.e., biotic) often require very rapid gene expression changes. In such conditions, it is advantageous for the cell to have the necessary transcription factors already synthesized, but present in an inactive form. This is often achieved by using post-translational modifications to control transcription factor activity. While certainly not the only post-translational modification that can affect transcription factor activity, phosphorylation/dephosphorylation is probably among the best studied. Transduction of the biotic or abiotic stress condition that ultimately culminates in the phosphorylation/dephosphorylation of a transcription factor often involves intracellular signaling pathways and cascades of enzymes involved in protein phosphorylation (**kinases**) or protein dephosphorylation (**phosphatases**) (see also Chapter 16).

Protein phosphorylation/dephosphorylation can modulate transcription factor activity in a number of different ways. For example, these modifications can determine the localization of the transcription factor in the cell (e.g., whether it will be nuclear or cytoplasmic, see Section 10.3). Phosphorylation/dephosphorylation can also control the stability of transcription factors: for instance, the phosphorylated form of the transcription factor gets degraded (see Chapter 17) more rapidly than the dephosphorylated form, or vice versa. A good example of this is provided by the *Arabidopsis* HY5 bZIP transcription factor, which promotes **photomorphogenesis** (i.e., developmental changes associated with growth in the light). HY5 exists in two forms, one phosphorylated and the other not. Only non-phosphorylated HY5 interacts with nuclear COP1, a negative regulator of photomorphogenesis (Figure 10.1). COP1 is one of a large group of enzymes, called the E3 ubiquitin ligases. As we discuss in more detail in Chapter 17, ubiquitin is a small protein (~80 amino acids) that gets covalently attached to many cellular proteins and marks them for degradation, through what is known as the proteasome pathway. The nuclear level of COP1 is negatively regulated by light, meaning that in the presence of higher light intensity, less COP1 is in the nucleus (Figure 10.1). Non-phosphorylated HY5 is

Plant Genes, Genomes and Genetics, First Edition. Erich Grotewold, Joseph Chappell and Elizabeth A. Kellogg.
© 2015 John Wiley & Sons, Ltd. Published 2015 by John Wiley & Sons, Ltd.
Companion Website: www.wiley.com/go/grotewold/plantgenes.

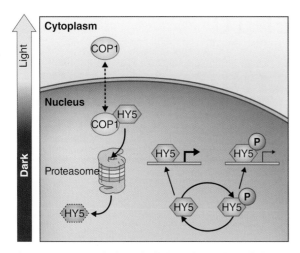

Figure 10.1 Control of HY5 by light and COP1 in *Arabidopsis*. In the dark, the level of COP1 in the nucleus is high. After light stimulus, the level decreases as COP1 moves into the cytoplasm. COP1 interacts with many proteins in addition to HY5 to facilitate protein degradation

the preferential target of COP1, which subsequently targets HY5 for degradation (Hardtke *et al.*, 2000). HY5 is only one of many other proteins that COP1 targets for degradation. Interestingly, and highlighting a remarkable feature of how biological systems are controlled, non-phosphorylated HY5 is also much more effective at activating transcription than phosphorylated HY5. The plant ensures that there is always a pool of phosphorylated HY5 in the dark, which can be rapidly converted into an active form by light. This example also highlights how phosphorylation can control protein–protein interactions, such as those described in Section 10.2.

Phosphorylation can modulate the DNA-binding activity of the transcription factor. As discussed in the previous chapter, the surface of the transcription factor that makes DNA contacts is usually basic (positively charged) to stabilize interactions with the negatively charged phosphate groups in the DNA. Therefore, phosphorylation of an amino acid involved in making DNA contacts is likely to decrease or abolish the interaction with the DNA because phosphate groups themselves tend to be negatively charged.

Proteins can be phosphorylated on serine, threonine, or tyrosine residues, and to a lesser extent on histidine. The enzymes responsible for these modifications are, respectively, known as serine/threonine protein kinases, tyrosine protein kinases, or histidine protein kinases. In plants, many biotic and abiotic

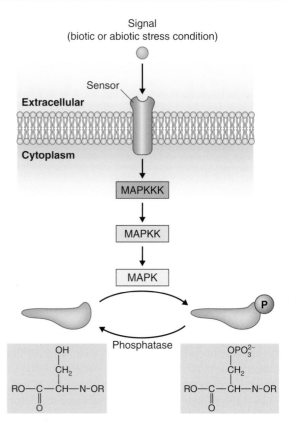

Figure 10.2 The MAPK cascade illustrating how the serine residue of a protein is ultimately phosphorylated

stress signals are conveyed by the mitogen-activated protein kinase (MAPK) cascade (Figure 10.2). The name MAPK comes from animals, where this cascade is stimulated by a number of factors that induce cell division (mitogen means that it induces mitosis). The cascade begins with a signal that activates a MAP kinase kinase kinase (MAPKKK), which phosphorylates a MAP kinase kinase (MAPKK), which in turns phosphorlyates a MAP kinase (MAPK). At the bottom of the cascade, a number of proteins end up phosphorylated on either serine or threonine. While a MAPK cascade can target many types of proteins for phosphorylation, transcription factors from a number of different families (e.g., R2R3 MYB, WRKY) have been identified as substrates for many of the plant MAPKs. However, MAPKs are not the only kinases that modulate transcription factor activity by phosphorylation; other families include the calcium-dependent protein kinases (CDPKs), the cyclin dependent kinases (CDKs, discussed also in Chapter 7 in the context of RNP-II) and the SNF1-related kinases (SnRKs).

10.2   Protein–protein interactions

As mentioned in the previous chapter, transcription factor activity is often controlled by protein–protein interactions. These interactions can expand the contact area of the complex with DNA, as is the case for basic helix-loop-helix (bHLH) or basic leucine-zipper (bZIP) transcription factors, which usually bind DNA only as homo- or heterodimers (see Chapter 9). In these cases, the short DNA contacts (3–4 base pairs) provided by the single basic (positively charged) helix of the transcription factor are expanded by bringing in close contact a second helix provided by the other subunit (Figure 9.7).

From the Experts

The (A)BC/E model of floral organ identity

Flowers are the reproductive organs of plants and arise from lateral meristems deriving from the shoot apical meristem (SAM). In many flowering plants, including the popular model *Arabidopsis thaliana*, four organ types are contained within the flower, arranged in four concentric circles, or whorls (Figure 10.3). The sepals are found in the outermost whorl, petals in the second whorl, stamens (male reproductive structures) in the third whorl and carpels (female reproductive structures) in the innermost whorl. However, considerable variation exists within angiosperms, especially in the outermost, non-fertile whorls of the flower. For example, lilies have petal-like organs (called tepals) in both the first and second whorls instead of distinct sepals and petals (Figure 10.3).

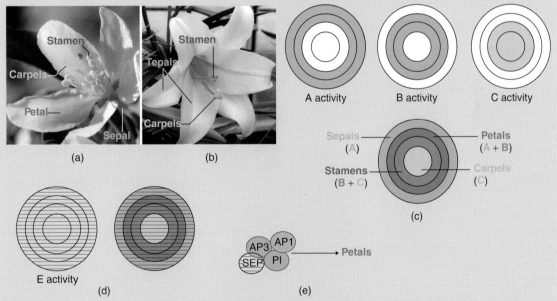

Figure 10.3 The (A)BC/E model for floral organ identity. (a) Most flowers, such as this cherry blossom, contain four organ types arranged in concentric whorls: sepals, petals, stamens, and carpels. (b) A common variation in floral composition is the presence of tepals in both the first and second whorls of the flower, such as is seen in this lily. (a, b) Photos courtesy of Matteo Citarelli. Reproduced with permission of Matteo Citarelli. (c) The classic ABC model visualized as floral meristems seen from above. Three gene activities, A, B, and C, act in two adjoining whorls each with overlapping patterns. In combination, these activities give each whorl a unique identity. In addition to the ABC genes, a group of SEP/E genes distinguish the flower from leaves and shoots. (d) The MADS box transcription factors encoded by the ABC and E genes act in multi-protein complexes with unique target specificities to specify organ identity. (e) In this example, the A protein AP1 acts in complex with SEP proteins and the B proteins AP3 and PI to specify petals

The floral organs arise from individual primordia within the floral meristem and acquire their identity from the combinatorial activity of a set of genes known as the floral organ identity genes or the (A)BC/E genes. These are homeotic genes, similar to those involved in segment identity in *Drosophila*; mutation in these genes causes one type of floral organ to be converted to another. The ABC model of floral development was articulated by Coen and Meyerowitz (1991) based on data from Arabidopsis and *Antirrhinum majus* (snapdragon). In this model, the floral meristem is divided into three overlapping concentric rings by the activity of the ABC genes (Figure 10.3), generating four unique regions. A activity alone specifies sepals, A+B petals, B+C stamens and C alone carpels. C genes also are essential to confer determinate growth on floral primordia.

B and C genes have been identified in diverse angiosperms and shown to function in floral organ identity as described above. In fact, these gene lineages predate flowering plants; B and C gene orthologs are expressed in gymnosperm cones (male cones and both male and female cones, respectively) (Theissen and Becker, 2004; Theissen *et al.*, 2000). The presence of tepals in some flowers has been demonstrated to correlate with expansion of the B expression domain into the first whorl (Kalivas *et al.*, 2007), suggesting that relatively simple modification of the basic program may underlie diverse morphologies across the angiosperms. However, it has become clear that A activity, as originally defined, is not conserved beyond Arabidopsis and its relatives. In other angiosperms, it appears that the orthologs of the Arabidopsis A gene *APETALA1* (*AP1*) function in floral meristem identity (the differentiation of a flower from a shoot) rather than in sepal and petal identity, while orthologs of the Arabidopsis A gene *APETALA2* (*AP2*) have diverse functions both within and without flowers. Therefore, additional mechanisms must exist to distinguish sepals from petals in flowering plants.

Subsequent to the discovery of the ABC genes, a fourth category of floral organ identity genes was identified, the SEPALLATA (SEP)/E genes (Pelaz *et al.*, 2000). Arabidopsis encodes four such genes (*SEP1–4*) and in aggregate they are expressed throughout the floral meristem. Thus, the E domain is superimposed over the ABC domains (Figure 10.3). Single mutants in any one *SEP* gene have little to no phenotype but mutant combinations change organ identity. *sep1 sep2 sep3* triple mutants exhibit phenotypes similar to a BC double mutant, indeterminate flowers consisting of sepal-like organs. Quadruple mutants display indeterminate flowers containing only leaf-like organs. Therefore, SEP activity is necessary for identity of all four organs in the flower. *SEP* orthologs are found across angiosperms and their function is conserved. For example, loss of function in the *SEP* ortholog *OsMADS1* in rice causes equivalent phenotypes to s*ep1 sep2 sep3* mutants in Arabidopsis (Agrawal *et al.*, 2005).

All of the (A)BC and E genes, except for the Arabidospis A gene *AP2*, encode MADS domain transcription factors. MADS proteins are known to act as dimers. In fact, the two B class transcription factors APETALA3 (AP3) and PISTILLATA (PI) act as obligatory heterodimers in Arabidopsis (Riechmann *et al.*, 1996a). The DNA binding specificity of the floral MADS box transcription factors are largely overlapping (Riechmann *et al.*, 1996a, 1996b), but the proteins function distinctly in floral organ identity. Therefore, an active area of research has been how specificity in target selection is conferred. This appears to be gained in part by the formation of higher order complexes of MADS proteins. Complexes of A and B gene products were first identified in snapdragon (Egea-Cortines *et al.*, 1999) and then the E class SEP proteins were shown to be in a complex with (A)BC class MADS proteins in Arabidopsis (Fan *et al.*, 1997; Honma and Goto, 2001) and several other angiosperms (Ferrario *et al.*, 2003; Leseberg *et al.*, 2008). For example, in the second whorl of the Arabidopsis flower where petals form, complexes containing the A protein AP1, the B heterodimer AP3/PI and the E protein SEP3 have been detected (Figure 10.3). These "quartets" of MADS proteins appear to have enhanced DNA binding affinity and transcriptional activation activity (Theissen and Saedler *et al.*, 2001). Each whorl of the flower will have unique MADS-containing complexes that regulate a unique suite of genes. Thus, regulation of protein–protein interactions between transcription factors is essential for proper floral morphogenesis.

By Rebecca Lamb

Protein–protein interactions also can expand the number of ways that the transcription factor complex can bind DNA. One example is provided by the interaction of a subgroup of R2R3 MYB proteins and a subgroup of bHLH transcription factors, which have been best described in the control of anthocyanin pigments in maize and other plants, in the differentiation of *Arabidopsis* leaf epidermal cells to hairs, or trichomes, and in the formation and positioning of root hairs (Feller *et al.*, 2011). Interactions between R2R3 MYB and bHLH factors occur through specific residues that are not involved in DNA binding; specifically, the solvent-exposed surface of the R3 repeat of the R2R3 MYB domain (Grotewold *et al.*, 2000: Zimmermann *et al.*, 2004) binds to an N-terminal region in the bHLH transcription factor (Goff *et al.*, 1992). This results in a protein complex with two distinct types of DNA-binding domains (the R2R3 MYB and the bHLH domains). The MYB-bHLH complex is thus multi-functional and its role in any particular cell is determined by how it binds to specific promoters. The active R2R3 MYB-bHLH complex can bind DNA either through the MYB domain, which usually recognizes a DNA sequence that fits the consensus $CC^T/_AAC$, or through the dimeric bHLH, which usually recognizes a DNA sequence with the consensus CANNTG, where N can be any nucleotide; the latter sequence is known as the E-box (or Enhancer box). Additional variation is provided by the dimerization of the bHLH protein, which may be hetero- or homodimeric. If it forms a homodimer, then the recognition sequence is a palindrome, such as CACGTG; this particular sequence is known as the G-box.

Additional regulation comes from proteins that disrupt the active R2R3 MYB-bHLH complex. For example, a group of small MYB proteins that contains just the R3 MYB repeat (Figure 10.4) can interact with the bHLH and compete with R2R3 MYBs for interaction with this essential bHLH co-regulator. Thus, the R3 MYB proteins block the formation of the transcriptionally active R2R3 MYB-bHLH complex, as shown in Figure 10.4.

Just as a transcription factor and inhibitor can form a heterodimer that cannot bind DNA, this can also be a feature of some regulatory proteins in that they need to be homo- or heterodimers to bind DNA. For example, in animals a group of at least four related bHLH factors that includes MyoD is required for cell differentiation processes that result in normal muscle

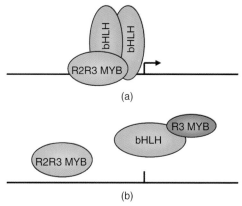

Figure 10.4 Interactions between R2R3-MYB and bHLH factors provide a good example of how transcription factor activity is controlled by protein–protein interactions. (a) The R2R3 MYB-bHLH complex is transcriptionally active, resulting in the activation of genes such as those involved in anthocyanin pigment biosynthesis or trichome formation. (b) Interaction of the bHLH factor with inhibitory R3 MYB proteins precludes the formation of the active R2R3 MYB-bHLH complex

development. MyoD binds DNA as a heterodimer with a member of the E2 family of bHLH proteins (such as E47). However, to prevent the cell differentiation process from taking place in the wrong cells or at the wrong time, MyoD and members of the E2 subgroup of bHLH factors interact with Id (inhibitor of differentiation), which contains the HLH motif but lacks the basic region. Hence, MyoD-Id heterodimers fail to bind DNA (Wright, 1992).

The response of plants to light is also controlled by inhibitory factors. Normally, a low ratio of red to far red light is interpreted by plants as being shaded by a neighboring plant, promoting cell elongation in a process that is known as **Shade Avoidance Syndrome**, or SAS. In the shade, the P_{FR} isoform of the phytochromes (PhyA through E in Arabidopsis) moves to the nucleus, where it interacts with a group of bHLH transcription factors (the PIFs, or Phytochrome Interacting Factors). The PIFs then trigger gene expression changes associated with SAS and with greening (de-etiolation). PIF3, PIF4, and PIF5 form homodimers and PIF3-PIF4 also form heterodimers, all exhibiting DNA binding (Toledo-Ortiz *et al.*, 2003; Hornitschek *et al.*, 2009). However, the related bHLH factor HFR1 (long hypocotyl in far red light) is induced in the shade and inhibits the shade avoidance response by interacting with PIF4 and PIF5 and forming heterodimers that will not bind to DNA (Hornitschek *et al.*, 2009). This creates a negative feedback loop that ultimately modulates to what extent plants respond to shade.

Protein–protein interactions are also important for integrating distinct signaling pathways. A good example is provided by the crosstalk between light and the brassinosteroid (BR) plant steroid hormone. BRs control many plant cellular processes including cell elongation, differentiation of the vascular and reproductive systems, photomorphogenesis, and stress responses. Most of the key players have been identified in the signal transduction pathway that results in BR cellular responses. BR is recognized in the cellular membrane by the BRI1 receptor (note that in animals, many steroid receptors are cytoplasmic, see Section 10.3), and this interaction indirectly leads to the dephosphorylation of the bHLH factors BZR1 and BZR2, resulting in their mobilization to the nucleus and regulation of gene expression. Cellular targets for BZR1 and BZR2 have been identified through the combination of genome-wide expression and ChIP-chip analyses. Such studies showed that the BZR1-repressed genes included many light signaling (e.g., *PhyB*) and light responsive genes. Moreover, BZR1 physically interacts with PIF4, one of the PIF proteins described earlier, and together they activate a large number of genes (Wang, Z.Y. *et al.*, 2012).

From the Experts

The plant circadian clock: interlocked feedback loops by interacting transcription factors

The circadian clock is a 24-h time-keeper that coordinates many molecular, physiological and metabolic processes to optimize the plant's health and survival in a changing environment. Clocks are found nearly universally, although across phyla the molecular components that underlie each system vary widely. The circadian system is comprised of multiple interlocked autoregulatory feedback loops consisting of activating and repressive transcriptional elements to sustain robust 24-h oscillations (McClung, 2011; Nagel and Kay, 2012). Most of the core components are themselves rhythmically expressed both at the mRNA and protein levels. These early-phased (morning) and late-phased (evening) genes often control the expression of each other, with morning transcriptional repressors inhibiting evening gene expression during the day, and evening complexes expressed at night which repress transcription of the morning genes (Pokhilko *et al.*, 2012).

An example of such a loop of reciprocal repression is based on two morning genes [*CIRCADIAN CLOCK ASSOCIATED 1* (*CCA1*) and *LATE ELONGATED HYPOCOTYL* (*LHY*)] inhibiting the early day expression of a late evening gene [*TIMING OF CAB EXPRESSION 1* (*TOC1*; *PRR1*)], thereby limiting its expression to very late in the day (night). With *TOC1* expression restricted to the night, it now acts to repress *CCA1/LHY* expression during that time, helping to define *CCA1/LHY* as morning-expressed genes.

TOC1 is the founding member of five closely related PSEUDO RESPONSE REGULATORS (PRRs: PRR9, PRR7, PRR5, PRR3). These proteins bind DNA through a conserved CCT domain at the carboxy terminus, but often work together with co-repressors (e.g., TOPLESS) and histone deacetylases (Gedron *et al.*, 2012; Wang, L. *et al.*, 2012). In addition to TOC1, the waveform of *CCA1/LHY* circadian expression depends on sequential inhibition by PRR9, PRR7 and PRR5. All four proteins are expressed at discrete times of the circadian cycle, with PRR9 accumulating early in the day followed by PRR7, PRR5 and TOC1 (Nakamichi *et al.*, 2010). Their protein expression patterns determine their occupancy of *CCA1* and *LHY* promoter regions, keeping *CCA1* and *LHY* transcription strongly repressed over most of the mid-morning and into late evening. However, control of another morning-phased gene, *PRR9*, relies on a very different repression complex. Here the transcription factor LUX ARRHYTHMO (LUX) complexes with EARLY FLOWERING 3 (ELF3) and 4 (ELF4) ("evening complex") to restrict *PRR9* expression to the morning (Helfer *et al.*, 2011; Nusinow *et al.*, 2011). Since CCA1/LHY activate *PRR9* expression, and PRR9 is a repressor of *CCA1/LHY*, their relationship defines an interacting loop of activation and repression.

The mechanism of CCA1/LHY-mediated repression of *TOC1* during the early day again involves very different factors. Here, the co-repressor DE-ETIOLATED1 (DET1) interacts with CCA1 and LHY at the *TOC1* promoter, most likely as a larger COP10-DET1-DDB1(CDD) complex (Lau *et al.*, 2011). *TOC1* (and *PRR5*) are also regulated by the CCA1/LHY-related MYB-transcription factor, REVEILLE8, which binds the *TOC1* and *PRR5* promoters, but as a positive activator (Hsu *et al.*, 2013).

Taken together, these studies highlight the complex transcriptional regulation of the clock. Current models are replete with transcriptional repressors, but have a relative dearth of known activators. Concomitant with these transcriptional control events are circadian oscillations in histone modifications, primarily methylation and acetylation (Malapeira *et al.*, 2012). These chromatin marks indicate that posttranslational mechanisms act at the nucleosomes, as well as on the transcription factors themselves (e.g., phosphorylation) to provide additional levels of control within the interlocking loops.

By David Somers

Thus protein–protein interactions between transcription factors and other regulatory proteins are most likely the norm, rather than the exception; the few cases that have been reported just highlight the need for more research in that area.

10.3 Preventing transcription factors from access to the nucleus

For a transcription factor to activate nuclear transcription, it obviously must be in the nucleus. Thus, after translation, the transcription factor needs to move across the nuclear envelope. As discussed in Chapter 16, this requires a particular protein sequence motif, the nuclear localization signal, or NLS. Thus, one way to control transcription factor activity is to prevent it from translocating to the nucleus.

The control of nuclear transport has been very well studied in animals, so we present two such examples here. The first (Figure 10.5a) is provided by the transcription factor NF-κB, which appears to be absent in the plant kingdom, yet is ubiquitous in animals. NF-κB is normally retained in the cytoplasm by the interaction with a factor known as IκB. IκB phosphorylation results in its degradation (through the proteosome pathway), freeing NF-κB to move to the nucleus, where it controls the expression of genes involved in a number of cellular functions, including the immune response. The case of NF-κB is in many ways similar to what happens with the phosphorylated forms of the BZR1 and BZR2 bHLH regulators described earlier, which in the absence of BR, are retained in the cytoplasm through the interaction with a 14-3-3 protein. 14-3-3 proteins are ubiquitous in all eukaryotes, and their odd name derives from how they are eluted in chromatographic purifications. What is relevant is that 14-3-3 proteins can recognize a number of proteins phosphorylated in Ser or Thr.

The second example (Figure 10.5b) is provided by a large group of transcription factors broadly known as the nuclear receptors. Similar to NF-κB, the nuclear receptors are specific to animals, and are responsible for binding a number of lipophilic hormones that include the steroid hormones progesterone, estradiol, and testosterone. The nuclear receptors that respond to these steroid hormones carry nuclear localization signals, yet are retained in the cytosol by interaction with a heat shock protein (HSP). Upon binding to the hormone ligand, this interaction is dissociated and the receptor translocates to the nucleus, where it regulates transcription of steroid hormone responsive genes.

An increasing number of plant transcription factors are being recognized as translocating between the cytoplasm and the nucleus when the cells are challenged by particular signals. Compared with what is known about NF-κB or nuclear receptors, the mechanisms remain in general less well understood. An exception to this, however, is a growing class of proteins known as membrane-bound transcription factors (Figure 10.5c). These transcription factors usually have an NLS, yet they are anchored to a cellular membrane, most often the membrane of the endoplasmic reticulum (ER), and hence are prevented from translocation to the nucleus. In the presence of a particular signal, the transcription

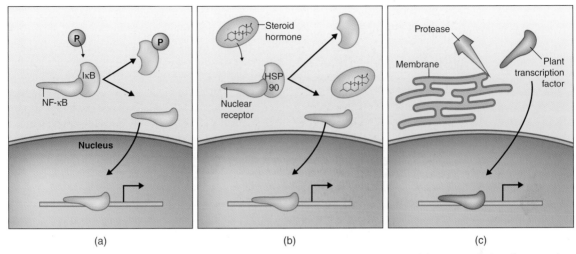

(a)   (b)   (c)

Figure 10.5 Control of transcription factor activity by modulating nuclear transport. (a) A transcription factor such as mammalian NF-κB is retained in the cytoplasm by the interaction with another protein, such as IκB. Phosphorylation of IκB results in the dissociation from NF-κB , allowing NF-κB to enter the nucleus and activate transcription. (b) Nuclear (steroid hormone) receptors are retained in the cytoplasm by interaction with HSP90. The binding to the steroid hormone releases the interaction with HSP90, permitting the nuclear receptor to translocate to the nucleus. (c) An increasing number of plant transcription factors are anchored to a membrane, such as the endoplasmic reticulum. When a protease cleaves the membrane attachment domain, the transcription factor can move to the nucleus

factor is released from the membrane, for example by the action of a protease that cleaves the ER anchoring region. It is estimated that about 10% of plant transcription factors (which represent 5–7% of all the protein-coding genes) are controlled by a mechanism like this. The first plant transcription factor identified as controlled by such a mechanism was Arabidopsis bZIP60, which is involved in the unfolded protein response (UPR), a cellular stress response that senses when proteins are not properly folded.

10.4 Movement of transcription factors between cells

Cell-to-cell communication by the direct movement of transcription factors appears to be common in many plant developmental processes (Kurata *et al.*, 2005; Wu and Gallagher, 2011). The maize KNOTTED1 (KN1) protein, a homeobox transcription factor, provided the first example of a plant transcription factor moving between cells. When KN1 is ectopically expressed in maize leaves, cells proliferate and form knot-like

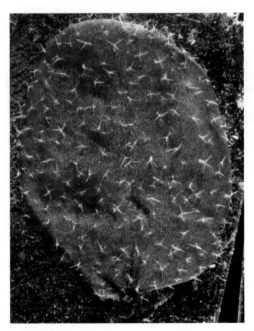

Figure 10.6 Characteristic pattern of trichome distribution on the adaxial surface of an Arabidopsis leaf. Such a pattern is in part provided by the lateral movement of small R3 MYB proteins. In Arabidopsis, trichomes are unicellular and usually have three branches. Photo courtesy of Marcelo Pomeranz. Reprouced with permission of Marcelo Pomeranz

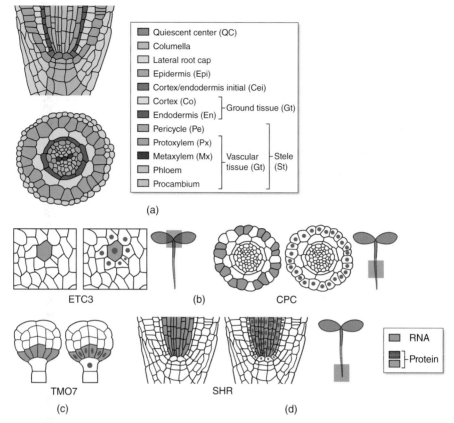

Figure 10.7 Intercellular movement of transcription factors. (a) Cellular organization of the Arabidopsis root shown in longitudinal (top) and transverse (bottom) section. Obtained from Miyashima *et al.*, 2011. (b–d) Schematic representations of the different distributions of the Arabidopsis mRNA (yellow) and protein (green) with the ability to move from one cell to another. (b) The diagram represents the localization of protein and mRNA for the small MYB proteins ETC3 and CPC involved in the control of trichome and root hair formation. (c) The diagram represents the distribution of TMO7 mRNA and protein in the globular stage embryos of Arabidopsis. (d) The diagram represents the distribution of the SHR transcription factor mRNA and protein in the meristematic region of the Arabidopsis root. (b–d) Adapted from Wu and Gallagher (2011). Reproduced with permission of Elsevier.

structures, hence the name (Hake and Freeling, 1986). KNOTTED1 is involved in establishment and maintenance of the plant SAM, the group of stem cells at the apex of the plant that remains in an undifferentiated stage and that ultimately will give rise to all aerial plant organs. The KN1 protein can move between cells. Interestingly however, although not directly relevant to protein movement, the KN1 protein participates also in the movement of the KN1 mRNA from one cell to another. Since these early studies with KN1, a number of other plant transcription factors have been shown to move between cells. Some other examples follow.

We described earlier in this chapter small R3 MYB proteins that block the interaction between specific bHLH and R2R3 MYB factors. Some of these small proteins can also move between adjacent cells. This becomes very important, for example, in the patterning of trichomes and root hairs in Arabidopsis. If one looks at the top (adaxial) side of an Arabidopsis leaf (Figure 10.6), there is a uniform distribution of trichomes, in general with 3–4 epidermal cells in between. This pattern is in part achieved by the movement of small R3 MYB proteins that include ETC3 (Figure 10.7b) and CAPRICE (CPC). The R3 MYB protein moves from a cell already committed to become a trichome to the adjacent cells to inhibit the formation of a trichome-inducing R2R3 MYB-bHLH complex, thus establishing the pattern.

Cell-to-cell movement of transcription factors is also central for root development (Figure 10.7a). The radicle is the first structure that emerges after germination from cells predetermined to have "root identity" in the *Arabidopsis* embryo. A key player in this process is a bHLH transcription factor called TMO7 (for TARGET OF MONOPTEROS7). TMO7 moves between embryonic cells (Figure 10.7b), and this is likely essential for root formation. The root also has a group of cells that remain undifferentiated and which, through cell divisions and differentiation, will result in formation of all the other root cells. This group of cells is known as the quiescent center (QC) and is located right above the root cap (Figure 10.7a). In the developing root, two transcription factors that belong to the GRAS family, SHORTROOT (SHR) and SCARECROW (SCR) are required for asymmetric division of QC cells and for specifying endodermis identity. The SHR mRNA is normally found in the stele (Figure 10.7d) while the SHR protein is also found in the adjacent cell layer. Wherever the SHR protein accumulates, it activates the expression of SCR. The SHR protein can be present both in the nucleus and the cytoplasm of stele cells. However, in cells that express SCR, SHR is only present in the nucleus and this prevents SHR from moving further into adjacent cell layers. In the different cell layers, SHR can work by itself or together with SCR to activate or repress specific sets of genes that ultimately lead to the correct radial patterning of the root.

10.5   Summary

While transcriptional regulation of transcription factors (by other transcriptional regulators) is often very important and results in the formation of gene regulatory networks, transcription factor activity can be modulated by a number of different mechanisms. We discuss in this chapter phosphorylation, protein–protein interactions, transport to the nucleus, and movement between cells. These mechanisms can activate or inhibit transcription factor activity, and also the biological activities of transcriptional regulators, expanding the functional repertoire of the limited number of transcription factors in a cell. The concepts described are applied to two very important biological processes: flower development and the circadian clock.

10.6   Problems

10.1 How does phosphorylation alter the charge of a protein?

10.2 Can phosphorylation alter the mobility of a protein in native polyacrylamide gel electrophoresis (PAGE)? And in denaturing PAGE? How?

10.3 What amino acids are most frequently found phosphorylated in plants?

10.4 In Chapters 9 and 10, we discussed protein–DNA and protein–protein interactions. Compare the chemical interactions (hydrogen bonds, hydrophobic, ionic, etc.) that might be at play in each case.

10.5 Would you expect that protein–protein or protein–DNA interactions evolve faster? Why?

10.6 Ubiquitin and SUMO are two small peptides that often modify plant proteins by covalent attachment. Investigate the main differences between them.

10.7 Why would there be a connection between the circadian clock and when a plant flowers?

10.8 Design two control circuits for a transcriptional factor regulating the biosynthesis of an essential amino acid. Design one circuit to be a positive control loop and the other to be a negative feedback loop mechanism.

References

Agrawal, G.K., Abe, K., Yamazaki, M. *et al.* (2005) Conservation of the E-function for floral organ identity in rice revealed by the analysis of tissue culture-induced loss-of-function mutants of the *OsMADS1* gene. *Plant Molecular Biology*, **59**, 125–135.

Coen, E.S. and Meyerowitz, E.M. (1991) The war of the whorls: genetic interactions controlling flower development. *Nature*, **353**, 31–37.

Egea-Cortines, M., Saedler, H. and Sommer, H. (1999). Ternary complex formation between the MADS-box proteins SQUAMOSA, DEFICIENS and GLOBOSA is involved in the control of floral architecture in *Antirrhinum majus*. *EMBO Journal*, **18**, 5370–5379.

Fan, H.Y., Hu, Y., Tudor, M. and Ma, H. (1997) Specific interactions between the K domains of AG and AGLs, members of the MADS domain family of DNA binding proteins. *Plant Journal*, **12**, 999–1010.

Feller, A., Machemer, K., Braun, E.L. and Grotewold, E. (2011) Evolutionary and comparative analysis of MYB and bHLH plant transcription factors. *Plant Journal*, **66**, 94–116.

Ferrario, S., Immink, R.G., Shchennikova, A. *et al.* (2003) The MADS Box Gene *FBP2* is required for SEPALLATA function in petunia. *Plant Cell*, **15**, 914–925.

Gendron, J.M., Pruneda-Paz, J.L., Doherty, C.J. *et al.* (2012) Arabidopsis circadian clock protein, TOC1, is a DNA-binding transcription factor. *Proceedings of the National Academy of Sciences USA*, **109**, 3167–3172.

Goff, S.A., Cone, K.C. and Chandler, V.L. (1992) Functional analysis of the transcriptional activator encoded by the maize *B* gene: evidence for a direct functional interaction between two classes of regulatory proteins. *Genes Development*, **6**, 864–875.

Grotewold, E., Sainz, M.B., Tagliani, L. *et al.* (2000) Identification of the residues in the Myb domain of maize C1 that specify the interaction with the bHLH cofactor R. *Proceedings of the Natiaol Academy of Sciences USA*, **97**, 13579–13584.

Hake, S. and Freeling, M. (1986). Analysis of genetic mosaics shows that the extra epidermal cell divisions in *Knotted* mutant maize plants are induced by adjacent mesophyll cells. *Nature*, **320**, 621–623.

Hardtke, C.S., Gohda, K., Osterlund, M.T. *et al.* (2000). HY5 stability and activity in *Arabidopsis* is regulated by phosphorylation in its COP1 binding domain. *EMBO Journal*, **19**, 4997–5006.

Helfer, A., Nusinow, D.A., Chow, B.Y. *et al.* (2011) *LUX ARRHYTHMO* encodes a nighttime repressor of circadian gene expression in the *Arabidopsis* core clock. *Current Biology*, **21**, 126–133.

Honma, T. and Goto, K. (2001) Complexes of MADS-box proteins are sufficient to convert leaves into floral organs. *Nature*, **409**, 525–529.

Hornitschek, P., Lorrain, S., Zoete, V. *et al.* (2009) Inhibition of the shade avoidance response by formation of non-DNA binding bHLH heterodimers. *EMBO Journal*, **28**, 3893–3902.

Hsu, P.Y., Devisetty, U.K. and Harmer, S.L. (2013) Accurate timekeeping is controlled by a cycling activator in *Arabidopsis. eLIFE*, **2**, e00473.

Kalivas, A., Pasentsis, K., Polidoros, A.N. and Tsaftaris, A.S. (2007) Heterotopic expression of B-class floral homeotic genes PISTILLATA/GLOBOSA supports a modified model for crocus (*Crocus sativus* L.) flower formation. *DNA Sequence*, **18**, 120–130.

Kurata, T., Okada, K. and Wada, T. (2005). Intercellular movement of transcription factors. *Current Opinion in Plant Biology*, **8**, 600–605.

Lau, O.S., Huang, X., Charron, J.B. *et al.* (2011) Interaction of *Arabidopsis* DET1 with CCA1 and LHY in mediating transcriptional repression in the plant circadian clock. *Molecular Cell*, **43**, 703–712.

Leseberg, C.H., Eissler, C.L., Wang, X. *et al.* (2008) Interaction study of MADS-domain proteins in tomato. *Journal of Experimental Botany*, **59**, 2253–2265.

Malapeira, J., Khaitova, L.C. and Mas, P. (2012) Ordered changes in histone modifications at the core of the Arabidopsis circadian clock. *Proceedings of the National Academy of Sciences USA*, **109**, 21540–21545.

McClung, C.R. (2011) The genetics of plant clocks. *Advances in Genetics*, **74**, 105–139.

Miyashima, S., Koi, S., Hashimoto, T. and Nakajima, K. (2011) Non-cell-autonomous microRNA165 acts in a dose-dependent manner to regulate multiple differentiation status in the *Arabidopsis* root. *Development*, **138**, 2303–2313.

Nagel, D.H. and Kay, S.A. (2012) Complexity in the wiring and regulation of plant circadian networks. *Current Biology*, **22**, R648–657.

Nakamichi, N., Kiba, T., Henriques, R. *et al.* (2010) PSEUDO-RESPONSE REGULATORS 9, 7, and 5 are transcriptional repressors in the *Arabidopsis* circadian clock. *Plant Cell*, **22**, 594–605.

Nusinow, D.A., Helfer, A., Hamilton, E.E. *et al.* (2011) The ELF4-ELF3-LUX complex links the circadian clock to diurnal control of hypocotyl growth. *Nature*, **475**, 398–402.

Pelaz, S., Ditta, G.S., Baumann, E. *et al.* (2000). B and C floral organ identity functions require SEPALLATA MADS-box genes. *Nature*, **405**, 200–203.

Pokhilko, A., Fernandez, A.P., Edwards, K.D. *et al.* (2012) The clock gene circuit in *Arabidopsis* includes a repressilator with additional feedback loops. *Molecular Systems Biology*, **8**, 574.

Riechmann, J.L., Krizek, B.A. and Meyerowitz, E.M. (1996a). Dimerization specificity of Arabidopsis MADS domain homeotic proteins APETALA1, APETALA3, PISTILLATA, and AGAMOUS. *Proceedings of the National Academy of Sciences USA*, **93**, 4793–4798.

Riechmann, J.L., Wang, M. and Meyerowitz, E.M. (1996b) DNA-binding properties of Arabidopsis MADS domain homeotic proteins APETALA1, APETALA3, PISTILLATA and AGAMOUS. *Nucleic Acids Research*, **24**, 3134–3141.

Theissen, G. and Becker, A. (2004) Gymnosperm orthologues of Class B floral homeotic genes and their impact on understanding flower origin. *Critical Reviews in Plant Sciences*, **23**, 129–148.

Theissen, G., Becker, A., Di Rosa, A. *et al.* (2000) A short history of MADS-box genes in plants. *Plant Molecular Biology*, **42**, 115–149.

Theissen, G. and Saedler, H. (2001) Plant biology. Floral quartets. *Nature*, **409,** 469–471.

Toledo-Ortiz, G., Huq, E. and Quail, P.H. (2003) The Arabidopsis basic/helix-loop-helix transcription factor family. *Plant Cell*, **15,** 1749–1770.

Wang, L., Kim, J. and Somers, D.E. (2012) Transcriptional corepressor TOPLESS complexes with pseudoresponse regulator proteins and histone deacetylases to regulate circadian transcription. *Proceedings of the National Academy of Sciences USA*, **110,** 761–766.

Wang, Z.Y., Bai, M.Y., Oh, E. and Zhu, J.Y. (2012) Brassinosteroid signaling network and regulation of photomorphogenesis. *Annual Review of Genetics*, **46,** 701–724.

Wright, W.E. (1992). Muscle basic helix-loop-helix proteins and the regulation of myogenesis. *Current Opinion in Genetics & Development*, **2,** 243–248.

Wu, S. and Gallagher, K.L. (2011) Mobile protein signals in plant development. *Current Opinion in Plant Biology*, **14,** 563–570.

Zimmermann, I.M., Heim, M.A., Weisshaar, B. and Uhrig, J.F. (2004). Comprehensive identification of *Arabidopsis thaliana* MYB transcription factors interacting with R/B-like BHLH proteins. *Plant Journal*, **40,** 22–34.

Chapter 11
Small RNAs

11.1 The phenomenon of cosuppression or gene silencing

In previous chapters we described how RNAs in general, and mRNAs in particular, are synthesized by the process of transcription, and how transcription is regulated. Over the past couple of decades, however, the RNA field has been revolutionized by the discovery of a completely new class of RNAs, known as small RNAs. This chapter describes their generation and function for the biology of plants

Before we plunge into small RNAs, we must introduce the phenomenon generally known as **cosuppression** or **gene silencing**, which has been very familiar to the plant molecular biology researcher since the early 1980s. Cosuppression occurs when a gene or another piece of genetic material is introduced into a plant, for example through transformation. Such a gene is normally referred to as a **transgene**. Transgenes integrate (in single or multiple copies, depending on the transformation method used) at variable locations in the plant genome through what is known as non-homologous recombination. The alternative is homologous recombination, in which the normal gene in the plant is precisely replaced by the transgene; this occurs at very low frequency in most land plants except for some mosses, so plant gene replacement is currently almost impossible. When

such randomly inserted transgenes are transcribed, researchers rapidly noticed that genes integrated in different parts of the genome showed very different expression levels; this was known as **positional effect**. However, if the sequence of the transgene contained a large portion of identity with another DNA sequence in the genome, then expression of both the transgene and the endogenous gene would often be reduced. This phenomenon was called cosuppression (Jorgensen, 1995). Gene cosuppression has by now been widely observed and is assumed to be near-universal in plants, but pigmentation patterns in the petunia flower (Figure 11.1) provided perhaps one of the best insights on the molecular bases of cosuppression, which ultimately contributed significantly to the discovery of small RNAs.

The bright flowers of petunia would seem like an unlikely place to look for novel mechanisms of RNA regulation, but studies of anthocyanin pigmentation in petunia (and other plants) provided surprising results. Anthocyanins are responsible for most of the red, orange, purple, and blue colors of flowers (Grotewold, 2006), which we discussed previously in the context of visible outputs for transposon activity (see Chapter 3). Rich Jorgenson and Carolyn Napoli, then at the DNA Plant Technology Corporation, in Oakland, CA, were attempting to increase production of anthocyanin genes in petunias to create flowers with deeper, richer colors. They used transgenic approaches to place the first gene in the anthocyanin pathway, chalcone synthase (CHS) into petunia. In some cases, they

Plant Genes, Genomes and Genetics, First Edition. Erich Grotewold, Joseph Chappell and Elizabeth A. Kellogg.
© 2015 John Wiley & Sons, Ltd. Published 2015 by John Wiley & Sons, Ltd.
Companion Website: www.wiley.com/go/grotewold/plantgenes.

Figure 11.1 Patterns of floral pigmentation from plants transformed with Petunia chalcone synthase (*CHS*) gene constructs that result in cosuppression. Que *et al.* (1997). Copyright 1997 American Society of Plant Biologists. Used with permission

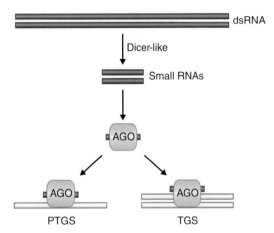

Figure 11.2 General pathway for the formation and action of small RNAs. Double-stranded RNAs (dsRNAs) are the substrates for different Dicer enzymes resulting in the formation of small RNAs that can then participate in transcriptional gene silencing (TGS) or post-transcriptional gene silencing (PTGS)

got the result they were expecting – more pigment. Surprisingly, they also got flowers with unusual color patterns and some with no pigment at all (Figure 11.1) (Napoli *et al.*, 1990; Que *et al.*, 1997). It appeared that introducing an extra copy of CHS (beyond the one that was in the plant already) caused both the introduced copy and the original copy to be turned off, or silenced. The mechanism was unknown. Subsequent studies resulted in the identification of two separable mechanisms for gene silencing: Transcriptional gene silencing (TGS) and post-transcriptional gene silencing (PTGS) (Figure 11.2).

11.2 Discovery of small RNAs

Shortly after the petunia studies described above, Andrew Fire and Craig Mello at the University of Massachusetts Medical School, who were working on the nematode worm *Caenorhabditis elegans* (*C. elegans*), discovered that, when injected into the worm, long double-stranded RNAs can silence endogenous genes that share significant regions of identity with the RNA introduced. This **RNA interference** (or RNAi) effect was maintained if a sense or an antisense RNA was introduced, and the silencing effect could be maintained over multiple generations (Fire *et al.*, 1998). The discovery of RNA interference resulted in Fire and Mello being jointly awarded the Nobel Prize in Physiology or Medicine in 2006. Several subsequent experiments demonstrated that the interfering RNA was somehow promoting degradation of the RNA. RNA interference appeared to occur post-transcriptionally (hence it is the plant equivalent of PTGS, Figure 11.2) in a multiplicative or catalytic fashion, since even very low quantities of the interfering RNA had a dramatic effect on the expression of the endogenous gene. *C. elegans* is very amenable to genetic analysis. Hence, it was possible to utilize mutagenesis to identify mutants that had a defective RNAi response. One of the first *C. elegans* mutants characterized corresponded to a gene similar to the just-identified *ARGONAUTE1* (or *AGO1*) gene from Arabidopsis, involved in the control of leaf development (Bohmert *et al.*, 1998). As we will discuss shortly, there are multiple AGO proteins in most organisms, each one playing specific roles in the molecular pathways that result in gene silencing.

In 1999, David Baulcombe's group (John Innes Centre, Norwich, UK), while trying to understand PTGS (the plant equivalent of RNAi), discovered in tomato small RNAs (~25 nucleotides long) that corresponded in sequence to plant genes targeted by PTGS (Hamilton and Baulcombe, 1999). These findings were followed by a rapid succession of publications in the fruit fly Drosophila showing that such small RNAs (now demonstrated to be 21–23 nucleotides long), called **small interfering RNAs** (or **siRNAs**), were derived from larger double-stranded RNAs by the action of an enzyme called Dicer. Moreover, the RNA-induced silencing complex (RISC) was identified as a nuclease responsible for associating with siRNAs, base pairing with the endogenous mRNA, and cleaving it (Matzke and Matzke, 2004) (Figure 11.2).

Were all these small RNAs just a consequence of the presence of the transgenes, perhaps part of an RNA-based cellular immunity? Probably not, since small RNAs had already been described in worms in 1993 in the absence of any transgene. The discovery of siRNA rekindled the interest in small RNAs, and during the first decade of the twenty-first century, a large number of publications reported the presence of such natural small RNAs (now known as microRNAs, or miRNAs) in metazoans and plants. Like siRNAs in RNA-based cellular immunity, miRNAs are key regulators of development in both plants and animals. Although siRNAs and miRNAs are

biochemically similar, they are distinct in the ways they form and function (Figure 11.3). Intriguingly, small RNA-mediated silencing signals can spread throughout the plant, helping explain many long-known, yet poorly understood, phenomena. In the next sections we discuss how small RNAs are produced, how they function, their intercellular and systemic movement and the importance that they have in the biology of plants.

11.3   Pathways for miRNA formation and function

MicroRNAs (miRNAs) are encoded by *MIR* genes transcribed by RNP-II (Figure 11.3) and are therefore subjected to all the various aspects of transcriptional control described in Chapters 9 and 10. The primary miRNA transcripts (pri-miRNAs) contain imperfect, self-complementary fold-back regions (Figure 11.4) that are processed by different types of proteins in animals and plants. In animals, initial processing is carried out by the Drosha nuclease (remember: nuclease is a general term used to describe enzymes that degrade nucleic acids) in association with an RNA-binding protein. The processed animal miRNA precursors (pre-miRNA) are exported from the nucleus to the cytoplasm, where they are processed into the mature miRNA (~22 nucleotides long) by another nuclease, Dicer. In plants, the Dicer-like protein DCL1 performs both the initial processing into the pre-miRNA as

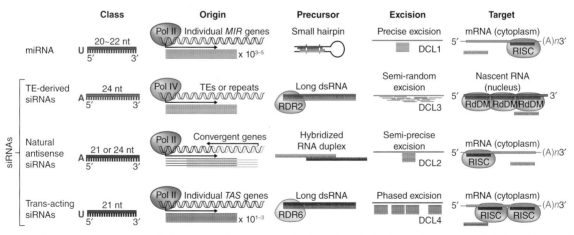

Figure 11.3 Classification and origin of the main classes of plant small RNAs that include miRNAs and different types of siRNAs. Adapted from Wang *et al.* (2011). Reproduced with permission of Elsevier

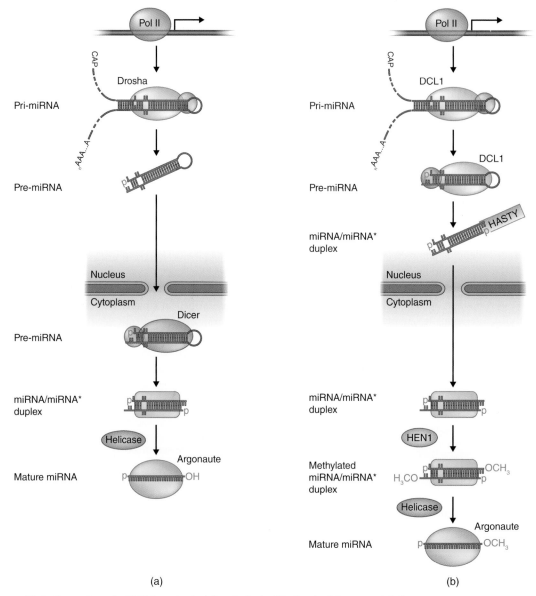

Figure 11.4 Processing of miRNA in animals (a) and plants (b). See text for more details

well as the final processing into the miRNA in the nucleus, resulting in the ~21 bp double-stranded duplex containing the mature miRNA as well as the complementary strand, called the passenger strand or miRNA*. The duplex is exported to the cytoplasm, where the miRNA strand is recognized by an AGO protein. In plants, both the miRNA and miRNA* are 2′-O-methylated by the HEN1 methyltransefrase (Yu *et al.*, 2005) (Figure 11.4). Note that

2′-O-methylation is an unusual nucleotide modification. Afterwards, the miRNA and miRNA* strands are separated by a helicase, prior to loading the miRNA strand onto an AGO protein.

miRNAs function by base-pairing with the target mRNA, resulting in either the inhibition of translation, or the cleavage of the target mRNA (Figure 11.5). There are clear differences between plants and animals in how they act on the corresponding mRNA

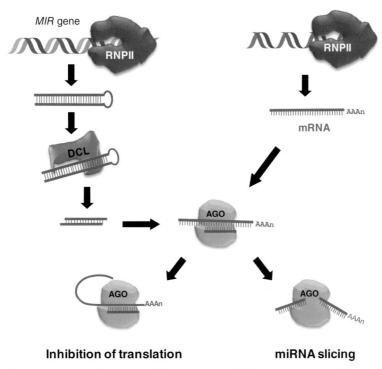

Table 11.1 Differences in miRNA function between plants and animals. The table reflects the most common characteristics; however, exceptions for pretty much every case have been found

	Plants	Animals
Region in the target mRNA	5'-UTR, coding sequence	3'-UTR
Complementarity between miRNA and target mRNA	Perfect	Limited
Consequences on target mRNA	Cleavage	Inhibit translation

(Table 11.1). In animals, miRNAs function by two possible mechanisms: if sequences are completely identical between the miRNA and the target mRNA, RISC (RNA-induced silencing complex) cleaves (slices) the target mRNA. However, if the sequences are not perfectly identical between the miRNA and the 3'-UTR of the target gene, then RISC recognizes the heteroduplex and blocks translation. The most frequent mechanism of miRNA action in plants is mRNA cleavage. In plants, miRNAs most often recognize the 5'-UTR or coding region of the target mRNA. Another difference between plant and animal miRNA targeting is that a plant miRNA usually binds a target mRNA at a single site, whereas animal miRNAs usually have multiple binding sites in a single mRNA target.

Different *MIR* genes can result, after processing, in the formation of identical (or nearly identical) miRNAs, which are usually grouped together into a family. Current evidence suggests that known miRNA families in plants and animals arose independently. The Arabidopsis genome harbors more than 600 *MIR* genes that can be grouped into more than 50 distinct families. The microRNA database (miR-Base, http://www.mirbase.org/) holds information on miRNAs from all species, and clear guidelines for nomenclature and annotation of miRNAs are available (Meyers *et al.*, 2008). Because high complementarity with the corresponding target mRNA is essential for miRNA function in plants, miRNAs evolve fairly

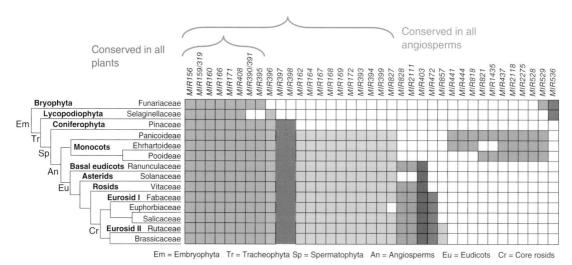

Figure 11.6 Conservation and divergence of *MIR* genes in the plant kingdom. Each column represents one MIR family and the colors represent the distribution of the MIR genes in the different taxonomic groups. Adapted from Cuperus *et al.* (2011). Copyright 2011 American Society of Plant Biologists. Used with permission

rapidly and estimates have suggested that in any given plant, a new miRNA appears or an existing miRNA disappears every one million years (Cuperus *et al.*, 2011). The consequence of this is that the vast majority of the *MIR* genes are lineage- or species-specific (Figure 11.6). Nevertheless, several groups of *MIR* genes are conserved among all plants (Figure 11.6). These *MIR* genes (and the resulting miRNAs) are often associated with metabolic processes or developmental pathways at the core of plant function. In Section 11.6, we describe the role that *MIR156* and *MIR166* play in plant development. In humans, mis-expression of miRNA provides one of the major hallmarks of cancer.

11.4 Plant siRNAs originate from different types of double-stranded RNAs

In contrast to miRNAs that are derived from mRNAs containing a hairpin-like structure, all siRNAs are produced from long double-stranded RNA molecules (Figure 11.3). Indeed, the ability of double-stranded RNAs to induce gene silencing has been conserved throughout evolution, leading to the hypothesis that a major function of double-stranded RNA-induced gene silencing is to be part of a surveillance mechanism for protection of the organism against unwanted nucleic acid 'invasions'. Such unwanted nucleic acids (often RNAs) can derive from transposons, viruses, or other parasitic DNA sequences; these ancient protective mechanisms are then activated in response to the new sort of invading RNA, the transgene. Transposable element-derived siRNAs constitute the vast majority of the siRNAs present in plants. They are typically 24 nucleotides long (Figure 11.3), but if formation of 24 nucleotide siRNA is blocked, then 21–22 nucleotide siRNAs direct TGS. The 21–22 nucleotide siRNAs are probably involved in initiation of transposable element silencing, while 24 nucleotide siRNAs play a major role in maintenance of TGS (Nuthikattu *et al.*, 2013). In Chapter 3, we discussed transposable elements and how the plant often controls their activities. If the double-stranded RNA-induced gene-silencing pathway is mutated, transposons are often mobilized, with consequent effects on genome instability. In metazoans, this often occurs through PTGS/RNAi while in plants this happens mainly through TGS (Figure 11.7), through the covalent modification of DNA and histones, processes that are described in

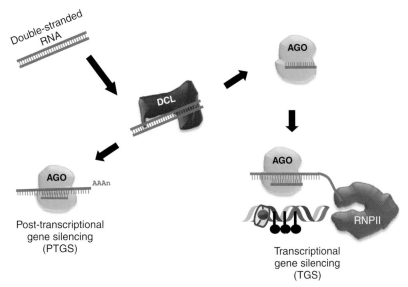

Double-stranded RNA

DCL

AGO

AGO

AGO
AAAn

Post-transcriptional
gene silencing
(PTGS)

RNPII

Transcriptional
gene silencing
(TGS)

Figure 11.7 Mechanisms for siRNA function in plants. PTGS results in mRNA slicing, while the effects of PTGS on histone modifications and chromatin structure will be discussed in detail in Chapter 12. RNPII corresponds to DNA-dependent RNA polymerase II, DCL to a Dicer-like protein and AGO to an ARGONAUTE protein, which is most often AGO4 in plants. Adapted from Williams (2013). Copyright 2013 American Society of Plant Biologists. Used with permission

more detail in Chapter 12. The signals that participate in RNA silencing mediated by transgenes can move from one cell to another and over long distances (**systemic movement**) in the plant. This movement is discussed in more detail in Section 11.6.

As with other pathogens, plants are involved in an arms race with viruses. Many plant viruses have RNA genomes that replicate through double-stranded RNA intermediates, which trigger the siRNA pathway. Pathways for antiviral silencing can be divided into three main stages: (i) sensing and processing viral RNAs into siRNAs specific to the virus; (ii) amplification of the siRNA signal (which involves RNA-dependent RNA polymerases, or RdRPs); and (iii) assembling RISC complexes specific for targeting the viral mRNAs.

While the RNAi pathway is selected to silence viral RNA expression, the virus in turn encodes suppressors of the RNAi pathway of the host cell. Targets of viral **silencing suppressors** include Dicer and DCL proteins, the RdRPs, and the AGOs (Burgyan and Havelda, 2011). Some of the most widely used viral suppressors include the P19 protein from tombusviruses and the *Tobacco etch virus* HC-Pro protein. Both of these viral suppressors target the silencing machinery by prevent-

ing the assembly of the RISC complex (Burgyan and Havelda, 2011). In addition to providing resistance to viruses, siRNAs have also been associated with bacterial pathogens. We mentioned before the challenge of keeping transgenes expressed in plants because of cosuppression. However, if a silencing suppressor, such as p19, is co-transformed and co-expressed with the transgene, then transgenes tend to have a much more elevated and stable expression.

Plants also generate other types of siRNAs that often play important regulatory roles. One good example is provided by a group of siRNAs known as *trans*-acting siRNAs, or tasiRNAs (Figure 11.3). tasiR-NAs are specific to the plant kingdom, and have even been found in the more basal plants such as mosses. They derive from specific genes in the plant known as *TAS* genes, which are transcribed by RNP-II. The produced *TAS* transcript is then cleaved by a miRNA, and the cleavage product serves as substrate for an RNA-dependent RNA polymerase (RdRP; RDR6 in Arabidopsis, Figure 11.8), resulting in formation of the double-stranded RNA that characterizes the formation of all siRNAs. This double-stranded RNA is then cleaved in a sequential fashion by a Dicer-like

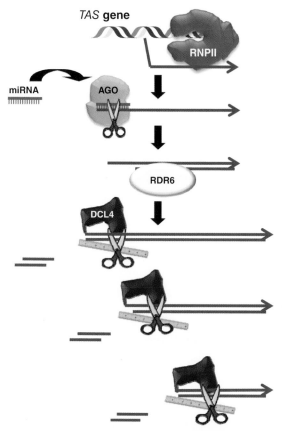

Figure 11.8 Formation and mechanism of action of *trans*-acting siRNAs (tasiRNAs). Adapted from Williams (2013). Copyright 2013 American Society of Plant Biologists. Used with permission

protein (DCL), resulting in multiple tasiRNAs that are 'phased', meaning that they are one adjacent to the next (Figure 11.8). In fact, the presence of phased 21 nucleotide small RNAs is indicative of a tasiRNA. tasiRNAs can be derived from either strand of the double-stranded RNA. So far, four *TAS* genes (*TAS1–4*) have been identified in Arabidopsis, although high-throughput analyses of small RNAs suggest that others might be present as well. Different from miRNAs, which are frequently very specific for a particular gene, *TAS* derived tasiRNAs target multiple genes, often corresponding to a gene family. For example, *TAS1* and *TAS2* tasiRNAs targets include pentatricopeptide repeat (PPR) genes (which regulate transcripts in organelles, see Chapter 5), TAS3 tasiRNAs target auxin response factor (ARF) transcription factors and TAS4 tasiRNAs target particular types of MYB transcription factors.

Another siRNA subfamily that plays an important regulatory role corresponds to natural antisense siRNAs (nasiRNA; sometimes also called natural *cis*-acting siRNAs) (Figure 11.3). nasiRNAs derive from transcripts that have overlapping regions, for example, resulting from mRNAs encoded by opposite DNA strands, but with convergent 3′ ends (Figure 11.3). Such overlapping genes are often involved in biotic and abiotic stress responses. They are clearly transcribed by RNP-II, but participation of RNP-IV was also proposed (Zhang *et al.*, 2012). nasiRNAs are not unique to plants and have also been described in mammals, Drosophila and yeast.

11.5 Intercellular and systemic movement of small RNAs

Intercellular communication is essential for the development and survival of multicellular organisms. Signaling molecules that move between cells include peptides, hormones, and transcription factors. However, by the late 1990s it became evident that small RNAs are also part of this plant mobile signaling repertoire (Figure 11.9). The movement of an RNA silencing signal helped explain a phenomenon observed almost a century ago: If the lower leaves of a tobacco plant are infected with an RNA virus, then these leaves show strong symptoms of infection. However, the upper leaves in the plant remained symptom-free, and more remarkably, they became resistant to subsequent infections by the same virus. Today, we understand that the RNA virus induced the formation of siRNAs in the lower leaves, which can then move through the phloem to the upper leaves. These siRNAs then interfere with any new viral RNA molecule that comes into these cells. Thus, these siRNA are part of the plant's innate immunity system.

Small RNA movement involves both movement through plasmodesmata (cytoplasmic connections between plant cells; see Figure 2) as well as systemic movement that involves the phloem (Figure 11.9). The current model is that small RNAs are largely transmitted as double-stranded (Dunoyer *et al.*, 2010). Systemic RNA interference is not unique to plants and has also been found in *C. elegans* and insects (Melnyk *et al.*, 2011; Molnar *et al.*, 2011).

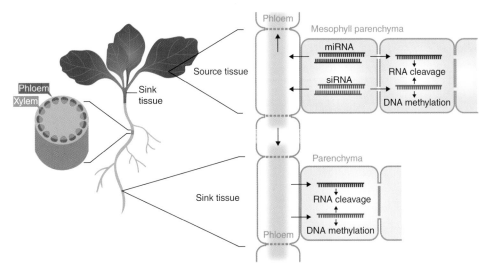

Figure 11.9 Movement of small RNAs in Arabidopsis. Small RNAs produced, for example, in leaves (source tissue) move to other tissues such as roots (sink tissue) through the phloem, the tissue responsible for transporting proteins, hormones, and nucleic acids. Parenchyma represents the bulk of the cell types in a particular tissue; the mesophyll parenchyma represents most of the photosynthetic tissues. Melnyk *et al.* (2011). Reproduced with permission of John Wiley & Sons

From the Experts

The control of transposable elements in plant germ cells by small RNAs

Transposable elements are fragments of DNA that can copy themselves and then the copies can move from one region of the genome to another (see Chapter 3). Transposable elements generate mutations and chromosome instability by inserting into DNA and generating chromosomal breaks. Small interfering RNAs (siRNAs) act throughout the plant lifecycle to repress the activity of transposable elements. However, at no point in the plant lifecycle is suppression of transposable element activity more critical than in the germ cells.

Transposable element transposition is most deleterious in germ cells or in a germline, since these induced mutations will be inherited by the next generation. Since plants do not set aside a germline early in development (as animals do), the only chance a plant transposable element can be assured to create a germinal transposition event is late in development in the gametes. In contrast to animals, plant gametes develop from the haploid products of meiosis only after a series of mitotic divisions that produce a multicellular haploid structure called a gametophyte. One or two cells in the gametophyte differentiate into gametes (sperm or egg). The male gametophyte of flowering plants is the pollen grain, which contains three cells: a large vegetative cell that controls pollen growth and fertilization, and two smaller sperm cells (SCs; Figure 11.10). The two SCs are embedded within the cytoplasm of the vegetative cell, which is controlled by the vegetative cell nucleus (VN), which does not pass its DNA to the next generation.

In 2009, a series of publications examined the chromatin state of the VN. These studies found that the VN loses normal chromatin condensation, resulting in the reactivation of transposable element transcription (Schoft *et al.*, 2009; Slotkin *et al.*, 2009). The decondensation of chromatin in nurse cells (the cells adjacent to gametes in plants or animals that supply nutrients to the gametes) is conserved, with examples from both the female and male lineages (Pillot *et al.*, 2010). In addition, transcriptional activation of transposable elements leads to post-transcriptional degradation of these transposable elements mRNAs into siRNAs (Slotkin *et al.*, 2009). Therefore, plant nurse cells are potent producers of transposable element siRNAs.

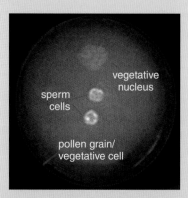

Figure 11.10 A mature wild-type Arabidopsis pollen grain stained with the DNA-binding fluorescent dye DAPI. The two sperm cells reside within the cytoplasm of the larger vegetative cell. The vegetative nucleus has decondensed chromatin (compared with the sperm cells), and is the site of transposable element transcriptional activation

The question remains why nurse cells such as the pollen vegetative cell undergo programmed reactivation of transposable elements. One hypothesis is that the plant purposely lets transposable elements reactivate in order to use the somatic nature of the VN to its advantage, "knowing" this DNA with transpositions and chromosome breaks will not be passed to the next generation. The purpose of this reactivation may be to generate transposable element siRNAs, which accumulate to high levels in pollen, from VN transposable element transcripts (Slotkin *et al.*, 2009). In addition, transposable element siRNAs concentrate in the SCs (Slotkin *et al.*, 2009), which are embedded in the cytoplasm of the vegetative cell and display visible cell-to-cell connections (McCue *et al.*, 2011). Therefore, the purpose of transposable element activity in the gametophyte may be to pre-load germ cells with transposable element siRNAs before fertilization (reviewed in Martinez and Slotkin, 2012). In addition, it was recently discovered that demethylation of VN transposable elements (and presumably transposable element activation) is necessary for proper methylation of the SC transposable elements (Ibarra *et al.*, 2012). Together, these data suggest that transposable element siRNAs produced from active transposable elements in the VN target SC methylation to efficiently silence transposable elements in gametes and to establish epigenetic marks for the next generation.

By R. Keith Slotkin

11.6   Role of miRNAs in plant physiology and development

As mentioned in previous sections, miRNAs play key roles in plant development and plant physiology. Table 11.2 lists a few examples of miRNAs and their key targets, with a description of the function that they have been shown to play.

Here, we will describe the role of miRNA156 (usually indicated as miR156) in the control of plant development, more specifically in the regulation of vegetative phase change. Similar to humans, in which childhood is a juvenile phase that transitions into adolescence and culminates in adulthood, plant development goes through different phases as well. After germination, but before flowering and entering reproduction, plants go through a period of vegetative growth, which can be further divided into a juvenile and an adult phase. Reproductive competence is

Table 11.2   Targets of some plant miRNAs. The colors correspond to the level of conservation, as shown in Figure 11.6

miRNA	Targets	Function
miR156	SPL transcription factors	Vegetative phase change
miR160	ARF transcription factors	Auxin hormone transport
miR165/166	HD-ZIPIII transcription factors	Development, polarity
miR172	AP2 transcription factors	Developmental timing, floral organ identity
miR390	TAS3	Auxin response, development
miR395	Sulfate transporter	Sulfate uptake
miR399	Protein ubiquitination	Phosphate uptake

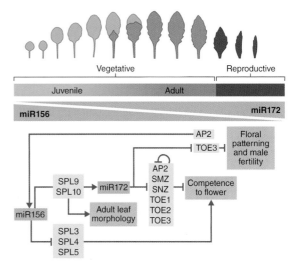

Figure 11.11 Regulation of phase change in Arabidopsis. See text for details. Redrawn with information from Huijser and Schmid, 2011 and *The Plant Cell* teaching tools in plant biology

usually acquired after the transition to the adult phase is complete (Huijser and Schmid, 2011). Depending on the species, morphology of juvenile and adult phases can appear as being different species or can be distinguishable only to the trained eye. Phase transition in Arabidopsis is easy to observe, so it has provided a very convenient system to investigate the molecular mechanisms associated with juvenile to adult transition (Huijser and Schmid, 2011). During early Arabidopsis development, miR156 levels are high (Figure 11.11), promoting the juvenile vegetative growth phase in seedlings. Arabidopsis juvenile leaves are almost round in shape and exhibit leaf hairs (trichomes, very large single-celled hairs) only on the upper surface of the leaf (the adaxial side). As the plant matures, the levels of miR156 steadily decrease, permitting mRNA accumulation for the transcription factors SPL3, SPL4, SPL5, SPL9, and SPL10 (all from the SPL family; SPL stands for Squamosa Promoter binding protein Like). SPL9 and SPL10 promote the appearance of adult leaf traits that include more elongated leaves with trichomes on the bottom of the leaf (abaxial side). Simultaneously, the SPL9 and SPL10 transcription factors bind to the regulatory region and induce the expression of *MIR172* genes. Increased levels of miR172 result in degradation of the mRNAs for six AP2-like transcription factors (AP2, SMZ, SNZ, TOE1, TOE2, TOE3; Figure. 11.11) that participate in the repression of flowering (degradation of repressors, hence promotion of flowering). This, in combination with the increased accumulation of SPL3, SPL4, and

SPL5 resulting from the decrease in miR156, have as consequence that the plant becomes competent for flowering. AP2 also plays a role in flower patterning, as described in Chapter 10.

11.7   Summary

Small RNAs are probably one of the most significant discoveries of the past decades. Much of the DNA that was previously considered to be "junk" (i.e., not coding for genes) is suddenly seen in a different way in the context of small RNA biology. This chapter describes two major types of small RNAs: miRNAs and siRNAs. Both of these small RNA classes are incredibly important in the biology of plants, controlling developmental processes and providing adaptive immunity to transposons and viruses. The mechanisms by which small RNAs are formed are well established, and involve specific members of the DICER and AGO protein families.

11.8   Problems

11.1 Arabidopsis and most characterized plants have several Dicer-like (DCL) proteins. In which small RNA pathway does each DCL protein participate?

11.2 How has the concept of "junk" DNA evolved after the discovery of small RNAs? Why?

11.3 How would you find a *MIR* gene in a genome? How would you test that it is a *MIR* gene?

11.4 What are characteristics that distinguish small RNAs from mRNA degradation products?

11.5 Describe the mechanisms by which some plant miRNAs function during plant development.

References

Bohmert, K., Camus, I., Bellini, C. *et al.* (1998) AGO1 defines a novel locus of *Arabidopsis* controlling leaf development. *EMBO Journal*, **17**, 170–180.

Burgyan, J. and Havelda, Z. (2011) Viral suppressors of RNA silencing. *Trends in Plant Science*, **16**, 265–272.

Cuperus, J.T., Fahlgren, N. and Carrington, J.C. (2011). Evolution and functional diversification of miRNA genes. *Plant Cell*, **23**, 431–442.

Dunoyer, P., Schott, G., Himber, C. *et al.* (2010) Small RNA duplexes function as mobile silencing signals between plant cells. *Science*, **328**, 912–916.

Fire, A., Xu, S., Montgomery, M.K. *et al.*, (1998) Potent and specific genetic interference by double-stranded RNA in *Caenorhabditis elegans*. *Nature*, **391**, 806–811.

Grotewold, E. (2006) The genetics and biochemistry of floral pigments. *Annual Review of Plant Biology*, **57**, 761–780.

Hamilton, A.J. and Baulcombe, D.C. (1999) A species of small antisense RNA in posttranscriptional gene silencing in plants. *Science*, **286**, 950–952.

Huijser, P. and Schmid, M. (2011) The control of developmental phase transitions in plants. *Development*, **138**, 4117–4129.

Ibarra, C.A., Feng, X., Schoft, V.K. *et al.* (2012) Active DNA demethylation in plant companion cells reinforces transposon methylation in gametes. *Science*, **337**, 1360–1364.

Jorgensen, R.A. (1995) Cosuppression, flower color patterns, and metastable gene expression states. *Science*, **268**, 686–691.

Martinez, G. and Slotkin, R.K. (2012) Developmental relaxation of transposable element silencing in plants: functional or byproduct? *Current Opinion in Plant Biology*, **15**, 496–502.

Matzke, M.A. and Matzke, A.J. (2004) Planting the seeds of a new paradigm. *PLoS Biology*, **2**, E133.

McCue, A.D., Cresti, M., Feijo, J.A., and Slotkin, R.K. (2011) Cytoplasmic connection of sperm cells to the pollen vegetative cell nucleus: potential roles of the male germ unit revisited. *Journal of Experimental Botany*, **62**, 1621–1631.

Melnyk, C.W., Molnar, A. and Baulcombe, D.C. (2011) Intercellular and systemic movement of RNA silencing signals. *EMBO Journal*, **30**, 3553–3563.

Meyers, B.C., Axtell, M.J., Bartel, B. *et al.* (2008). Criteria for annotation of plant microRNAs. *Plant Cell*, **20**, 3186–3190.

Molnar, A., Melnyk, C. and Baulcombe, D.C. (2011) Silencing signals in plants: a long journey for small RNAs. *Genome Biology*, **12**, 215.

Napoli, C., Lemieux, C. and Jorgensen, R. (1990) Introduction of a chimeric chalcone synthase gene into petunia results in reversible co-suppression of homologous genes in trans. *Plant Cell*, **2**, 279–289.

Nuthikattu, S., McCue, A.D., Panda, K. *et al.* (2013) The initiation of epigenetic silencing of active transposable elements is triggered by RDR6 and 21-22 nucleotide small interfering RNAs. *Plant Physiology*, **162**, 116–131.

Pillot, M., Baroux, C., Vazquez, M.A. *et al.* (2010) Embryo and endosperm inherit distinct chromatin and transcriptional states from the female gametes in *Arabidopsis*. *Plant Cell*, **22**, 307–320.

Que, Q., Wang, H.Y., English, J.J. and Jorgensen, R.A. (1997) The frequency and degree of cosuppression by sense chalcone synthase transgenes are dependent on transgene promoter strength and are reduced by premature nonsense codons in the transgene coding sequence. *Plant Cell*, **9**, 1357–1368.

Schoft, V.K., Chumak, N., Mosiolek, M. *et al.* (2009). Induction of RNA-directed DNA methylation upon decondensation of constitutive heterochromatin. *EMBO Reports*, **10**, 1015–1021.

Slotkin, R.K., Vaughn, M., Borges, F. *et al.* (2009) Epigenetic reprogramming and small RNA silencing of transposable elements in pollen. *Cell*, **136**, 461–472.

Wang, X., Laurie, J.D., Liu, T. *et al.* (2011) Computational dissection of *Arabidopsis* smRNAome leads to discovery of novel microRNAs and short interfering RNAs associated with transcription start sites. *Plant Genomics*, **97**, 235–243.

Williams, M.E. (2013). The small RNA world. Teaching tools in plant biology: lecture notes. *The Plant Cell* (online), doi/10.1105/tpc.110.tt10210.

Yu, B., Yang, Z., Li, J. *et al.* (2005). Methylation as a crucial step in plant microRNA biogenesis. *Science*, **307**, 932–935.

Zhang, X., Xia, J., Lii, Y.E. *et al.* (2012) Genome-wide analysis of plant nat-siRNAs reveals insights into their distribution, biogenesis and function. *Genome Biology*, **13**, R20.

Chapter 12
Chromatin and gene expression

12.1 Packing long DNA molecules in a small space: the function of chromatin

In this chapter, we describe another major component of gene regulation: how DNA is packed into chromatin, and how modifications of this chromatin can influence gene expression. But first, let us briefly review what was discussed in Chapter 4 regarding the packaging of DNA into chromatin. Recall that chromatin is a complex mixture of DNA and proteins, and many researchers would probably argue that RNA is part of chromatin as well. In particular, the DNA forms a complex with histone proteins to create the basic structural unit of chromatin known as a nucleosome (Figure 4.3). In the nucleosome, the DNA is wound around a complex of eight histones, assembled as two tetrameric complexes, each of which includes histones H2A, H2B, H3 and H4. In addition to these core histones, there are also a number of histone variants that serve more specialized functions (e.g., CENH3, H2A.Z, H2A.X). In this chapter, we will focus on the core histones. Histone H1 then stabilizes the histone–DNA interaction by interacting with the DNA as it enters the nucleosome. Nucleosomes are separated by stretches of unbound DNA to create the "beads-on-a-string" structure known as the 10 nm fiber (Figure 4.4). These are then further packaged,

with the help of histone H1, to create a slightly denser 30 nm fiber (Figure 4.5).

It is easy to imagine, based on what has been described in previous chapters of this book, that the assembly of DNA into chromatin would have a significant impact on allowing transcription factors and other regulatory proteins to bind DNA and ultimately to control gene expression. The rest of this chapter describes what is known in about how chromatin structure and modifications influence biological processes such as gene expression and recombination.

12.2 Heterochromatin and euchromatin

The chromatin structure is characterized by two main forms, **heterochromatin** and **euchromatin**, which can be easily distinguished cytologically based on using reagents that stain DNA. Regions of the chromatin that are always heterochromatic, referred to as **constitutive heterochromatin**, include the centromeres and telomeres, discussed in Chapter 4. Other heterochromatic regions can be characteristic of some cells, or of cells grown under particular conditions. Such chromatin regions are known as **facultative heterochromatin**. As we will discuss in subsequent sections, heterochromatin is usually associated with genomic regions that are silent, that is, do not express any genes.

Plant Genes, Genomes and Genetics, First Edition. Erich Grotewold, Joseph Chappell and Elizabeth A. Kellogg.
© 2015 John Wiley & Sons, Ltd. Published 2015 by John Wiley & Sons, Ltd.
Companion Website: www.wiley.com/go/grotewold/plantgenes.

12.3 Histone modifications

Since the early 1960s, it has been known that histone proteins can be post-translationally modified. It was not until the late 1990s that it became evident that, similar to the regulatory code embedded within DNA, the pattern of histone modifications in a particular region of the genome provides another level of regulation.

Histones can be modified in several different ways (Bannister and Kouzarides, 2011), but the main modifications (histone 'marks') that we will discuss here include **acetylation** and **methylation**, and to a lesser extent, **phosphorylation**. Histone acetylation and methylation occur most frequently on lysine residues (K in Figure 12.1), although arginines (R in Figure 12.1) can also be modified. Methylation can involve the addition of one, two or three methyl groups on a single lysine residue. Phosphorylation, a

very common mechanism by which protein function is modified, occurs primarily on serine or threonine (S or T, respectively, in Figure 12.1) residues, and from the perspective of histone modification, has been studied mainly in the context of regulation of the cell cycle or in response to DNA damage. All covalent histone modifications are concentrated in the histone N-terminal tails that stick out from the nucleosome and, among other protein–protein interactions, may participate in making contacts with neighboring nucleosomes (Figure 12.1; see also Figure 4.15).

When the structure of the nucleosome was first established, it was assumed that histone modifications would participate in stabilizing or destabilizing (depending on the particular modification) the nucleosome, for example by changing the positive charge of histones and affecting the interactions with DNA or other histones, and therefore impacting

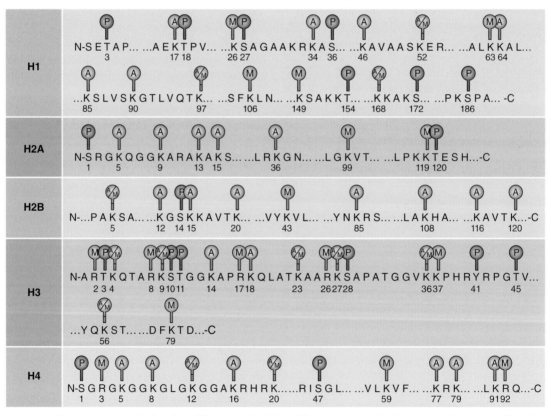

Figure 12.1 Histone post-translational modifications. The figure illustrates several of the major histone modifications so far described in each of the five core histones. The numbers under the amino acid sequence represent the position in the protein (the N-terminus corresponds to 1). The following is the code used for the modifications: A, acetylation; M, methylation; P, phosphorylation. Adapted from Portela and Estellar (2010). Reproduced with permission from Macmillan Publishers Ltd

chromatin structure. Today, it is clear that, while histone modifications do affect nucleosome structure, their more important function is to mediate the interaction of chromatin with a large number of proteins that participate directly or indirectly in the regulation of gene expression. These proteins are generally known as histone '**readers**', while histone '**writers/erasers**' are the enzymes that introduce or remove particular **histone marks**. Before we describe what is known about readers and writers/erasers, let us look into how particular histone modifications affect gene expression.

12.4   Histone modifications affect gene expression

None of the histone modifications have universal effects on gene expression, but certain histone marks correlate better than others (Table 12.1). Before we discuss the effect of particular histone modifications, it is important to clarify commonly used nomenclature for histone modifications. For example, H3K9 means the histone 3 protein (H3) is modified at the lysine (K) residue, 9 amino acids into the protein from the amino terminus. Acetylation is usually symbolized by "ac" and methylation by "m". Since amino groups

Table 12.1 Some common histone modifications and their effects on gene expression. Histone acetylation (ac) is generally associated with higher gene expression than found in the absence of this histone mark (hence reflected in activation), while histone methylation (m) can have activating or repressing consequences, depending on the particular histone residue and position in the genome

Histone mark	Effect on gene expression
Acetylation	
H3K9ac	Activation
H3K14ac	Activation
H3K27ac	Activation
Methylation	
H3K4m1	Activation/repression
H3K4m2	Activation/repression
H3K4m3	Activation
H3K9m2	Repression
H3K27m3	Repression
H3K36m3	Activation

can be mono-, di- or trimethylated, we will use the nomenclature m1, m2 and m3, respectively.

Acetylation alters the charge of histones, by neutralizing the (positive) charge of lysine with the (negative) charge of the acetyl group. In general, this reduces the interaction of the acetylated histone with the DNA, providing the open chromatin structure that is generally associated with the expression of the genes that carry histones with this mark (Table 12.1). For example, the acetylated histone H3K9 (H3K9ac) is found close to the transcription start site of expressed genes, and associated with regions where transcription factors bind.

In contrast, histone **methylation** does not change the overall charge of the nucleosome, and the effects of histone methylation are much more gene- and context-dependent. Moreover, single, double or triple methylation of the same histone residue can have very different effects on gene regulation (e.g., H3K4 methylation in Table 12.1).

Histone modifications are variably present in different genes, and when present, they can be differentially distributed across the gene (Figure 12.2). For example, methylation of H3K4 is found in nearly all expressed Arabidopsis genes, as evidenced from ChIP-chip experiments (Figure 9.5) using antibodies that recognize specifically H3K4m1, H3K4m2 or H3K4m3. Within genes, the distribution patterns of H3K4m1, H3K4m2 or H3K4m3 are different (Figure 12.2). H3K4m3 is primarily found in the 5′ and regulatory regions of genes, while H3K4m1 is predominantly found in transcribed regions (introns and exons). Indeed, H3K4m1 correlates well with the localization of DNA methylation marks (Zhang *et al.*, 2009) (Figure 12.2). H3K4m2 is not a very frequent mark in Arabidopsis genes. The relationship between DNA methylation and histone modifications is further discussed in Section 12.8.

12.5   Introducing and removing histone marks: writers and erasers

A dedicated group of enzymes is responsible for modifying histones post-translationally, primarily once the histones are already assembled in the chromatin as part of nucleosomes. Some enzymes introduce specific

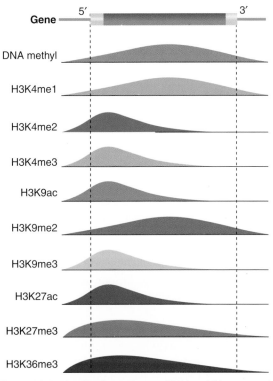

Figure 12.2 Distribution patterns of DNA and histone modifications in genes. Figure obtained from He *et al.* (2011)

marks and are therefore known as 'writers', while other enzymes called 'erasers' remove marks or entire nucleosomes from chromatin.

12.5.1 Histone acetylation

Examples of writers include **histone acetyltransferases** (HATs), which catalyze the addition of an acetyl group from the donor acetyl-CoA onto specific lysine (K) residues in histones. In Arabidopsis, the lysines that are acetylated by the action of dedicated HATs include H3K9, H3K14, H3K18, H3K23, H3K27, H4K5, H4K8, H4K12, H4K16, and H4K20. HATs can be classified into two major families depending on whether they acetylate newly synthesized free histones in the cytoplasm, or whether they work as part of larger protein complexes to target histones already incorporated into chromatin. Based on protein sequence conservation and the presence of particular domains, this latter group of HATs is further classified into four major families known as GNAT, MYST, CBP/p300 and TAF1/TAF$_{II}$250. (Remember that we discussed the

TAFs or TBP associated factors in Chapter 8.) While originally described in animals and yeast, members of these HAT families have been characterized in plants (Berr *et al.*, 2011). All these HATs function as transcriptional co-activators in activating gene expression under particular conditions.

Histone acetylation is removed by a group of enzymes known as **histone deacetylases**, or HDACs. Histone deacetylation restores the positive charge of the lysine residue, potentially stabilizing nucleosome structure. Not surprisingly, most HDACs function as transcriptional repressors. Three major families of HDACs have been described in plants, RPD3/HDA1, SIR2 and HD2, with the HD2 family likely being plant-specific.

12.5.2 Histone methylation

The addition of one, two or three methyl groups to lysine residues, or one or two to arginine residues is carried out by a group of enzymes collectively known as **histone methyltransferases** (HMTs). All lysine HMTs contain a catalytic SET domain (named for the first three proteins discovered to contain it), which is conserved from plants to animals. In contrast, arginine HMTs fall into two distinct classes that are structurally different from the lysine HMTs. In the rest of this section, we will refer mostly to SET-containing lysine HMTs. Highlighting the complexity of histone methylation, the Arabidopsis genome harbors at least 47 genes encoding SET-containing HMTs. Unlike acetylation, methylation does not change the overall charge of the histone. Similar to acetylation, histone methylation can result in either gene activation or repression (Table 12.1). Methylation of histones is a very dynamic process. Histone demethylation is carried out by histone demethylases, which belong to two major families. Members of the LSD (lysine specific demethylase) family target demethylation of H3K4m1/m2 and H3K9m1/m2. LSDs can also target other proteins, as demonstrated for the demethylation of specific methylated lysine residues in the vertebrate tumor suppressor protein p53. In plants, *LSD* is a small gene family with around four members. A key member of this family in Arabidopsis is *FLOWERING LOCUS D* (*FLD*), which helps establish when a plant flowers (see Section 12.10). The second group of histone demethylases belongs to the Jumonji (JMJ) family (the name comes from the Japanese meaning "cross-shaped", reflecting the effect that the mutant in one of the *JMJ* genes has on the neural plates of mice). There are at least 21 JMJ proteins in Arabidopsis, but specific substrates and functions are known for only a handful of these proteins.

12.6 'Readers' recognize histone modifications

Histone modifications, in particular acetylation, can directly influence chromatin structure by decreasing the characteristic positive charge of histones. But the most important role of histone modifications is to provide surfaces for proteins to recognize chromatin, as proposed by the '**histone code**' hypothesis. Proteins that recognize particular histone marks are generally known as chromatin 'readers'. Chromatin readers then interact with other proteins that allow them to execute specific functions, such as additional histone modifications or chromatin remodeling. The latter proteins are therefore known as effectors.

At least three major types of domains [bromodomain, Royal family protein domain, and PHD domain (PHD stands for plant homeodomain)] recognize particular histone marks. For example, the bromodomain binds acetylated histone lysines, and bromodomains have been found in several chromatin remodeling proteins as well as in HATs. This ability of bromodomains to recognize acetylated histones and recruit HATs results in the propagation of histone acetylation signals, in which a primary acetylated histone recruits additional HATs resulting in large neighboring chromatic regions being acetylated.

The 'Royal family' of chromatin readers includes chromodomains Tudor, PWWP, and plant Agenet domains. Members of the Royal family recognize methylated H3 and H4. An example of a plant member of this family is the LIKE HETEROCHROMATIN PROTEIN 1 (LHP1), which recognizes H3K27m3 with very high affinity, and promotes gene silencing. In contrast, animal HETEROCHROMATIN PROTEIN 1 (HP1) specifically binds H3K9m3.

The PHD domain is composed of a 50–80 amino acid domain that contains one or more zinc-finger structures (Figure 9.7). When binding to chromatin, PHDs recognize primarily the methylation state of H3K4 and to a lesser extent the methylation state of H3R2 and acetylation state of H3K14. In some PHD proteins, a single zinc finger can recognize two distinct histone marks. In other PHD proteins, each of the multiple zinc fingers recognizes a different histone modification, providing opportunities for the combinatorial reading of multiple histone marks.

12.7 Nucleosome positioning

Nucleosome positioning is defined as the location of nucleosomes with respect to the genomic sequence (Struhl and Segal, 2013) (see also Chapter 4). The position typically changes frequently, but molecular tools are currently available to determine with some certainty where all the nucleosomes are located (on average) in a particular cell. Nucleosomes are typically excluded (or are found significantly less often) at promoters, enhancers and transcriptional terminator regions. In addition, an array of ordered nucleosomes flanks the transcription start sites (TSSs) of genes, and the order decreases as the distance from the TSS increases. This structured nucleosomal distribution is important for the recruitment of components of the transcriptional machinery. The picture that has emerged over the past decade is that nucleosomes are distributed in a non-random fashion in the genome.

Nucleosome positioning is strongly affected by the sequence of DNA. In particular, DNA sequences that bend easily are more likely to be part of nucleosomes. So, what affects DNA bending? Sequences formed by tracts of A-T or G-C are intrinsically less bendable (more rigid); not surprisingly, many promoters have A-T-rich tracts and hence fewer nucleosomes, and in many organisms, the presence of A-T tracts can influence gene expression. But DNA sequence is only one contributor to nucleosome positioning; enzymes that remodel nucleosomes or that compete with transcription factors significantly contribute to establishing nucleosome patterns. Four types of **chromatin remodelers** are known, all of which use ATP hydrolysis (energy) to change the composition, structure and position of the nucleosomes. Nucleosome remodelers can be recruited to particular chromatin regions by transcription factors, or by specific histone modifications.

12.8 DNA methylation

So far we have discussed only histone modifications as major participants of chromatin structure and gene expression. However, in plants and other eukaryotes, DNA methylation also provides a mark that can be inherited and that contributes to marking chromatin that will become transcriptionally inactive, or silent (heterochromatin). In mammals, for example, DNA methylation is central to keeping one of the two X chromosomes in females inactive (remember that females have two X chromosomes, while males are XY). Defects in DNA methylation in mammals are indeed embryonic lethal.

Methylation of DNA occurs at cytosines on the carbon numbered 5 within the pyrimidine ring (m^5C). Arabidopsis has approximately 6% of the cytosine residues methylated, while maize has over 25% reflecting the significantly larger amount of heterochromatic DNA associated with transposable elements and other highly repetitive sequences that are characteristic of the maize genome. Historically, patterns of DNA methylation were studied by using pairs of restriction enzymes that recognize identical sequences, yet in which one is inhibited by DNA methylation. A methylated site would not be cut by the methylation-sensitive enzyme and thus would produce a single DNA fragment, whereas the same site would be cut by the methylation-insensitive enzyme, producing two fragments. A classic example is the *Hpa*II-*Msp*I pair of restriction enzymes. Both enzymes recognize the DNA sequence 5'-CCGG-3', but *Hpa*II is unable to cut 5'-C^{m}CGG-3'. Hence comparing restriction patterns obtained with these two enzymes, the level of DNA methylation can be estimated for a particular gene, or genome-wide. Figure 12.3 shows an example of how such results are interpreted (Mathieu *et al.*, 2007). Wild-type (WT) Arabidopsis rDNA genes have a high degree of methylation, such that *Hpa*II shows a ladder of mostly very large fragments (WT, on left). In contrast, if the same DNA is cut with *Msp*I (which is DNA-methylation insensitive), the ladder is significantly displaced towards the bottom, that is, smaller fragments, reflecting that both methylated and non-methylated DNA is similarly restricted (Figure 12.3; WT, middle lane). Now, if we look at the *met1* mutant (described later in this section), which is significantly reduced in DNA methylation activity, we see that irrespective of which of the two enzymes

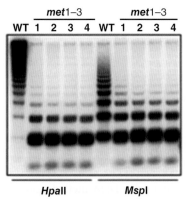

Figure 12.3 Restriction enzyme pairs can be used to map the methylation status of a gene, or of the entire genome. This figure shows a Southern blot of Arabidopsis genomic DNA obtained from wild-type (WT, non mutant) or *met1* mutant plants cut with the enzyme pair *Hpa*II or *Msp*I and hybridized with a 5S rDNA fragment. The numbers 1–3 indicate successive generations of propagation of the *met1* mutant. Remember that agarose gels are usually shown such that the negative electrode is on top and the positive on the bottom, so larger DNA fragments are closer to the top and smaller ones towards the bottom. See the text for an explanation of the results. Mathieu *et al.* (2007). Reproduced with permission of Elsevier

is used, the DNA is almost completely digested, producing ladders that are collapsed into a few bands of small molecular weight. The patterns are very similar, indicative that most of the DNA methylation has disappeared.

Today, a technique called **bisulfite sequencing** (see Figure 1.5) can establish genome-wide patterns of DNA methylation. In this technique DNA is treated with sodium bisulfite, which converts all unmethylated C, but not m^5C, residues to U. When amplified by polymerase chain reaction, these U residues are replaced by T residues by a uracil tolerant DNA-dependent DNA polymerase (see Figure 1.5). Hence comparing the genome sequence before and after the bisulfite treatment, each C that was not replaced by T corresponds to m^5C, since methylation prevents bisulfite from reacting with the cytosine. In contrast, every replacement of C by T indicates the presence of a C base that was not methylated.

In plants, DNA methylation occurs on cytosine residues in three different contexts: in the cytosine residue of a 5'-CG-3' pair, of 5'-CHG-3' and of 5'-CHH-3' (where H is A, T, or C). These types of DNA methylation are commonly known as CG,

CHG, and CHH, respectively. As can be deduced from drawing the complementary strand, only CG and CHG methylation occur in symmetric sites. The replication of these symmetrically methylated sites results in a "hemi-methylated" substrate that is the target of certain methyltransferase enzymes, ensuring inheritance of these symmetrically methylated sites. In contrast, CHH methylation is lost during DNA replication since the daughter strands would not carry this DNA mark.

DNA methylation can happen *de novo* (i.e., in the absence of a prior DNA methylation mark; by its nature, all CHH DNA methylation is *de novo*), or can be maintained, based on the presence of a methylation mark on the other strand. Different plant enzymes are involved in maintenance or *de novo* DNA methylation.

DNA methylation is carried out by a group of enzymes called DNA methyltransferases that use *S*-adenosyl methionine (AdoMet) as a methyl donor (the methyl group is transferred from AdoMet to the cytosine residue). Maintenance DNA methylation of CG sites is carried out in plants by MET1, a DNA methyltransferase that can function either during or after DNA replication. MET1 is homologous to animal DNMT1.

CMT chromomethylases (e.g., CMT2 and CMT3 in Arabidopsis) carry out *de novo* and maintenance DNA methylation of non-CG sites (i.e., CHG and CHH). Indeed, CMTs have almost the same activity on hemimethylated (one strand methylated, but not the other) as on unmethylated DNA substrates. CMTs are plant-specific and the absence of CMT-like genes in mammals explains the lack of CHG methylation in these species. CMTs have a chromodomain; as discussed in the previous section, this domain (Figure 12.4) allows for example CMT3 to be recruited to chromatin regions that contain the H3K9m2 mark. (In this section, please do not confuse histone methylation and DNA methylation; these are interlinked events, yet they are directed by different enzymes). The *de novo* DNA methyltransferase activity of CMTs depends on the presence of methylated H3K9 marks, while maintenance DNA methylation does not require H3K9 methylation (Du *et al.*, 2012). Highlighting

the relationship between histone marks and DNA methylation (which will be further expanded in the following section), the H3K9 HMT KRYPTONITE (KYP) binds directly to m^5CHG (methylated DNA). Thus, KYP binding to DNA results in H3K9 methylation, which in turn results in recruitment of CMT3, further methylating DNA. Hence, KYP and CMT3 participate in a reinforcing loop; this loop is commonly used for silencing transposable elements. CMTs are partially redundant in the methylation of non-CG sites with the DOMAINS REARRANGED METHYLASE (DRM). CHH methylation is by nature asymmetric and methylation cannot be maintained in a semi-conservative fashion, as happens for CG or CHG. There are likely multiple pathways for inducing CHH methylation, one of which is triggered by RNA-mediated gene silencing in what is known RNA-dependent DNA methylation (RdDM), a process that we described briefly in the previous chapter.

Methylation marks can be removed from DNA if the DNA replicates and maintenance methylation does not restore the marks present in the original strand. In addition to this passive removal of m^5C marks, specific DNA glycosylase enzymes remove the methylated cytosine base from the DNA, which is then subjected to repair through what is known as base excision repair. In Arabidopsis, well-described groups of DNA glycosylases include DEMETER (DME) and REPRESSOR OF SILENCING1 (ROS1). DME functions during gametogenesis to control the expression of genes derived specifically from one parent. This phenomenon is called **imprinting**. Understanding how imprinting works is an active area of research in both animals and plants.

12.9   RNA-directed DNA methylation

In Chapter 11, we discussed how small interfering RNAs (siRNAs) could affect gene expression post-transcriptionally (PTGS) or transcriptionally (TGS). Chromatin modifications triggered by siRNAs are observed in many eukaryotes, not just in plants, and are a hallmark of TGS. In organisms such as the fruitfly *Drosophila melanogaster*, which has little or no DNA methylation, modification of histones is the major cause of TGS.

Figure 12.4 Schematic representation of the CMT3 protein structure

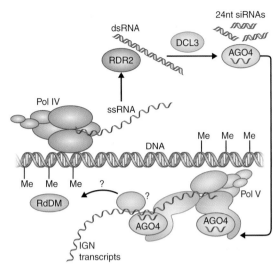

Figure 12.5 Simplified model of RNA directed DNA methylation (RdDM). Single-stranded RNAs corresponding to transposons and repetitive DNA elements are transcribed by RNP-IV. An RdRP such as RDR2 generates the double-stranded RNA (dsRNA), which is then processed as described in Chapter 11. RNP-V is likely to transcribe intergenic non-coding (IGN) regions, which serve as scaffolds to recruit AGO4 loaded with the 24 nucleotide siRNA. This complex then recruits an RdDM which targets for methylation DNA regions corresponding to both the IGN and siRNAs. ssRNA, single-stranded RNA; siRNAs, small interfering RNAs; Me, DNA methylation marks; IGN, intergenic non-coding region. Adapted from Law and Jacobsen (2010). Reproduced with permission of Macmillan Publishers Ltd

As we described in the previous chapter, transposable elements can result in the production of single stranded RNA transcripts (ssRNA in Figure 12.5) by RNP-IV (RNA polymerase IV) (Law and Jacobsen, 2010). The RNA-dependent RNA polymerase RDR2 then generates double-stranded RNAs (dsRNA, Figure 12.5) from these ssRNAs. These dsRNAs are processed by DCL3 (DICER-LIKE3) into 24 nucleotide siRNAs. These 24 nucleotide siRNAs are then incorporated into the AGO4 (ARGONAUTE4) complex. AGO4 loaded with the siRNA then interacts with RNP-V (RNA polymerase V), which is actively transcribing intergenic regions, including transposable elements. These non-coding RNAs generated by RNP-V then serve as scaffolds to recruit the siRNA-AGO4 complex by base-complementarity. This then results in DNA methyltransferases (such as DRM2 in Arabidopsis; other plants surely have equivalent factors, but the names might be different) to be recruited to the complex, and methylate the genomic DNA, ultimately resulting in regions in the genome that are homologous to the sequence carried by the siRNA being methylated (Figure 12.5). In organisms such as fission yeast (*Schizosaccharomyces pombe*), which lack the plant specific RNP-V, but which also experience RNA-mediated DNA methylation, heterochromatic DNA is transcribed by RNP-II.

12.10 Control of flowering by histone modifications

The control of flowering time is a fundamental plant process that is regulated by histone and DNA modifications. Many plant species need to reach a particular size or age before they can flower, while in other plants environmental factors such as length of day or temperature determine when the switch to flowering will happen. Thus, the transition from a vegetative to a reproductive phase is critical for the reproductive success of a plant.

Flowering in Arabidopsis is controlled by both day length and temperature. Most Arabidopsis species are winter annuals (like winter wheat), meaning that long days (characteristic of spring/summer) induce flowering, while short days have no effect (hence the plant continues to grow vegetatively). In most cases, Arabidopsis plants also must pass through a cold period followed by warming temperatures, before they can flower. This is called **vernalization**, a process that allows Arabidopsis (and similar overwintering plants) to delay flowering until the warmer days of spring arrive. By characterizing mutants that show abnormal flowering patterns (e.g., that do not require vernalization, or that can flower in short days), a large number of genes have been characterized that help control the transition to flowering (He, 2009). In this section, we focus on an example that highlights how chromatin or DNA modifications play a key role in this process.

FLOWERING LOCUS C (*FLC*) is a MADS box transcription factor (see Chapter 9) that inhibits the transition to flowering. FLC expression is positively controlled by FRIGIDA (FRI) (i.e., the more FRIGIDA RNA, the more FLC, so flowering is blocked). Vernalization represses FLC in a FRI-independent fashion (Amasino, 2004), permitting the transition to flowering to happen. A period of low temperatures induces proteins that include a PHD protein (VERNALIZA-TION INSENSITIVE3, or VIN3) that introduces a number of repressive histone marks (H3K9m2/m3, H3K27m2/m3) into the *FLC* locus, switching it from being highly expressed to being repressed. VIN3 is only one out of more than 30 factors that participate in chromatin modifications at *FLC* (He, 2009). Repressive marks such as H3K27m3 are recognized by LHP1 (see Section 12.8), which remains bound even after the plant continues to grow in the higher temperatures of spring and summer, keeping *FLC* off. This provides a 'molecular memory' of winter. As mentioned above, FRI induces *FLC* expression (hence repressing flowering). FRI accomplishes recruiting proteins to the FLC promoter that cause chromatin modifications that activate expression. Indeed, FRI encodes a plant-specific scaffold protein, suggesting that its main function is to work through protein–protein interactions, rather than itself introducing chromatin changes (He, 2012).

Another important player is FLOWERING LOCUS T (FT). FT encodes '**florigen**', an elusive factor that centuries of physiological studies had proposed to exist, but which was only recently identified. Expression of FT is repressed by FLC, but is activated in long days. As described above for FLC, FT activation and repression are controlled by chromatin modifications (He, 2012). FT is expressed in vascular cells and moves to the shoot apical meristem, where it interacts with other proteins and activates genes required for flower formation.

12.11   Summary

We have delayed until this late in the chapter to introduce the term **epigenetics**, in the hope that, by having described all the possible ways in which changes in gene expression can happen without changes in the underlying DNA sequence, the mysticism associated with epigenetics is mostly eliminated. The term epigenetics was first used by Conrad Waddington in 1942 in the context of an 'epigenetic landscape' to explain how gene function may be influenced by the environment. Today, epigenetics is used in a more restricted way to describe a "stably heritable phenotype resulting from changes in a chromosome without alterations in the DNA sequence" (Berger *et al.*, 2009). Based on what we described in this chapter, it is easy to see that there are three main types of processes that are

often associated with epigenetic phenomena: DNA methylation, histone modifications, and nucleosome positioning. In many instances, DNA methylation and histone modifications are strongly coupled, for example by siRNAs derived from transposons and other repetitive sequences, resulting in the accumulation of repressive marks that ultimately can establish heterochromatic regions, or domains, in the genome.

12.12   Problems

12.1   Draw illustrations of all the different types of histone modification.

12.2   What are the main consequences of histone modifications?

12.3   Draw illustrations for the most common modification of DNA.

12.4   How can the presence of DNA methylation be determined?

12.5   What is the difference between epigenetics and Mendelian genetics?

12.6   Why are histone modifications considered epigenetic marks?

12.7   What is the relationship between siRNAs, histone modifications, and DNA methylation?

References

Amasino, R. (2004). Vernalization, competence, and the epigenetic memory of winter. *Plant Cell*, **16**, 2553–2559.

Bannister, A.J. and Kouzarides, T. (2011) Regulation of chromatin by histone modifications. *Cell Research*, **21**, 381–395.

Berger, S.L., Kouzarides, T., Shiekhattar, R. and Shilatifard, A. (2009) An operational definition of epigenetics. *Genes Development*, **23**, 781–783.

Berr, A., Shafiq, S. and Shen, W.H. (2011) Histone modifications in transcriptional activation during plant development. *Biochimica et Biophysica Acta*, **1809**, 567–576.

Du, J., Zhong, X., Bernatavichute, Y.V. *et al.* (2012). Dual binding of chromomethylase domains to H3K9me2-containing nucleosomes directs DNA methylation in plants. *Cell*, **151**, 167–180.

He, G., Elling, A.A. and Deng, X.W. (2011) The epigenome and plant development. *Annual Review of Plant Biology*, **62**, 411–435.

He, Y. (2009) Control of the transition to flowering by chromatin modifications. *Molecular Plant*, **2**, 554–564.

He, Y. (2012) Chromatin regulation of flowering. *Trends in Plant Science*, **17**, 556–562.

Law, J.A. and Jacobsen, S.E. (2010) Establishing, maintaining and modifying DNA methylation patterns in plants and animals. *Nature Reviews Genetics*, **11**, 204–220.

Mathieu, O., Reinders, J., Caikovski, M. *et al.* (2007) Transgenerational stability of the Arabidopsis epigenome is coordinated by CG methylation. *Cell*, **130**, 851–862.

Portela, A. and Esteller, M. (2010) Epigenetic modifications and human disease. *Nature Biotechnology*, **28**, 1057–1068.

Struhl, K. and Segal, E. (2013) Determinants of nucleosome positioning. *Nature Structural & Molecular Biology*, **20**, 267–273.

Zhang, X., Bernatavichute, Y., Cokus, S. *et al.* (2009) Genome-wide analysis of mono-, di- and trimethylation of histone H3 lysine 4 in *Arabidopsis thaliana*. *Genome Biology*, **10**, R62.

Part III

From RNA to Proteins

Chapter 13
RNA processing and transport

13.1 RNA processing can be thought of as steps

In the first part of this book, we addressed current concepts of genes, their molecular architecture and organization into genomes, and in Part 2 the processes associated with the developmental and regulated expression of these genes were discussed. In Part 3, we will be reviewing the processing of RNA transcripts from the time a messenger RNA (mRNA) originates from transcription to when the mRNA becomes engaged in the translation process of protein synthesis, or when the mRNA is degraded (Figure 13.1). Although all these processing steps represent essential modifications to each mRNA species and are presented here as successive events in the life cycle of an mRNA, it is probably more accurate to envision several of them as occurring in a continuum. For instance, an mRNA may undergo capping and splicing reactions simultaneously.

Another reason for presenting each processing step as a discrete event is that each also represents a point for possible regulation and control of gene expression. Alternative splicing, for instance, obviously generates diverse mRNAs from expression of a single gene, and thus provides a mechanism for generating multiple, diverse proteins. As we have emphasized throughout this book, gene expression is the means by which a gene contributes to a cellular phenotype, which in turn leads to a physiological impact of a cell's functioning in a tissue or organ. And whole plant traits and characteristics are the integrated functioning of tissues and organs. Hence, all the mechanisms that regulate when, how and to what extent a gene is expressed are as important as the gene itself in contributing to a plant's growth, development and physiology.

Another concept to keep in perspective is the timing of events. The time it takes RNA polymerase to transcribe DNA into RNA varies from 10 to 400 nucleotides per second. Assuming an average primary transcript of 2000 bases in length, an average gene can be transcribed in ~1 min (2000 bases of a typical gene divided by ~50 nucleotides polymerized per second by RNA polymerase). In contrast, many of the mRNA processing steps take considerably longer (Figure 13.2).

13.2 RNA capping provides a distinctive 5′ end to mRNAs

Almost all mRNAs are chemically modified at their 5′ terminus by the addition of a guanosine nucleotide added in a non-template dependent fashion (meaning that the information is not encoded in the coding strand of the DNA) to the first (5′) nucleotide of the mRNA, followed by up to three methylations to generate the type 0, 1 and 2 cap structures (Figure 13.3). Type 0 caps dominate and account for the modifications of 40% or more of all the mRNAs, with type 1 and 2 caps each occurring in up to 30% of the mRNAs. Ribosomal RNA (rRNA) and transfer RNAs (tRNAs)

Plant Genes, Genomes and Genetics, First Edition. Erich Grotewold, Joseph Chappell and Elizabeth A. Kellogg.
© 2015 John Wiley & Sons, Ltd. Published 2015 by John Wiley & Sons, Ltd.
Companion Website: www.wiley.com/go/grotewold/plantgenes.

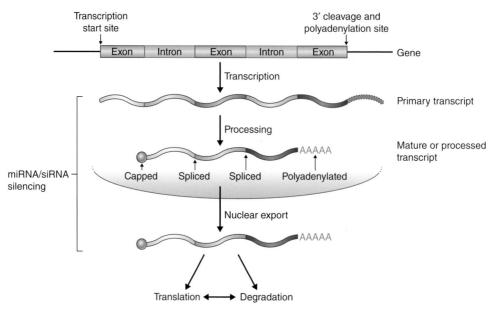

Figure 13.1 A schematic representation of the steps in mRNA processing and maturation; steps that contribute to the regulation and specificity of gene expression

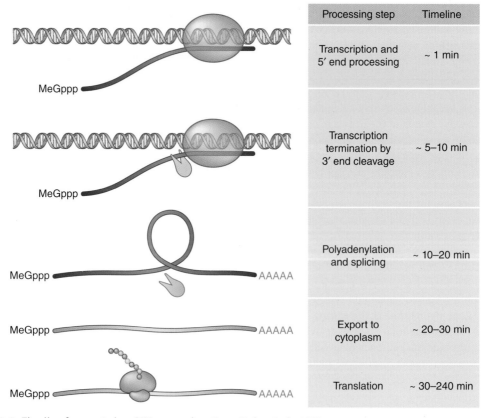

Processing step	Timeline
Transcription and 5' end processing	~ 1 min
Transcription termination by 3' end cleavage	~ 5–10 min
Polyadenylation and splicing	~ 10–20 min
Export to cytoplasm	~ 20–30 min
Translation	~ 30–240 min

Figure 13.2 Timeline for events in mRNA processing. From Krebs *et al.*, 2010

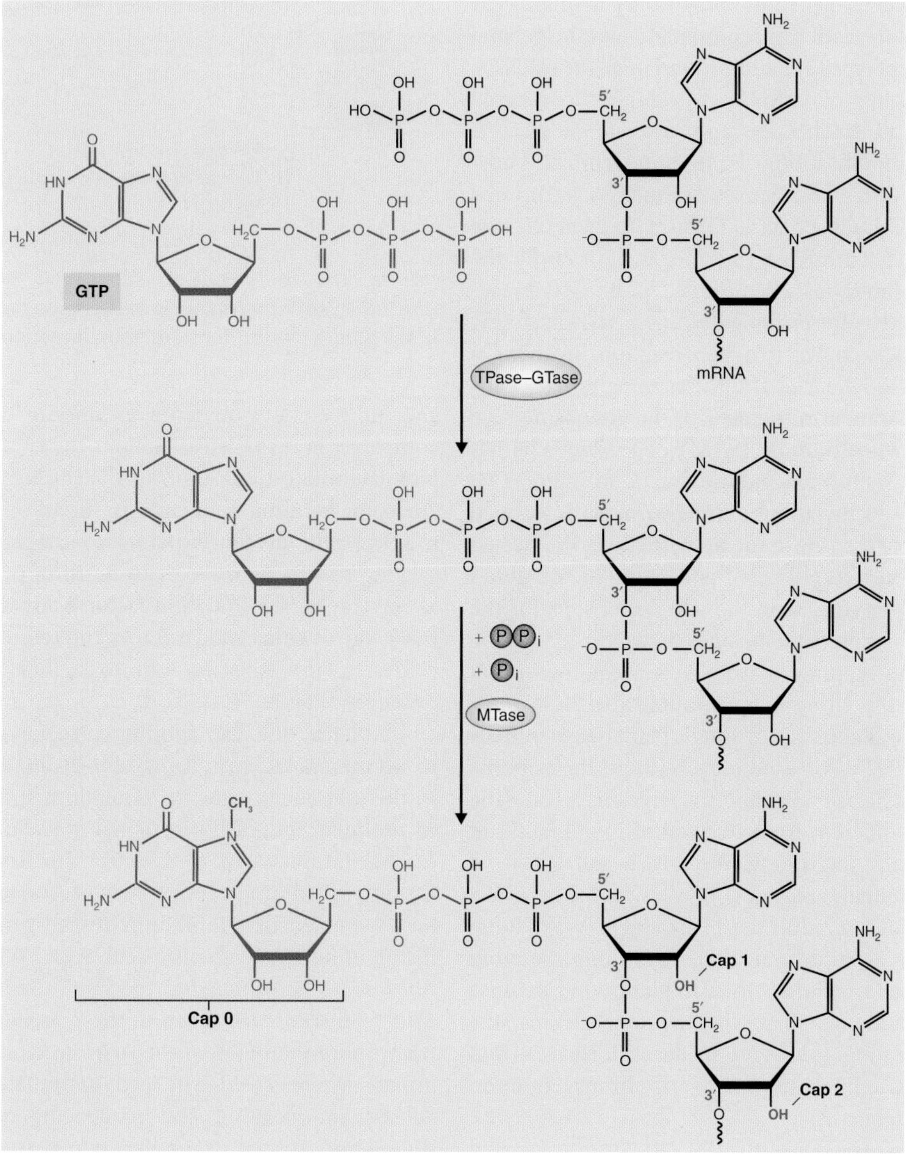

Figure 13.3 The steps in capping a mRNA consist of adding a guanosine nucleotide to the 5′ terminus of the mRNA, catalyzed by a fusion enzyme consisting of a terminal phosphatase (TPase) and guanosine transferase (GTase), followed by nitro-methylation at the 7 position of guanosine base by a methyltransferase (MTase) using S-adenosylmethionine (SAM) as the methyl donor to generate the initial and dominant capped mRNA structure. Subsequent methylation at the 2′ hydroxyl of the second and third ribose sugars yields equally abundant cap 1 and cap 2 structures

are not capped and there is no evidence, at least up to this point in time, for specificity of the type of 5′ cap associated with the expression of any one gene. That is, there is no evidence yet for whether an mRNA arising from a particular gene consists of only a single type of cap structure, or if it is representative of the general distribution of types 0, 1 and 2 found in all the mRNAs.

The capping of mRNA is a highly conserved process in all eukaryotes, and consists of discrete sequential modifications. First, the terminal phosphate of the 5′ triphosphate end of mRNA is trimmed off by an exo-phosphatase (TPase) to generate the diphosphorylated mRNA intermediate. Secondly, the intermediate mRNA is then capped with GMP, a reaction catalyzed by the enzyme RNA guanylyltransferase (GTase) in a complex two-step reaction involving a GMP–enzyme intermediate. The capping process ends with methylation at nitrogen 7 of the guanosine base by S-adenosylmethionine (SAM) dependent methyltransferase (MTase) to yield the "cap 0" structure. Two additional O-methylations of the hydroxyls positioned at carbon 2 of the ribose sugars of the two 5′ terminal nucleosides yield cap forms 1 and 2, and are catalyzed by different enzymes than the one responsible for the initial methylation of guanosine base.

While the capping of mRNA is ubiquitous in all eukaryotes, the enzymes responsible for these reactions behave differently in fungi, plants and animals (Shuman, 2002). For example, in fungi, the triphosphate phosphatase appears to rely on a catalytic mechanism different from that found in animals and plants. Equally interesting, the TPase and GTase of plants and animals are encoded by a single gene with a chimeric structure, while the TPase and GTase of fungi are found as separate genes. Because fungi are more closely related to animals than to plants, and because plants and animals share capping mechanisms, the condition in fungi is likely to be derived, whereas that in plants and animals appears to arise from a common ancestral origin.

The 5′ cap of eukaryotic mRNA has been associated with several cellular functions including splicing, nuclear export, and mRNA half-life stabilization. The role of mRNA caps in contributing to these processes is perhaps best appreciated by considering their role in translation initiation. First, consider the function of the Shine-Dalgarno sequence, which resides 7–10 bases upstream of the methionine initiation codon in prokaryotic mRNAs. This consensus sequence is also referred to as the ribosome-binding

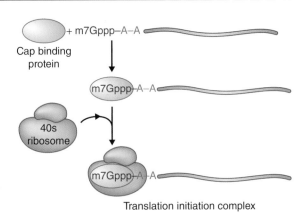

Figure 13.4 The 5′ cap of mRNA brings the mRNA into association with the ribosome and assures that it is bound in the proper register for translation initiation

site and facilitates mRNA association with ribosomal complexes and spatially positions the ribosomal complex to initiate translation of the mRNA. No similar ribosome-binding sequence is found in eukaryotic mRNAs and evidence points to the cap structure serving this role. In fact, **cap-binding proteins** such as the translation initiation factor 4, eIF4E, lock onto the 5′ cap of eukaryotic mRNAs and recruit the bound mRNA to the 40S ribosome to facilitate translation initiation (Figure 13.4).

In plants, the cap binding proteins also appear to serve regulatory roles (Kuhn *et al.*, 2008). *Abh1* is the dominant gene in Arabidopsis that appears to modulate an early hormonal signal transduction cascade (Hugouvieux *et al.*, 2001). The *Abh1* gene was first identified in a genetic screen of Arabidopsis based on a hypersensitive inhibition of seed germination to the plant hormone abscisic acid, ABA. Without ABA, *Abh1* seeds germinated at the same time and rate as wild type seeds. However, in the presence of 0.3 μM ABA, more than 90% of the *Abh1* seeds failed to germinate, whereas wild-type seeds germinated normally. Subsequent isolation and sequencing of the *Abh1* gene demonstrated its similarity to yeast and human homologs of a subunit of a heterodimeric nuclear cap-binding complex, which was experimentally confirmed. Further, DNA chip-microarray analysis demonstrated that 18 genes showed significantly lower expression levels in the *Abh1* genetic background, while 13 genes were significantly up regulated. Many of the aberrantly regulated transcripts in *Abh1* corresponded to genes known to be involved in ABA signaling, such as regulatory phosphatases and kinases.

Equally interestingly, mutants in the *Abh1* gene closed their stomata abnormally early if they were treated with ABA or exposed to drought, and hence were more tolerant of water stress conditions, representing the first evidence for a cap binding protein contributing to a direct physiological phenotype in plants.

13.3 Transcription termination consists of mRNA 3′-end formation and polyadenylation

Transcription initiation, as described in Chapter 8, is well studied because it provides an obvious means for regulating differential gene expression patterns, and thus the specialization of cells and cellular functions in tissues and organs. Upon first consideration, it is less obvious how transcript termination might also influence or contribute to the accumulation of particular mRNAs in different cells or conditions. However, recent studies have documented how important **3′-end formation and polyadenylation** are, as discussed below in regard to how 3′-end formation for a particular class of genes contributes to the timing of flowering.

Almost all mRNAs are polyadenylated, distinguishing this class of RNA from all other classes, including rRNAs, tRNAs and small ($\leq$100 bases) RNAs. Notable exceptions to the mRNA polyadenylation rule are mRNAs coding for canonical histone proteins, which instead contain a unique 3′-end structure. Polyadenylation also appears to be important for transport of mRNAs from the nucleus to the cytoplasm, stabilization of the RNA (affecting the turnover rate), and engagement of the mRNA into the protein translational machinery (Hunt, 2008). But then all of this seems dependent upon where the nascent mRNA chain is cleaved and polyadenylated (Figure 13.5). That is, as RNA polymerase transcribes a gene and the nascent mRNA is extruded from the transcriptional machinery, another complex of proteins surveys the RNA for distinct *cis*-elements (sequences enriched for specific nucleotide patterns), which dictate site-specific cleavage. The cleavage complex then introduces a single-strand specific break in the mRNA molecule and adds adenosine residues

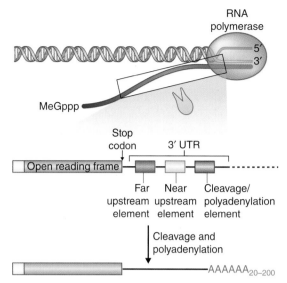

Figure 13.5 Overview of mRNA 3′-end formation and polyadenylation. Adapted from Hunt (2008)

in a template independent fashion. The number of adenosine residues added to any one mRNA is not precise and typically an mRNA species can contain from 20 to more than 100 terminal adenosine residues. This variation in the number of A's is known as microheterogeneity, to distinguish it from other forms of variation described below.

Early experiments in mammals and systems other than plants identified a putative recognition sequence AAUAAA for cleavage and polyadenylation of mRNAs, normally known as the polyadenylation sequence, since the polyA tail is usually found 10–30 nucleotides 3′ (downstream) of such a sequence. However, this sequence is not conserved in plant mRNAs, suggesting that additional elements might contribute to marking the site for mRNA cleavage. To date, two such elements have been defined – **the far-upstream (FUE) and near-upstream (NUE) elements** – by compiling and comparing 3′UTR domains, and also by experiments to document the specific role of distinct sequences in the formation of mature mRNAs. As their name implies, these elements are found relatively close to the ultimate cleavage site, but like the cleavage site itself, they are not defined by absolute conservation of their sequences. The FUEs are positioned between 30 and 150 nucleotides upstream (5′) to the cleavage site and are enriched with uridine (U) residues. The NUEs lie between 10 and 30 bases upstream from the cleavage site and have a compositional bias for

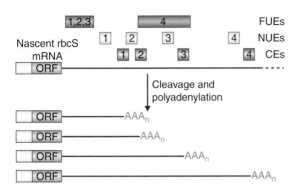

Figure 13.6   An example of the multiple 3′ ends and polyadenylation sites observed for the pea *RbcS* mRNA. The far upstream (green boxes, FUE) and near upstream elements (light green boxes, NUE) that coordinate cleavage site (red boxes, CE) selection, identified by detailed molecular dissection studies (Hunt, 2008), are noted above the nascent mRNA

adenosine (A) residues. The cleavage site itself appears to be highly enriched in U residues.

Experiments to define the role of these sequence elements in 3′-end formation have revealed a surprising level of **alternative polyadenylation site-selection** in mRNAs, an observation that is now widely recognized as applicable to all eukaryotic organisms and a significant means for generating additional diversity and heterogeneity to mRNAs. For example, as illustrated in Figure 13.6, the mRNA arising from the pea rubisco small-subunit gene (*RbcS*) can have four different termination/polyadenylation sites and therefore four distinct cleavage elements (red boxes). Identification of the sequence elements associated with the selection of each of these termination sites has involved experiments that delete or reposition sequences, leading to the identification of one FUE (1, 2, 3) (dark green) that operates in combination with separate NUEs 1, 2 and 3 (light green) and a second FUE (4) that appears to function in combination with a single NUE and cleavage element or site (CE).

13.3.1   Alternative polyadenylation creates additional mRNA diversity

Alternative polyadenylation refers to distinct cleavage site selection as noted for the pea RbcS mRNA in Figure 13.6., and is very different from the single nucleotide differences of 3′-end formation that might

occur within a single cleavage site (microheterogeneity; see above). In contrast to microheterogeneity, alternative 3′-end formation represents a significant means for dramatically altering the resulting mRNA itself. Alternative polyadenylation of mRNA creates alternative 3′UTR sequences in the mature mRNA, sequences that can influence the stability of the mRNA species, as well as their translatability. Alternative polyadenylation also includes 3′-end formation across an entire mRNA that can actually alter the coding sequence of the mRNA. Alternative polyadenylation sites have been documented within introns, for instance. Alternative polyadenylation in these circumstances means inclusion and exclusion of particular exon elements, thus creating different proteins from the differentially terminated mRNA species. This diversity in 3′-end formation has direct parallels to alternative intron splicing, another mechanism for creating alternatively encoded protein products, which is discussed later in this chapter.

The overall importance of alternative polyadenylation site selection is still being evaluated, but appears to be significant. About half or more of all mammalian genes appear subject to alternative polyadenylation site selection. Although there are increasing efforts to document, quantify and qualify the significance of these alternative mechanisms in plants (Wu *et al.*, 2011), notable examples include the *S* locus glycoprotein and its corresponding *S*-receptor kinase mRNAs (loci involved with self-incompatibility recognition between pollen and stigma) (Giranton *et al.*, 1995), the mRNAs for the ethylene receptor-like protein, ETR1, in peach (*Prunus persica*) (Bassett *et al.*, 2002), the mRNAs encoding the receptor-like protein containing TIR motifs in Arabidopsis (Meyers *et al.*, 2002), an mRNA that encodes both lysine-ketoglutarate reductase and saccharopine dehydrogenase in cotton (Tang *et al.*, 2002), and an ascorbate peroxidase-encoding mRNA in spinach (Ishikawa *et al.*, 1997). One example of alternative polyadenylation in plants has been particularly well documented for the gene *FCA*, which is associated with flowering time in Arabidopsis (Terzi and Simpson, 2008).

The biochemical function of 3′-end selection, cleavage and polyadenylation is mediated by a complex of interacting proteins that provide both specificity and selectivity of the complex as a whole. Figure 13.7 illustrates our current understanding of the proteins associated with the 3′-end formation and polyadenylation complex. This illustration is derived from many

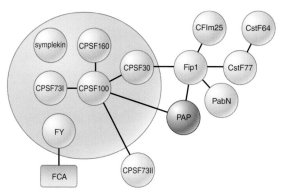

Figure 13.7 A cartoon depiction of some of the proteins associated with the mRNA cleavage and polyadenylation complex. Typically, the complex recognizes the sequences like the FUE, NUE and CE (illustrated in Figure 13.2) of the growing mRNA, introduces a break in the sugar–phosphate backbone at the CE, then adds in a non-template dependent fashion a chain of up to several hundred adenosine bases to the 3′ end of the mRNA. Some of the proteins making up the complex physically bind to one another, as shown by the interconnecting lines, and some of these proteins have the ability to bind directly to RNA (light red color) based on inferences from their amino acid sequence as well as experimental evidence. The proteins within the green circle make up a distinct complex of polyadenylation factors or proteins that have been readily isolated from the nuclei of plants. The nomenclature for these proteins for the most part does not relate to any specific biochemical function like 3′ cleavage. Instead, a role of CstF64 in the actual cleavage reaction has been suggested from a variety of corollary observations with yeast and mammalian experimental systems. The polyadenylation polymerase or PAP, which adds the polyA tail to each mRNA, is shown in bright red. Some of the interactions of the complex with other proteins provide a means for possible regulation of the 3′ cleavage and polyadenylation reaction, such as the FCA interaction (noted in blue). Hunt (2008). Reproduced with permission of Springer Science+Business Media

experiments to evaluate the interaction between the various protein components and RNA, as well as their catalytic roles in the cleavage of the nascent mRNA and the addition of a polyA tail. Hence, the nomenclature indicates those protein components (**s**pecificity **f**actors) associated with **c**leavage and **p**olyadenylation activities (**CPSF**) as well as other proteins like the **p**olyA **p**olymerase (**PAP**), which actually adds the polyA tails to the ends of the mRNAs. Figure 13.7 also illustrates which protein factors can interact with or bind to RNA (red color), as well as interact amongst themselves (indicated by interconnecting lines).

Not all the interactions of the various proteins in the complex are necessarily physical (i.e., there may not be direct binding between the proteins) and all these interactions should be considered dynamic. That is, these interactions may vary and change depending on the developmental stage of a cell and in response to various environmental cues. The *FY* gene encodes a protein that illustrates this point. *FY* is associated with the so-called autonomous pathway controlling flowering, a set of innate mechanisms controlling flowering time that are different from mechanisms that monitor environmental cues to control the flowering program. Specific mutations within the *FY* gene resulted in Arabidopsis plants that flowered later than wild type plants, but a complete knockout of the *FY* gene never yielded seeds that were homozygous for the mutant allele and able to germinate. The mutants with delayed flowering resulted from carboxy-terminal deletions of FY and loss of protein domains that mediate protein–protein interactions. While FY is a well-documented component of the CPSF complex, the exact mechanism of its interactions with the other CPSF proteins was not clear. Nonetheless, the carboxy-terminal domains of FY have been shown to be important for FY's interactions with FCA, another protein associated with the autonomous flowering program.

FCA gene expression is subject to several layers of post-transcriptional regulation, but alternative polyadenylation appears to be a dominant and important mechanism. At least three forms of *FCA* transcripts can be found in wild type plants (Quesada *et al.*, 2003). *FCA* gamma is the conventional and fully processed form of the transcript, *FCA* alpha has all the introns spliced out properly except for intron 3, and *FCA* beta is created by 3′-end formation and polyadenylation within the third intron of the transcript (Figure 13.8). Mutants that alter the formation of the beta form affect flowering. If little of the beta form accumulates, more of the gamma form does, and flowering is normal to early in these plants. If more of the beta form accumulates, and less of the gamma form accumulates, then flowering is delayed. Subsequent work has shown that *FCA* is self-regulating. *FCA* encodes an RNA-binding protein that specifically binds to the region within intron 3 of the mature *FCA* mRNA. The interaction between FCA and FY recruits the cleavage and polyadenylation machinery to intron 3 and thus yields the alternative polyadenylated form, FCA beta. Mutants in either FY or FCA

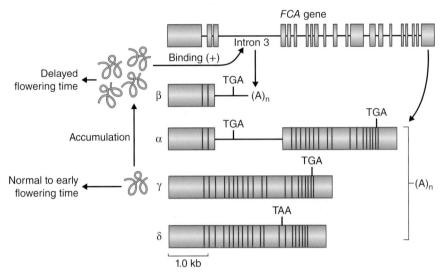

Figure 13.8 Depiction of how alternative 3′-end formation can regulate flowering time. *FCA* is a negative regulator of another factor of the autonomous flowering program. That is, *FCA* promotes flowering by suppressing the activity of another protein that tends to delay flowering. Hence, the normally processed gamma *FCA* mRNA leads to normal flowering time. The encoded FCA protein (blue curly symbol) is also self-regulatory and, when too much of it accumulates, it binds to intron 3 of the primary FCA transcript and promotes alternative 3′-end formation and polyadenylation (the plus symbol). This results in more of the beta form of *FCA* mRNA and lower amounts of the gamma form accumulating. The net result is that the time to flowering is delayed. This is because less of the FCA protein is around to counteract the effects of the flowering suppressor factor. Adapted from Quesada *et al.* (2003). Reproduced with permission of John Wiley & Sons

that alter the ability of these proteins to interact with one another result in more full-length FCA gamma mRNA accumulating in cells and thus these plants flower earlier than normal wild type plants.

13.4   RNA splicing is another major source of genetic variation

Almost all prokaryotic genes are co-linear with their protein products. That is the gene is transcribed into an mRNA that is directly translated into a protein based on reading 3 nucleotide codons one at a time and ratcheting together the amino acids that constitute the final protein product. But the case in eukaryotes, plants included, is quite different because most nuclear genes are interrupted by intervening sequences of DNA that do not end up in the final mRNA and hence do not contribute to the final protein product (Figure 13.1). The intervening sequences of DNA, commonly referred to as introns, are transcribed

into the primary transcript but are processed out, bring together the exons into the mature or fully processed transcript that is finally exported to the cytoplasm for translation. The process of removing the intronic sequences from the primary RNA is known as **RNA splicing**. While we do not fully appreciate why eukaryotic genes consist of introns and exons, we do understand that, unless this processing occurs with exacting precision, aberrant and mutant RNAs might arise. It might be this possibility of generating such genetic diversity at the RNA level, rather than at the DNA level, that gave rise to this processing system.

Figure 13.9 illustrates some of the more common processing specifics for intron removal that occur for a nuclear gene. There are examples of organellar transcripts in chloroplast and mitochondria having sequences spliced out, but these are much less typical than for nuclear genes. Perhaps the first layer of the splicing puzzle is what defines an intron and how it is recognized distinct from an exon. Interestingly, when homologous genes (genes coding for identical or similar protein functions) from distant plant species are compared, the sizes of their exons tend to be conserved but not so for their introns. Intron size

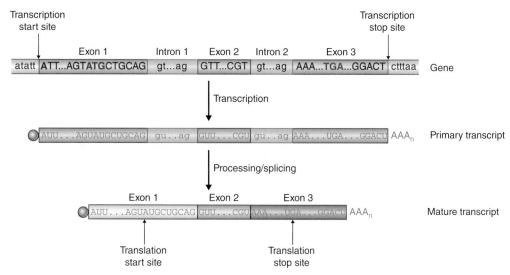

Figure 13.9 Illustration of how intron sequences are spliced out of primary mRNA transcripts to yield a mature mRNA containing a contiguous open-reading frame encoding a protein product

(numbers of base pairs) seems much more variable than that for exons. But what does seem conserved for introns are the sequences at their boundaries with exons.

Brown and Simpson (1998) defined the initial step in RNA splicing as the assembly of a ribonucleoprotein complex, the so-called **spliceosome**, with RNA poised to undergo processing. This event is followed by *trans*-acting factors, including small nuclear protein-RNA complexes (snRNPs) (Ru *et al.*, 2008), recognizing the intron sequences and defining the 5′ and 3′ intron splice junctions, introducing a cleavage at the 5′ junction allowing for lariat formation between the 5′ end of the intron with another nucleotide within the intron near the 3′ end, followed by cleavage at the 3′ splice junction and ligation of the neighboring exons into a contiguous RNA. While conserved sequence elements within the intron elements are thought to mediate these snRNP interactions, no sequence-specific elements, other than the splice site junctions, have yet been defined. The common consensus sequence for the 5′ splice site in plant is AG/GTAAGT with the frequency of the designated nucleotides at each position of 62, 79, 100, 99, 70, 58, 49%, and TGCAG/G for the 3′ splice site junction with a frequency bias of 64, 42, 95, 100, 100, 57% (/ represents splice site junctions). The intron border sequences are therefore not invariant with the strongest conservation being only for the immediate splice site nucleotides, GT at the 5′ end and AG at the 3′ end.

Forms of alternative splicing are now well recognized as a significant means for generating additional layers of genetic diversity beyond a genome's DNA sequence (Figure 13.10). Alternative splicing refers to the situation where, for example, exon 1 can be ligated directly to exon 3 rather than to exon 2. When exon 1 is ligated to exon 3, then a significant amino acid coding sequence found in the initial or primary transcript will be missing from the final mature mRNA species. There are also examples of so-called cryptic intron deletion. This is where splicing leads to the deletion of a region of an exon that is normally found in the mature, fully processed mRNA. Alternative splicing also refers to cases where an intron sequence is not removed during process. Inclusion of intronic sequences within mRNA can have profound effects on the translation product. Pre-mature stop codons can yield small, truncated protein products or if the intron sequence does continue the open-reading frame, then relatively large amino acid insertions into the protein can result. Alternative splicing can be quite significant and examples of upwards of 50 different splice variants of a single primary mRNA have been documented. In fact, recent sequencing projects comparing the sequence of RNA transcripts relative to their respective genomes have determined that 22% of the total transcriptomes of Arabidopsis and rice represent alternative or variant spliced mRNA species (Wang and Brendel, 2006). In summary, alternative splicing events are categorized in four cases: Alternative 5′ splice site

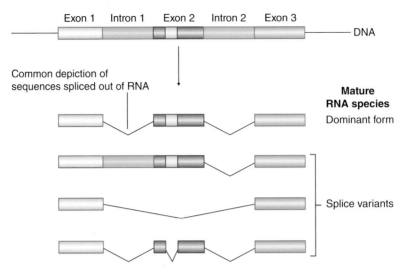

Figure 13.10 Illustrations of possible alternative RNA splicing events occurring during the processing of a primary transcript. Adopted from Zhang and Gassmann, 2007

choice, alternative 3′ splice site choice, cassette-exon inclusion and skipping, and intron retention (Nilsen and Graveley, 2010).

Documentation of alternative splicing raises interesting questions about the possible roles of splicing for genome evolution and gene expression. One school of thought has suggested that exons may represent selectable, inheritable evolutionary units. That is, exons could represent "functional" units within their protein products that can serve to enhance or alter the activity of an enzyme, for example. This same notion applies to how different splice variants might be expressed in plants grown under different environmental conditions. For instance, variant 1 might be expressed under non-water stress conditions and variant 2 would be expressed under water stress conditions. Under the water stress conditions, the protein encoded by variant 2 might provide a selective advantage to the plant (Richardson *et al.*, 2011). Another interesting case is the necessity of splicing for the generation of phenotypic traits. Expression of the genetic loci encoding the receptors of specific pathogens are examples of this wherein expression of an intronless form of the gene does not appear sufficient to confer resistance responses to the transgenic plants, but gene constructs harboring introns do (Zhang and Gassmann, 2007). The interesting conundrum arising from these diverse observations is how alternative splicing becomes fixed into the molecular genetic machinery of a plant to ensure that the alternative splicing event occurs reproducibly, or how these events get captured into a more stable form (a new allelic gene form) within the genome.

13.5 Export of mRNA from the nucleus is a gateway for regulating which mRNAs actually get translated

In plants, like all eukaryotes, the genome is sequestered in the nucleus, an organelle surrounded by a double-layered membrane called the nuclear envelope. As we have discussed up to this point, transcription of a gene, transcript termination, polyadenylation and intron splicing are all processes occurring exclusively in the nucleus. But the ultimate manifestation of gene expression is when the mRNA is translated into protein in the cytoplasm. This necessitates the movement of the transcript from the nucleus to the cytoplasm by yet another process referred to as **nuclear export**, a process that is not very well described but one that we know imposes another level of regulation on gene expression because of its gate-keeping function (Chinnusamy *et al.*, 2008).

Figure 13.11 illustrates a working model of the nuclear transport of mRNA from the nucleus in plants,

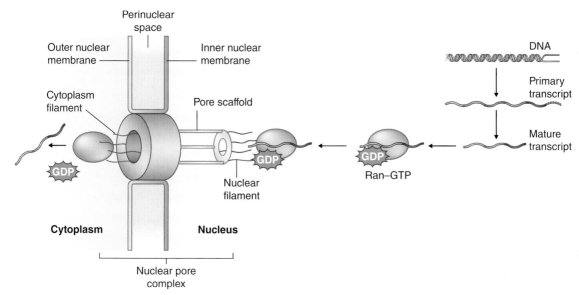

Figure 13.11 A cartoon depiction of how RNAs are exported out of the nucleus and released into the cytoplasm (Rout et al., 2000)

much of which is based on parallel observations with yeast and mammalian systems. While small molecules, less than a few thousand daltons are probably freely diffusible into and out of the nucleus via the nuclear pores, proteins and mRNAs are not. Their transport is an active process requiring energy and the export of mRNAs utilizes GTP, which is a cue for its export. As a primary transcript is processed to its mature form, a number of proteins form a complex with the RNA and ferry it to the nuclear pore. One conspicuous protein in these complexes is RAN, a relatively small protein known for its role in chaperoning proteins and RNA into and out of the nucleus, and its ability to release energy upon hydrolysis of GTP to GDP and inorganic phosphate. Once interactions between the newly transcribed and processed RNA and transport proteins are complete, the RNA-protein complex finds it way to the nuclear filaments of the nuclear pore, the RNA-protein complex is tracked through the pore complex and upon hydrolysis of the GTP, the RNA is released into the cytoplasm. Ran and other RNA binding proteins are then recycled back into the nucleus to re-initiate another round of RNA export. Many proteins make up the nuclear pore complex, upwards of 30, and are collectively referred to as nucleoporins.

All the machinery associated with mRNA export suggests that there are many potential places in this process where specificity and regulation must come into play. For instance, how is assembly of the nuclear export proteins with fully processed and mature mRNA distinguished from the complex with primary transcripts? While no specific sequence motifs within mRNA have been identified for mediating the binding with the nuclear export proteins, abnormal mRNAs, mRNAs derived from mutant or engineered genes having defects in 3'-end formation or splicing, tend to accumulate in the nucleus and not be exported. This has led to the suggestion that some nuclear export proteins are recruited to mRNA by the processing and splicing machinery proteins. Hence, if an mRNA transcript is not actively processed, it is less likely to be exported from the nucleus. While this may be the case for some or even most mRNAs, it is not sufficient to account for all the mRNA export specificity. There are genes which do not have any introns and their primary transcripts are not subject to the splicing process, yet these mRNAs appear to be efficiently exported from the nucleus.

The export process has also been shown to be involved with a variety of physiological responses including innate hormonal signals as well as external biotic (i.e., pathogens) and abiotic (i.e., temperature adaptation) stress (Gong et al., 2002, 2005;Dong et al., 2006). Plants in particular certainly face unique challenges in being able to adapt to very broad temperature extremes. Chilling tolerance, for instance, is where plants subject to repeated cycles of cold temperatures adapt their physiological machinery so that they are functioning sufficiently under optimal and cold conditions. Some genetic mutants that do not exhibit cold acclimation map to defects in nuclear RNA export. In the normal or wild type plants, cold adaptation appears to include reduced export of transcripts from the nucleus. This reduced transport response is lost in the mutants and the causes have been traced to defects in specific proteins associated with the mRNA export machinery.

13.6 Summary

This chapter is focused on the maturation events and movement of mRNA from the nucleus to the cytoplasm. Transcription of a gene generates the primary transcript, which then undergoes a number of processing steps that occur more or less simultaneously. First, a chemical cap is appended to the 5′ end of the newly synthesized mRNA and then the 3′ end of the transcript is generated by a complex assemblage of proteins that cleave the nascent RNA chain, then adds a polyA tail in a non-template dependent fashion. Equally important, the intervening sequences, segments of RNA that do not appear in the final mRNA, are spliced out to yield a final mature mRNA. But that mRNA must be exported from the nucleus and this process too involves a whole host of proteins that serve to track the mRNA movement to nuclear pores and ultimate release into the cytoplasm. The complexity in each of these processing steps suggests that each could serve as another kind of checkpoint providing another type of mRNA proof reading as well as another means for controlling gene expression. Much of what we know about these regulatory functions of the RNA processing events come from an examination of genetic mutants in discrete steps of these events.

13.7 Problems

13.1 mRNA capping consists of methylation at 3 positions generating forms known as Cap 0, 1 and 2. What are all the possible different methylation patterns using these three sites? Given all the different possible forms, why do you think only three, Cap 0, 1 and 2, are observed in nature?

Can you derive a mechanistic model for such a preference?

13.2 Calculate the percentage of time it takes for each of the processing steps from transcription initiation to the time of nuclear export for an average mRNA. For instance, transcription takes x% of the total time. Assuming that the most complicated processing steps would take the longest, list the processing steps in increasing order of complexity. Then comment about what makes each processing step more complex than the previous one.

13.3 Develop a regulatory network as in Figure 13.8 except for the gene encoding for the enzyme hexokinase, an enzyme that mediates the utilization of sugars for respiration and energy production. Assume the HEX gene has one intron and that the network would want to regulate hexokinase activity based upon the availability of sugar for building biomass and energy requirements. Integrate the network with photosynthesis that occurs during the day/night cycle.

References

Bassett, C.L., Artlip, T.S. and Callahan, A.M. (2002) Characterization of the peach homologue of the ethylene receptor, PpETR1, reveals some unusual features regarding transcript processing. *Planta*, **215**, 679–688.

Brown, J.W.S. and Simpson, C.G. (1998) Splice site selection in plant pre-mRNA splicing. *Annual Review of Plant Physiology and Plant Molecular Biology*, **49**, 77–95.

Chinnusamy, V., Gong, Z. and Zhu, J.K. (2008) Nuclear RNA export and its importance in abiotic stress responses of plants. *Current Topics in Microbiology and Immunology*, **326**, 235–255.

Dong, C.H., Hu, X.Y., Tang, W.P. *et al.* (2006) A putative Arabidopsis nucleoporin, AtNUP160, is critical for RNA export and required for plant tolerance to cold stress. *Molecular and Cellular Biology*, **26**, 9533–9543.

Giranton, J.L., Ariza, M.J., Dumas, C. *et al.* (1995) The S locus receptor kinase gene encodes a soluble glycoprotein corresponding to the SRK extracellular domain in *Brassica oleracea*. *Plant Journal*, **8**, 827–834.

Gong, Z.Z., Dong, C.H., Lee, H. *et al.* (2005) A DEAD box RNA helicase is essential for mRNA export and important for development and stress responses in Arabidopsis. *Plant Cell*, **17**, 256–267.

Gong, Z.Z., Lee, H., Xiong, L.M. *et al.* (2002) RNA helicase-like protein as an early regulator of transcription factors for plant chilling and freezing tolerance. *Proceedings of the National Academy USA*, **99**, 11507–11512.

Hugouvieux, V., Kwak, J.M. and Schroeder, J.I. (2001) An mRNA cap binding protein, ABH1, modulates early abscisic acid signal transduction in Arabidopsis. *Cell*, **106**, 477–487.

Hunt, A.G. (2008). Messenger RNA 3′ end formation in plants. *Current Topics in Microbiology and Immunology*, **326**, 151–177.

Ishikawa, T., Yoshimura, K., Tamoi, M. *et al.* (1997). Alternative mRNA splicing of 3′-terminal exons generates ascorbate peroxidase isoenzymes in spinach (*Spinacia oleracea*) chloroplasts. *Biochemical Journal*, **328**, 795–800.

Kuhn, J.M., Hugouvieux, V. and Schroeder, J.I. (2008) mRNA cap binding proteins: effects on abscisic acid signal transduction, mRNA processing, and microarray analyses. *Current Topics in Microbiology and Immunology*, **326**, 139–150.

Krebs, J.E., Goldstein, E.S. and Kilpatrick, S.T. (2010) Lewin's Essential Genes, Jones and Bartlett, Boston.

Meyers, B.C., Morgante, M. and Michelmore, R.W. (2002) Tir-X and Tir-NBS proteins: two new families related to disease resistance Tir-NBS-LRR proteins encoded in Arabidopsis and other plant genomes. *Plant Journal*, **32**, 77–92.

Nilsen, T.W. and Graveley, B.R. (2010) Expansion of the eukaryotic proteome by alternative splicing. *Nature*, **463**, 457–463.

Quesada, V., Macknight, R., Dean, C. and Simpson, G.G. (2003) Autoregulation of FCA pre-mRNA processing controls Arabidopsis flowering time. *EMBO Journal*, **22**, 3142–3152.

Richardson, D.N., Rogers, M.F., Labadorf, A. *et al.* (2011) Comparative analysis of serine/arginine-rich proteins across 27 eukaryotes: insights into sub-family classification and extent of alternative splicing. *PLOS ONE*, **6**, e24542. doi:10.1371/journal.pone.0024542.

Rout, M.P., Aitchison, J.D., Suprapto, A. *et al.* (2000) The yeast nuclear pore complex: composition, architecture, and transport mechanism. *Journal of Cell Biology*, **148**, 635–651.

Ru, Y., Wang, B.B. and Brendel, V. (2008) Spliceosomal proteins in plants. *Current Topics in Microbiology and Immunology*, **326**, 1–15.

Shuman, S. (2002) What messenger RNA capping tells us about eukaryotic evolution. *Nature Reviews Molecular Cell Biology*, **3**, 619–625.

Tang, G.L., Zhu, X.H., Gakiere, B. *et al.* (2002) The bifunctional LKR/SDH locus of plants also encodes a highly active monofunctional lysine-ketoglutarate reductase using a polyadenylation signal located within an intron. *Plant Physiology*, **130**, 147–154.

Terzi, L.C. and Simpson, G.G. (2008) Regulation of flowering time by RNA processing. *Current Topics in Microbiology and Immunology*, **326**, 201–218.

Wang, B.B. and Brendel, V. (2006). Genomewide comparative analysis of alternative splicing in plants. *Proceedings of the National Academy of Sciences USA*, **103**, 7175–7180.

Wu, X.H., Liu, M., Downie, B. *et al.* (2011) Genome-wide landscape of polyadenylation in Arabidopsis provides evidence for extensive alternative polyadenylation. *Proceedings of the National Academy of Sciences USA*, **108**, 12533–12538.

Zhang, X.C. and Gassmann, W. (2007) Alternative splicing and mRNA levels of the disease resistance gene *RPS4* are induced during defense responses. *Plant Physiology*, **145**, 1577–1587.

Chapter 14

Fate of RNA

14.1   Regulation of RNA continues upon export from nucleus

The export of RNA from the nucleus appears at first glance to represent the last of the control points regulating the expression level of a gene. That is, once the mRNA appears in the cytoplasm, it engages with the translational machinery to produce the encoded protein product. But many observations suggest this is far from the actual case. Messenger (mRNA) and other RNA molecules are subject to yet another range of mechanisms modulating when and how an mRNA is ultimately engaged with the ribosomes and translated into protein. Several of these mechanisms control the half-life of the mRNA in the cytoplasm by regulating the degradation rate of the mRNA, and others appear associated with the selective movement of RNA between cell types and even longer distance movement, thus providing yet another means for imposing cell-specific gene expression patterns.

14.2   Mechanisms for RNA turnover

In retrospect, it seems obvious that RNAs and especially mRNAs must have mechanisms for their degradation, more technically referred to as

their turnover rate, which influences the **half-life of mRNA**. Otherwise, cells would not have a way to modulate what mRNAs are expressed during the lifetime of the cell, nor would cells responding to stimuli be able to revert to their normal metabolite state, nor would they be able to handle the occasional damaged mRNA. Hence, various mechanisms for how cells dispose of mRNAs have been elucidated. These mechanisms have been roughly divided between those that control the turnover and degradation of normal mRNAs during the typical life cycle of an mRNA, and those that serve as protective mechanisms controlling the deletion of damaged mRNA species (Figure 14.1). There are obvious overlaps in these mechanisms. At least in plants, much more attention has been given to the special conditions for handling abnormal mRNAs because these have been easier to study experimentally.

It is important to recall that the level of any particular mRNA is a consequence of its synthesis rate plus its degradation rate. Because it is typically easier to measure the abundance of mRNAs by techniques like Northern blot hybridization, reverse transcription polymerase chain reaction or even advanced transcriptomic sequencing methodologies (i.e., RNA-Seq), the steady-state level of mRNA is often reported. However, as emphasized here, the amount of any one mRNA in a cell at any particular time in the life cycle of the cell is the sum total of the transcription rate of the gene coding for the mRNA and a variety of mechanisms responsible for the degradation of that mRNA.

At least two major mechanisms appear to be responsible for the typical turnover of mRNAs in plant cells,

Plant Genes, Genomes and Genetics, First Edition. Erich Grotewold, Joseph Chappell and Elizabeth A. Kellogg.
© 2015 John Wiley & Sons, Ltd. Published 2015 by John Wiley & Sons, Ltd.
Companion Website: www.wiley.com/go/grotewold/plantgenes.

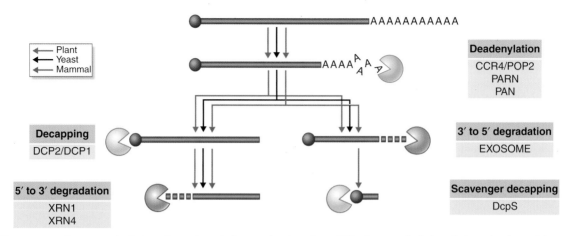

Figure 14.1 A cartoon depiction of two degradation mechanisms for mRNA, the so-called 5′ and 3′ mechanisms. Adapted from Chiba and Green (2009). Reproduced with permission of Springer Science+Business Media

much like in any other eukaryotic cells. These have been described as the 5′ to 3′ and the 3′ to 5′ degradation mechanisms and a list of the relevant proteins associated with these processes is shown in Table 14.1. Both however are dependent on the initial loss of the polyA tail at the 3′ end of the mRNA molecule. This is considered the crucial key step, because cytosolic polyA binding proteins, PABPs, are known to bind to the polyA tails and block access of the mRNA to degrading nucleases. (Remember that a nuclease is any enzyme that degrades a nucleic acid; enzymes that degrade RNA are generally known as RNases.)

In the **5′ to 3′ degradation mechanism**, deadenylation specific RNases (CCR4/POP2, PARN and PAN)

remove the majority of the polyA tail, which leads to a cascade of events resulting in mRNA degradation. Once the polyA tail sequence is removed, PABP can no longer bind to the mRNA. Once PABP is no longer bound to the mRNA, 5′ cap binding proteins are also released, thus exposing the 5′ terminus of the mRNA to the decapping enzymes, DCP1 and DCP2, enzymes that remove the terminal methylated nucleotide. The resulting unprotected mRNA lacking the 5′ cap and 3′ polyA tail can then be degraded by general RNases (XRN 1 and 4), which appear to exist in multiple copies in plant genomes.

The **3′ to 5′ degradation mechanism** differs from the 5′ to 3′ in that the deadenylated mRNA can associate with a complex of proteins known as the **exosome**. The exosomes characterized in yeast and mammalian systems include several to many different exonucleases, enzymes that can cleave nucleotides from the 3′ terminus, and these complexes might even be associated with cytoplasmic inclusions known as P-bodies (see Section 14.4). The 3′ to 5′ degradation mechanism has the potential to degrade the mRNA down to a few residues, including the 5′ cap nucleotide, which must be further metabolized by yet another possible class of decapping nucleases, although these enzyme(s) have not been confirmed in plants to date.

While the 5′ to 3′ and 3′ to 5′ mechanisms are thought to operate as described above in plants, the evidence is largely taken from yeast and mammalian studies and plant-specific evidence is still rather limited. The current best evidence is based on the presence of genes in plant genomes that are

Table 14.1   Proteins associated with RNA degradation

Protein	Role
PABP	PolyA tail binding proteins, serve to protect RNA from the initial events of RNA degradation
CCR4/POP2, PARN, PAN	RNases that remove the polyA tail found on the 3′ end of mRNAs
DCP1, DCP2	Enzymes that can remove the 5′ terminal nucleotide of mRNAs, thus poising the mRNA for 5′ degradation
XRN1 and XRN4	General RNases that can degrade mRNA in the 5′ to 3′ direction
Exosome	Complex of proteins that can degrade RNA in the 3′ to 5′ direction

homologous to those that have been studied in yeast and mammalian systems. Putative genes encoding homologs of the deadenylating enzyme, for example, are readily apparent in genomic sequence data for plants. Over-expression of the pepper CAF1 gene, a homolog of POP2, in tomato stimulates overall growth and disease resistance traits, while silencing this gene in pepper shows the converse characteristics. However, direct demonstration of an exonuclease or deadenylating enzyme activity with the CAF1 protein is missing. In contrast, an Arabidopsis PARN-like deadenylase was shown to degrade the polyA tail of RNA and mutants of this gene were associated with alterations in embryo development with embryo-specific transcripts accumulating with longer polyA tails.

Similar data exist for the decapping and the exoribonuclease steps of mRNA turnover. Several homologs of the *DCP* genes exist in Arabidopsis and a T-DNA insertional mutant in one of these, *AtDCP2*, showed a seedling lethality phenotype. Initial growth of these mutant seedlings was accompanied by a significant elevation in the level of over 140 specific mRNAs. Arabidopsis also contains homologs of the yeast/mammalian exoribonuclease Xrn2p, and when three of these genes (AtXRN2, 3 and 4) were expressed in yeast, 5′ to 3′ exonuclease activities were observed. Interestingly, a mutant having a non-functional AtXRN4 gene did not have defects in overall growth and development as one might expect. The AtXRN4 mutant did, however, reveal possible roles for this exonuclease in the phenomena of transgene-dependent gene silencing and how the signaling cascade associated with the gaseous growth regulator ethylene is transduced. The AtXRN4 mutation may in fact have minimal effect because the other two gene products continue to function.

Evidence for the 3′ to 5′ exosome complex in plants is also somewhat circumstantial. The exosome complex best studied is that in yeast, consisting of ten specific core proteins plus additional proteins that seem to associate based on the intracellular location of the exosome. Two homologs of these core components have been identified in Arabidopsis and characterized. Both *AtRRP4* and *AtRRP41* encode proteins with exonuclease activity and both appear to be associated with high molecular weight complexes isolated from Arabidopsis, which might be the plant complement to the yeast exosome. Interestingly, these two Arabidopsis genes appear to affect different stages of plant development. Mutation of the *AtRRP41* gene results in

alterations in female gametogenesis, while the *AtRRP4* mutant affects embryogenesis. This may be an example of subfunctionalization, as described in Chapter 2; in this case, AtRRP41 and AtRRP4 might have divided up an ancestral role in both the female gametophyte and the embryo. Further support for the existence of exosome-like complexes comes from studies of **process bodies** (PBs) and **stress granules** (SGs) (see Section 14.4).

Overall, plants seem to have all the component parts for both the 5′ to 3′ and the 3′ to 5′ degradation pathways, yet the details for how these pathways actually operate remain to be determined.

14.3   RNA surveillance mechanisms

Given that plants do possess the machinery for RNA turnover, what specific mechanisms control or regulate this process? Certainly there are mechanisms to degrade and recycle the chemical components of RNA just in the course of normal metabolism, otherwise there would be no way for a cell to change its metabolism or developmental fate. Deciphering such a general mechanism poses unique experimental challenges, the greatest being that mutations in such machinery are likely to have severe phenotypic consequences including lethality. Inroads have hence focused more on select or specific mechanisms that lend themselves to experimental dissection (Figure 14.2). **Nonsense mediated decay** (NMD) is the best studied of these systems across all organisms, including mammals, **Drosophila**, yeast and plants. In this circumstance, a point mutation that inserts into the coding region of a gene either via some natural process or introduced experimentally results in conversion of a codon originally calling for the insertion of an amino acid into a growing polypeptide chain into a stop codon, causing the premature termination of translation. If not corrected or repaired, such a mutation would have serious consequences for the organism, generating a truncated protein that may be deleterious for its proper functioning. Lots of incomplete, non-functional protein could accumulate and lots of energy would be wasted in this process. NMD is the mechanism investigators think evolved to handle this situation.

Figure 14.2 illustrates how the NMD mechanism is thought to operate. As the protein translational

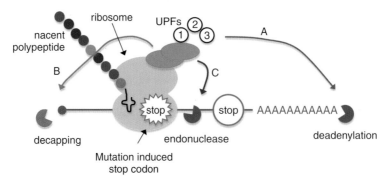

Figure 14.2 An illustration of the NMD mechanism. When the ribosome pauses to terminate translation at the aberrant stop codon, the free 5′ and 3′ ends of the mRNA molecule can interact with accessory proteins like UPF 1–3 or others not yet identified (designated in blue), to initiate 3′ deadenylation (A) or 5′ decapping (B) or internal endonuclease cleavage (C), all leading to the complete degradation of the mRNA

machinery moves down an mRNA molecule and meets with an unusually placed stop codon in the middle of the mRNA molecule, accessory proteins UPF1-3 and perhaps other less well-defined proteins (blue designations) mediate interactions with exonucleases that can degrade the mRNA from the 5′ or the 3′ end, or even possibly endonucleases that cleave the mRNA into smaller fragments that then go on to complete degradation by general ribonucleases. Hence, pausing the translational machinery along the mRNA instigates RNA degradation. Some investigators have suggested that the length of exposed RNA at the 5′ and 3′ ends of the mRNA at the time of the translation pause is sufficient for the NMD mechanism to kick into action. Others have noted that the insertion point of the nonsense codon relative to exon–exon junctions is also important, suggesting the involvement of other proteins associated with the splicing machinery. Regardless, when the translational machinery pauses, the exposed 5′ and 3′ ends of the mRNA have sufficient opportunity to physically interact with accessory proteins like UPFs 1–3 and degradation is initiated.

Investigators have described a couple of parallel surveillance-like mechanisms that might be considered more rare than NMD, but nonetheless exhibit some parallels to the NMD mechanism. Read-through of the mRNA into the 3′ untranslated region may induce degradation via a mechanism referred to as **non-stop mRNA decay** (NSD), as would translational pausing caused by secondary structure in the RNA, a hairpin structure for example, in a process known as **no-go decay** (NGD).

14.4 RNA sorting

As noted above, tracking RNA from the nucleus to the translational versus degradation machinery is one level of regulation. However, there is yet another level of RNA sorting that constitutes storage of mRNA in the cytoplasm, its release to the translational machinery, or its diversion to degradation (Balagopal and Parker, 2009) (Figure 14.3). The evidence for these sorting mechanisms comes from microscopic observations, in which RNA-binding proteins are bound to antibodies, or the genes coding for these RNA-binding proteins are fused to fluorescent tags. Under some conditions, these proteins are found diffusely throughout the cytoplasm. However, under other conditions where cells are differentiating or induced by environmental conditions like heat stress or oxygen deprivation, RNA-binding proteins localize in discrete foci in the cytoplasm (Weber *et al.*, 2008; Pomeranz *et al.*, 2010). The RNA sequestered by RNA-binding proteins results in cytoplasmic inclusions known as PBs or SGs. The PBs and SGs share some proteins in common, but also have distinctly different protein members, which might serve specific physiological functions. PBs and SGs thus appear to serve as mechanisms controlling the fate of RNA. RNA can be stored in PBs and SGs for variable lengths of time (minutes to days) to mitigate a particular situation, then released later to either re-initiate translation under the right developmental or environmental cues, or directed to the RNA degradation machinery. One example of the possible role of PBs and SGs in regulating gene expression relates to their formation in seeds (Bogamuwa and

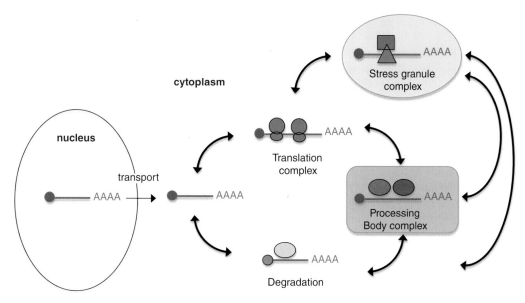

Figure 14.3 RNA in the cytoplasm is subject to multiple layers of regulation. Processing bodies and stress granules can sequester RNA, later to be released for either translation or degradation

Jang, 2013). Protein composition of these storage/sorting complexes and their existence *in planta* are sensitive to two key plant growth hormones, abscisic acid (ABA) and gibberellic acid (GA). Both ABA and GA are also known for their roles in regulating seed germination. However, how formation of PBs and SGs contributes to the ABA and GA regulatory process of seed germination remains to be determined.

14.5 RNA movement

One other fate of mRNA that was not considered until recently is the movement of mRNA from the cell where it is synthesized to another cell where it is translated. The issue of how small RNAs move locally and systemically was considered in Chapter 11, but longer RNAs can move as well. Such a possibility becomes more obvious when considering virus movement in plants. Entire virus particles (viral genome wrapped together with coat protein and other viral proteins) rarely move from cell to cell in plants. Instead, viruses move via their genomes, their DNA or RNA components, as well as transcripts derived from their genomes. This does make some sense in the case of viral pathogenesis. Intact virus particles could not move easily between cells, because the virus would have to enter a host cell, replicate its genome, express all the viral proteins necessary for packaging (i.e., coat protein), and then lyse or be released from the host cell. In this case, viral spread would be rather local. However, plant viruses have evolved the capability to move between cells via the plasmodesmata, physical interconnections between cells (Figure 2). Viral movement under these circumstances can include movement of the viral genome as well as encoded mRNAs.

The movement of mRNA from its site of synthesis in one cell to that in a neighboring cell is best described for the Knotted and Knotted-like messenger (Figure 14.4). The **Knotted/KN**-like genes (already mentioned in Chapter 10) encode transcription factor proteins that orchestrate gene expression programs by selective interactions with genomic DNA in the nucleus. Interestingly, the KN1 protein can also bind its own mRNA, facilitating its movement through plasmodesmata to neighboring cells, wherein translation of the KN1 mRNA would yield the KN1 protein, and that in turn could regulate cell specific gene expression in the neighboring cell type as well.

Messenger RNA movement is not restricted to just local movement. There are many examples of long-distance movement as well, wherein mRNA moves from cell-to-cell via the plasmodesmata until the message gets near the phloem, the vascular system responsible for the transport of nutrients, hormones,

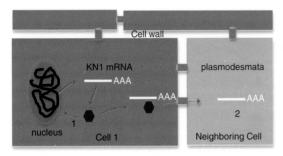

Figure 14.4 An illustration of how the KN1 mRNA might migrate to neighboring cells, where the Knotted protein could influence what genes are expressed because of its role as a transcription factor. Adapted from Bolduc *et al.* (2008)

proteins and nucleic acids throughout the entire plant (see Introduction). In a process thought to resemble that of localized movement, the mRNA probably associates with specific proteins that chaperone the messenger through specialized plasmodesmata into the sieve element cells of the phloem and it is thus transported to distal points within the plant. Some of these transported mRNAs are noted for the long distance traveled and their physiological importance for the recipient cells. For instance, Banerjee *et al.* (2006) demonstrated that leaf specific expression of a BEL1-like transcription factor gene in potato leaves led to accumulation of the corresponding mRNA in stolon (root) tips (Figure 14.5). Because this mRNA encodes a transcription factor that coordinates tuber formation, increased tuber production was also observed.

RNA movement also includes the movement of small RNAs, RNAs associated with gene silencing, and this special topic of RNA metabolism is discussed elsewhere (see Chapter 11).

14.6   Summary

While issues associated with the translation of mRNA will be dealt with in the next chapter, this chapter focused on the mechanisms sorting mRNAs for cytoplasmic storage or degradation. The central issue is the longevity of an mRNA species, and this has important temporal and spatial ramifications. Very simply, the amount of a protein produced from the translation of a transcript is somehow proportional to how long that mRNA exists in the cytoplasm and how available it is to the translational machinery. We really do not

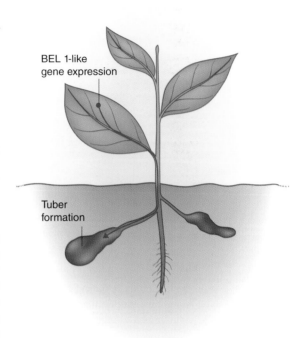

Figure 14.5 Leaf specific over-expression of the BEL 1 transcription factor gene in leaves leads to detection of the corresponding mRNA in roots and the increased production of tubers. From Lin *et al.*, 2013

understand what mechanisms regulate the half-life of any one mRNA, but have relied on studies to elucidate the mechanisms at play to degrade abnormal mRNA species, largely by documenting the genes encoding these RNA degradation enzymes and proteins. The mechanisms include 5′ and 3′ processes, processes that initiate at the head or tail of an mRNA and include exo- and endonuclease activities. The question of how regulation of gene expression might be imposed at the level of mRNA availability to the translational machinery was noted by the selective transport of mRNA from the cells where it is synthesized to different cells where the mRNA is actually translated.

14.7   Problems

14.1 Using the model shown in Figure 14.2, can you propose a way in which RNA secondary structure might be recognized and used to differentiate between short-lived versus long-lived mRNAs?

14.2 What does the broadly conserved existence of RNA surveillance mechanisms in all eukaryotic organisms suggest about the fidelity of gene transcription? Is it possible that transcriptional infidelity could be another means for molecular evolution? But, then how would a plant "capture" such an event in subsequent generations? How does this infidelity get incorporated back into the genome?

14.3 Make a list of the different mechanisms, proteins and factors that you would expect to be involved in RNA transport. How might these be similar or different for short range transport (between cells) and long range transport (movement from organ to organ)?

Further reading

Belostotsky, D.A. (2008) State of decay: an update on plant mRNA turnover. *Current Topics in Microbiology and Immunology*, **326**, 179–199.

Bolduc, N., Hake, S. and Jackson, D. (2008) Dual functions of the KNOTTED1 homeodomain: sequence-specific DNA binding and regulation of cell-to-cell transport. *Science Signaling*, **1**, pe28. doi: 10.1126/scisignal.123pe28.

Carlsbecker, A., Lee, J.Y., Roberts. C.J. *et al.* (2010) Cell signaling by microRNA165/6 directs gene dose-dependent root cell fate. *Nature*, **465**, 316–321.

Chiba, Y. and Green, P.J. (2009) mRNA degradation machinery in plants. *Journal of Plant Biology*, **52**, 114–124.

David-Schwartz, R., Runo, S., Townsley, B. *et al.* (2008) Long-distance transport of mRNA via parenchyma cells and phloem across the host-parasite junction in *Cuscuta*. *New Phytologist*, **179**, 1133–1141.

Kehr, J. and Buhtz, A. (2008) Long distance transport and movement of RNA through the phloem. *Journal of Experimental Botany*, **59**, 85–92.

Kim, M., Canio, W., Kessler, S. and Sinha, N. (2001) Developmental changes due to long-distance movement of a homeobox fusion transcript in tomato. *Science*, **293**, 287–289.

Nyiko, T., Sonkoly, B., Merai, Z. *et al.* (2009) Plant upstream ORFs can trigger nonsense-mediated mRNA decay in a size-dependent manner. *Plant Molecular Biology*, **71**, 367–378.

Rymarquis, L.A., Souret, F.F. and Green, P.J. (2011) Evidence that XRN4, an Arabidopsis homolog of exoribonuclease XRN1, preferentially impacts transcripts with certain sequences or in particular functional categories. *RNA*, **17**, 501–511.

Walley, J.W., Kelley, D.R., Nestorova, G. *et al.* (2010) Arabidopsis deadenylases AtCAF1a and AtCAF1b play overlapping and distinct roles in mediating environmental stress responses. *Plant Physiology*, **152**, 866–875.

References

Balagopal, V. and Parker, R. (2009) Polysomes, P bodies and stress granules: states and fates of eukaryotic mRNAs. *Current Opinion in Cell Biology*, **21**:403–408.

Banerjee, A.K., Chatterje, M., Yu, Y. *et al.* (2006) Dynamics of a mobile RNA of potato involved in a long-distance signaling pathway. *Plant Cell*, **18**, 3443–3457.

Bogamuwa. S. and Jang, J.-C. (2013) The Arabidopsis tandem CCCH zinc finger proteins AtTZF4, 5 and 6 are involved in light-, abscisic acid- and gibberellic acid-mediated regulation of seed germination. *Plant Cell & Environment*. **36**: 1507–1519.

Lin, T., Sharma, P., Gonzalez, D.H., Viola, I.L., Hannapel, D.J. (2013). The impact of the long-distance transport of a *BEL1*-like messenger RNA on development. *Plant Physiology* **161**: 760–772.

Pomeranz, M.C., Hah, C., Lin, P.-C. *et al.* (2010) The Arabidopsis tandem zinc finger protein AtTZF1 traffics between the nucleus and cytoplasmic foci and binds both DNA and RNA. *Plant Physiology*, **152**,151–165.

Weber, C., Nover, L. and Fauth, M. (2008) Plant stress granules and mRNA processing bodies are distinct from heat stress granules. *Plant Journal*, **56**, 517–530.

Chapter 15
Translation of RNA

15.1 Translation: a key aspect of gene expression

Gene expression in a technical sense spans everything from transcription of a gene in the nucleus to the functioning of the encoded protein in a cell, including its longevity. It is not the "functioning" of genes per se, but the activity or role of their encoded protein products that creates cells with unique biochemical abilities, giving them their phenotypes, traits, and characteristics. Hence, it is not unexpected that plants have evolved, like all organisms, mechanisms to control or regulate all the steps in the gene expression process. In this chapter, we review the many facets of messenger RNA (mRNA) translation common to all organisms, then focus on a few features of translation that give plants a unique ability to contend with special environmental conditions like variable light, water, salinity, and oxygen availability.

Once mRNA makes it way to the cytoplasm, it proceeds through the three phases of translation, known as **initiation**, **elongation** and **termination**. While it is helpful to think of each phase as a distinct and independent activity, they actually occur simultaneously and in concert with one another. It is equally important to keep the overall process and the various component players in this finely orchestrated ensemble in some perspective, so we tend to segregate the events into phases for ease of discussion (Browning, 1996; Futterer and Hohn 1996).

All mRNAs are translated on the basis of consecutive groups of three bases, codons, being interpreted by the translational machinery (Figure 15.1). Each codon in turn calls for a specific amino acid, which the ribosomes bring together via a peptide bond (Figure 15.2), forming a polypeptide chain and ultimately the mature protein. Which codons call for the incorporation of what amino acids is the basis of the genetic code (Figure 15.3). The genetic code is redundant, in that there are 64 possible three-letter combinations of four nucleotide bases. But it is not ambiguous, each codon specifies only one amino acid. Three of the codons are reserved for termination of translation (stop codons), leaving 61 possible codons divided amongst the 20 naturally occurring amino acids. With the exception of methionine (Met) and tryptophan (Trp), all other amino acids have multiple codon possibilities, hence the degeneracy of the code.

Many diverse proteins and RNAs are involved in translation of mRNA. First is the mRNA itself, which is the template "read" and translated into a protein product. Ribosomes are very large complexes of RNA and protein that "read" the mRNA and translate it into a protein. The 80S ribosome is actually composed of two individual complexes, the 40S and 60S ribosomal subunits (see Table 6.1). The nomenclature of ribosomes and their subunits is derived from their sedimentation rate upon centrifugation (the higher the number, the larger the subunit is and the more quickly it sediments) and is indicative of their overall mass. The 40S subunit consists of a single RNA species containing 1870 bases

Plant Genes, Genomes and Genetics, First Edition. Erich Grotewold, Joseph Chappell and Elizabeth A. Kellogg.
© 2015 John Wiley & Sons, Ltd. Published 2015 by John Wiley & Sons, Ltd.
Companion Website: www.wiley.com/go/grotewold/plantgenes.

codon

..caaca aug gga aag cua ugg caauugg.. mRNA

↓ translation

Met-gly-lys-leu-trp .... nascent protein

Figure 15.1 The overall process of protein synthesis is to translate the nucleotide sequence of mRNA into protein using the order of three nucleotides as codons calling for the incorporation of specific amino acids into the growing peptide chain

(actual size is species dependent and variable by a few bases) and upwards of 33 distinct proteins. The 60S subunit consists of 3 RNA species, the 5S RNA of approximately 120 bases, the 5.8S RNA with upwards of 160 bases and the 28S consisting of over 4700 bases. The 60S subunit also contains upwards of 50 different proteins. When the 40S and 60S ribosomal subunits combine, they form a structure that sediments as an 80S complex (Figure 15.4). The ribosomes have two transfer RNA (tRNA) binding sites, P and A, which provide the means for tRNAs to bring amino acids into the translation machinery (A site) and the transfer of that amino acid to the nascent polypeptide chain held in the P site by a second tRNA molecule.

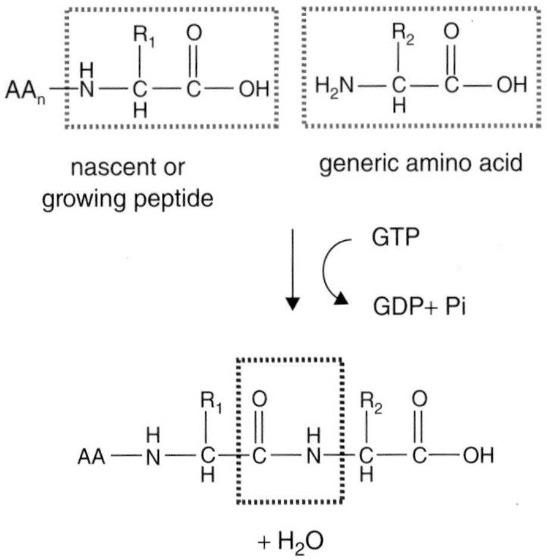

Figure 15.2 Peptide bond formation is a very unique chemical reaction in protein synthesis. Translation of mRNA consists of an mRNA's codons calling for specific amino acids to be brought together in a very specific order to create a protein. In order for the translational machinery to bring amino acids together in a stable form, the machinery catalyzes formation of covalent linkage between amino acids (red boxes) known as a peptide bond (blue box). Peptide bond formation requires significant energy input, which is derived from the cleavage of the terminal phosphate from GTP

		Second nucleotide				
		U	C	A	G	
U		Phe	Ser	Tyr	Cys	U
		Phe	Ser	Tyr	Cys	C
		Leu	Ser	STOP	STOP	A
		Leu	Ser	STOP	Trp	G
C		Leu	Pro	His	Arg	U
		Leu	Pro	His	Arg	C
		Leu	Pro	Gln	Arg	A
		Leu	Pro	Gln	Arg	G
A		Ile	Thr	Asn	Ser	U
		Ile	Thr	Asn	Ser	C
		Ile	Thr	Lys	Arg	A
		Met	Thr	Lys	Arg	G
G		Val	Ala	Asp	Gly	U
		Val	Ala	Asp	Gly	C
		Val	Ala	Glu	Gly	A
		Val	Ala	Glu	Gly	G

First nucleotide (left). Third nucleotide (right).

Figure 15.3 The genetic code – specific sequences of three nucelotides in mRNAs (codons) call for the incorporation of the designated amino acids (three-letter abbreviations) into the growing peptide chain. Stop codons are used to terminate mRNA translation and to dissociate the ribosomes and protein synthesis machinery

Translation of mRNA also requires other RNAs and proteins to help complete the job of protein synthesis. Forty-two tRNAs are the vehicles which ferry specific amino acids to the ribosomes engaged in translating an mRNA. Each tRNA has a 3′ region that is conjugated with a specific amino acid based on an anticodon sequence (base pair complement to the codon for which it serves in protein synthesis) found in a loop in the middle of the tRNA molecule. Only 42 tRNAs are required and not 61 tRNAs as predicted

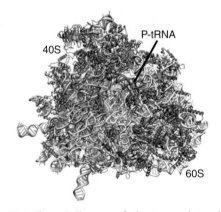

Figure 15.4 The 80S ribosome of wheat germ determined by a combination of cryro-electron microscopy and molecular modeling. The rRNA and protein components are shown in yellow and orange for the small (40S) subunit, and gray and blue for the large (60S) subunit, respectively. A tRNA in the P site is also shown (green). Taken from Armache *et al.*, 2010

by the genetic code. This is because of the wobble effect associated with the anticodon of several tRNAs. The third base in anticodons of several tRNAs can base pair with multiple codons differing only in the third base of the codon. This promiscuity is limited to those tRNAs that base pair with degenerate codons for amino acids like alanine. The degenerate codons for alanine are GCU, GCC, GCG, or GCA (Figure 15.3). Three termination tRNAs also exist. These tRNAs do not carry amino acids to the growing peptide chain, but instead terminate translation causing the release of the newly synthesized protein and disassembly of the translational machinery.

Many different proteins are associated with each phase of the translation process. While some are explicitly identified below, the roles for others are not yet fully defined and, hence, will not be discussed further here. Nomenclature for the various proteins is standardized. Proteins associated with the initiation process are referred to as **eukaryotic initiation factors**, or eIFs, and are numbered and sub-lettered consecutively (i.e., eIF4E) according to their role in the process and to designate their physical associations, respectively. Proteins associated with elongation are likewise labeled as eEFs (**eukaryotic elongation factors**), while those associated with termination and release of the protein product are referred to as **eukaryotic release factors** (eRFs).

15.2   Initiation

Translation initiation takes place when the 5′ cap of an mRNA is recognized by a suite of initiation factors or eIFs (Figure 15.5). Initiation factor 4 (eIF4) and several related proteins bind to the 5′ terminus of the mRNA, eliminating any secondary structure in the larger 5′ region of the mRNA and recruiting a 40S ribosomal subunit with a methionine charged tRNA in the P site of the ribosome along with other eIFs. The resulting 43S to 48S initiation complex (size of the complex depends on how many eIFs are associated) then scans 3′ down the mRNA in search of the first AUG start codon. Upon pausing at the first start codon, many of the eIF proteins fall off the mRNA-ribosome complex allowing a 60S ribosomal subunit to associate in forming the 80S initiation complex. The 80S complex is thus poised to translate an open-reading frame, a contiguous stretch of codons, into the biosynthesis of a protein product.

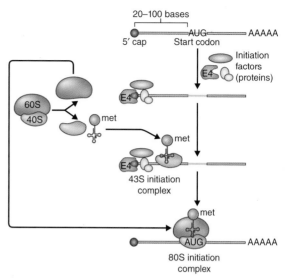

Figure 15.5 The initiation steps of mRNA translation. Initiation proteins, especially E4, recognize the 5′ terminal cap of an mRNA, and serve to recruit the small ribosomal subunit, the 40S subunit, along with the first tRNA carrying the methionine amino acid. This initiation complex then scans down the mRNA until it reaches the first start codon (AUG), which recruits a large ribosomal subunit, a 60S subunit, into the complex. The complex is now ready to initiate the elongation step of protein synthesis

Translation initiation in prokaryotes occurs by a different mechanism than that in eukaryotes. In prokaryotes, the small ribosomal unit binds to the mRNA at a specific nucleotide sequence, the Shine-Delgarno sequence. This sequence is found in virtually all prokaryotic mRNAs and positions the ribosomal small subunit at the start codon and poised for assembly with the ribosomal large subunit. One other distinguishing feature of prokaryotic mRNA translation is that it occurs simultaneously with transcription of the corresponding gene, largely because prokaryotes do not have intracellular compartments like the nucleus that can segregate the steps of transcription from translation.

15.3   Elongation

Once the 80S complex is formed at the start codon, the ribosome is set for the elongation phase (Figure 15.6). The ribosomal translation complex consists of the P site occupied by the initial methionine-charged tRNA or

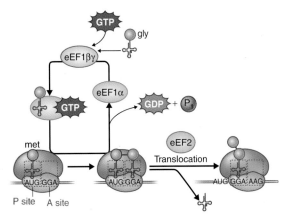

Figure 15.6 The elongation steps of mRNA translation. With the 80S initiation complex in position, elongation factor 1a facilitates the docking of the next amino acid charged tRNA into the A site. Following GTP hydrolysis and release of eEF1a, peptide bond formation between the nascent peptide chain in the P site is catalyzed with the amino acid in the A site. Elongation factor 2 (eEF2) then facilitates a translocation of the ribosome complex down the mRNA to reposition the A site over the next codon and to release the free tRNA from the P site. The cycle continues with eEF1a being reloaded with another aminoacyl-charged tRNA and GTP coordinated by a couple of additional eEFs

tRNA with a growing peptide chain (peptidyl tRNA), and the A site is unoccupied. Elongation factor eEF1 with GTP bound then helps position the next amino acid-charged tRNA into the A site such that the tRNA's anticodon can base pair with the codon of mRNA. The anticodon to codon base pairing in the A site in combination with eEF2 and GTP hydrolysis to GDP and inorganic phosphate (GTP is the high energy currency of protein synthesis, and functions in a way similar to the way that ATP provides energy to other reactions in cells) results in the release of energy and release of eEF1 from the ribosome complex. The released eEF1 is then reloaded by other eEFs with GTP and amino acid-charged tRNA. The released energy is also used to reposition the growing peptidyl-tRNA in the P site in proximity to the amino acid-charged tRNA in the A site. Concurrent with these conformational changes, the peptidyl transferase activity associated with the 60S ribosome catalyzes the transfer of the polypeptide associated with the tRNA in the P site to the amino acid bound to the tRNA in the A site. This transfer results in the formation of a peptide bond wherein the carboxy terminal end of the peptide in the P site is condensed with the amino terminal end of the amino acid in the

A site. The net result is that the growing peptide chain is now tethered to the tRNA in the A site. A translocation event then ratchets the peptidyl-tRNA from the A site to the P site, in turn releasing the free tRNA from the P site and repositioning the empty A site over the next codon to re-initiate the elongation cycle. The rate of the elongation cycle is estimated at 2–10 amino acids incorporated per second per translating ribosome, and elongation is generally not considered a rate-limiting step. Instead, initiation of translation is thought to be slower and limiting for overall protein synthesis rates.

15.4   Termination

When the 80S-nascent polypeptide complex reaches a stop codon, protein synthesis ceases and the polypeptide is released into the cytoplasm (Figure 15.7). The termination process, like initiation and elongation, is facilitated by several eRFs. eRF1 is a structural mimic of an amino acid-charged tRNA, but in fact does not contain an amino acid for transfer to the peptidyl-tRNA positioned in the P site. The eRF1 can also bind to all three stop codons, hence only one eRF1 form is necessary. The relative abundance of eRF1 in cells is not as large as some of the other translation factors, nonetheless this step in the termination process appears to occur quite readily and is not limiting. The complete dissociation of the ribosomal complex from the mRNA is also facilitated by a second releasing factor, eRF3,

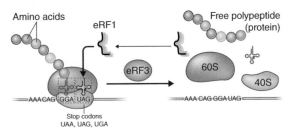

Figure 15.7 Termination of mRNA translation. When the 80S complex reaches a termination codon, instead of an aminoacyl-tRNA binding to the A site, the eukaryotic releasing factor (eRF1) binds to the stop codon. Because eRF1 does not contain an amino acid, the growing polypeptide chain is cleaved from the tRNA and released into the cytoplasm. With the assistance of another releasing factor (eRF3), the ribosomal complex dissociates releasing the tRNA, mRNA and ribosomal subunits to reinitiate another round of protein synthesis

which yields free tRNA, 40S and 60S ribosomal subunits to be recycled for another round of translation, and free polypeptide, which will have already initiated folding to its final three-dimensional conformation.

15.5   Tools for studying the regulation of translation

Many experimental studies seek to determine at what stage of gene expression a process is controlled. Is it via selective transcription or translation? For these reasons, investigators have sought and developed a variety of tools. For instance, Chapter 7 described the use of *α*-**amanitin** to inhibit RNA polymerase II, a tool for demonstrating the essentiality of mRNA synthesis in a biological process. The same is true for protein synthesis. **Cycloheximide** and **puromycin** are two chemical reagents that have been used extensively to inhibit protein synthesis in plants. Cycloheximide blocks the elongation step of protein synthesis, while puromycin acts as a tRNA mimic and cause premature chain termination (Michels *et al.*, 2000). Interestingly, these reagents are still being used to corroborate the importance of translational control on the expression of select genes during a developmental process or in response to an environment cue.

Another tool used to demonstrate the differential translation of mRNAs has been to determine the amount of free versus polysome bound forms of specific mRNAs (Bailey-Serres, 1999). Cytoplasmic mRNA does not have to be associated with any translational machinery and hence is known as free mRNA. Polysomes occur when an mRNA has successively initiated translation and thus has multiple ribosomes bound per mRNA. The exact mechanism regulating

polysome formation is not well understood, but it has been associated with the phosphorylation of specific ribosomal proteins in response to stress conditions like heat. The differential recruitment of mRNA into polysomes has also been reported for roots challenged with nitrogen-fixing, symbiotic bacteria versus non-challenged roots (Reynoso *et al.*, 2013).

15.6   Specific translational control mechanisms

Initiation of translation offers many opportunities for controlling the biosynthesis of a protein. As noted above, initiation commences with initiation factors recognizing the 5′ end of the mRNA and recruiting 40S ribosome binding. Hairpin structures within the 5′ UTR do in fact reduce the translation efficiency of mRNA, supporting the notion that the initiation complex serves to disrupt secondary structural elements in the mRNA and facilitates scanning of the mRNA for the start codon (Figure 15.8). In maize, the *Lc* gene is a member of the *R/B* transcription factor gene family important for expression of the anthocyanin biosynthetic pathway and the purple pigmentation in kernels. Simple sequence analysis of the normal *Lc* mRNA predicts two features not common in most mRNAs. First is a palindromic sequence that can form a hairpin structure near the 5′ end of the mRNA. Since double-stranded RNA is a more favorable form than single-stranded (See Chapter 6), such hairpin structures are predicted to form spontaneously and the relative likelihood of a hairpin existing is based on the energy input necessary to disrupt the hydrogen bonding of the base pairs within the palindrome (Figure 15.8). Using the tools of molecular genetics to

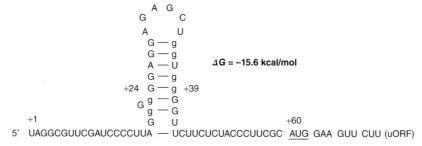

Figure 15.8   Illustration of the predicted palindrome that forms at the 5′ end of the maize *Lc* mRNA. Wang and Wessler (2001). Copyright 2001 American Society of Plant Biologists. Used with permission

Two levels of translational repression

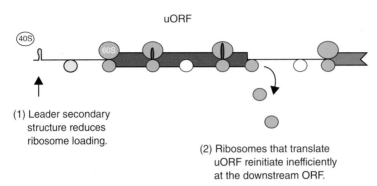

uORF

(1) Leader secondary
structure reduces
ribosome loading.

(2) Ribosomes that translate
uORF reinitiate inefficiently
at the downstream ORF.

Figure 15.9 A depiction of the 235 bases of the 5′ leader sequence of the *Lc* mRNA and the possible mechanisms controlling translation of the mRNA. The Lc protein is encoded by the open-reading frame depicted in green and is preceded by a hairpin palindrome at the 5′ end of the mRNA and a short ORF designated as uORF (upstream ORF). The role, if any, for the protein product derived from the uORF remains unknown. However, both the hairpin and uORF impact the relative translation efficiency of the *Lc* ORF. Wang and Wessler (2001). Copyright 2001 American Society of Plant Biologists. Used with permission

introduce mutations within test constructs of the putative *Lc* hairpin structure, elimination of this structure resulted in a 2-fold and almost 10-fold increase in the translation efficiency of the mRNA when tested *in vitro* and *in vivo*, respectively (Wang and Wessler, 2001).

A second feature of the *Lc* 5′ leader sequence thought to be important for translation efficiency is the short, upstream ORF (uORF) preceding the ORF for the Lc transcription factor (Figure 15.9). While the short protein encoded by the uORF has no known function, this ORF has been shown to reduce the amount of Lc protein synthesized by the ribosomes scanning down the *Lc* mRNA. Removal of the uORF from the 5′ leader sequence leads to a 2-fold increase in the *in vitro* translation of the downstream ORF, and greater than a 10-fold enhancement of the downstream translation product when mutant forms of mRNA are introduced into plant cells (Wang and Wessler, 1998).

While the examples above for the *Lc* 5′ leader sequence illustrate **suppressive translational control mechanisms**, other 5′ leader sequences and those particular to viral RNAs have been shown to enhance translation efficiency. The 68 nucleotides leader sequence of the TMV RNA (the omega sequence) and the 36 nucleotides leader of the alfalfa mosaic virus RNA were the first such **translation enhancer sequences** identified. When either of these sequences is appended to the 5′ end of many different mRNAs,

greater accumulation of the encoded proteins was observed in both *in vitro* and *in vivo* studies.

The 5′ leader sequences that stimulate and suppress expression appear to be structurally distinct from one another. Elements that suppress translation efficiency tend to possess secondary structure elements like inverted repeats that can form palindromes. Those 5′ elements that stimulate expression tend to consist of sequence specific elements. The TMV omega sequence consists of three copies of an 8 nucleotide repeat and a 25 nucleotide CAA-rich element. This suggests that these leader sequences might somehow stimulate translation via their ability to bind factors (proteins), which in turn stimulate overall translation efficiency.

Several other distinctive models for regulating the translation of mRNA are worth noting as well. The sequence 3′ to the stop codon of an mRNA (the 3′ UTR) can also influence the efficiency of translation. How does a 3′ tail influence events at the 5′ end of an mRNA? One possibility is that the 3′ tail loops back and interacts with the 5′ end of the mRNA, probably via the proteins binding these sequences (Figure 15.10). These RNA binding proteins might serve to recruit the actual translational machinery or could facilitate how quickly the machinery scans down an mRNA to find the first start codon and initiate translation.

A second somewhat unusual translational control mechanism has been reported for the α-amylase mRNA in barley. α-Amylase is produced in the

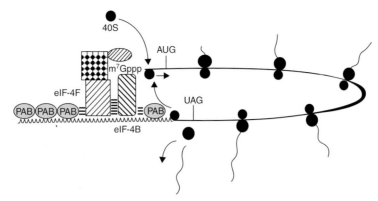

Figure 15.10 The co-dependent model for how the 5′ and 3′ UTR regions of an mRNA can influence the translation efficiency of an mRNA. In this example, proteins binding to the 3′ end, like polyA binding proteins (PABs), coordinate and interact with proteins binding in a sequence specific manner to the 5′ leader sequence, resulting in a more efficient recruitment of the translational machinery or scanning of the leader to the start codon (AUG). Adapted from Gallie (1996). Reproduced with permission of Springer Science + Business Media

aleurone cells of cereal grains during imbibition and secreted into the endosperm to mobilize the starch reserves to the growing embryo. Interestingly, the 5′ leader sequence of the barley α-amylase mRNA appears to stimulate this mRNA's translation only in aleurone cells (Gallie and Young, 1994).

Several other translational control mechanisms appear to be specific to viruses, including **frameshift skipping** and **stop codon suppression**. In frameshift skipping, the translational machinery will skip one or two bases forward or backward, thus changing the open reading frame. Stop codon suppression appears to be another example of this skipping phenomenon, wherein the translational machinery can skip over the stop codon and continue to read through the RNA. Neither of these skipping mechanisms has been reported for a nuclear encoded mRNA, but this could be because so few studies have sought this type of information.

One final factor that influences translation efficiency is **codon usage and bias**. Given the redundancy in the genetic code (Figure 15.2), more than one codon can call for the incorporation of the same amino acid into the growing peptide chain. Such synonymous codons are not equally represented in the open reading frames of a genome, hence there is a bias in their codon use. These biases correlate with the relative abundance of the respective tRNA species and the enzymes responsible for charging these tRNAs with their appropriate amino acids. Perhaps the best example of such codon usage biases in plants has been

reported for the expression of foreign genes in transgenic plants. In general, genes introduced from other kingdoms often have different codon biases than plant genes do and thus are poorly expressed. For instance, a human acetylcholinesterase gene (without introns) with human codon composition was introduced into a plant and its expression level compared with that of the same gene but with the codon composition optimized for dicots; expression of the codon optimized gene was 5- to 10-fold greater (Geyer *et al.*, 2007). This clearly demonstrates the potential for how codon usage could be utilized as a translational control mechanism. However, codon bias has only been identified as an important control mechanism in gender specific genes, genes expressed only in maternal versus paternal tissues. More specifically, genes expressed only in maternal or paternal tissues show distinctive codon usage biases (Whittle *et al.*, 2007). This is not exactly evidence for a codon bias mechanism controlling translation, but suggests that such a mechanism could be present and operating.

15.7   Summary

This chapter considered the major features of mRNA translation into protein and specifically the mechanisms of translation initiation, elongation and termination. With a basic understanding of the overall process, we then considered specific mechanisms that have been demonstrated to regulate translation. The importance of the 5′ and 3′ UTRs of mRNA as control points were noted because of both structural features (i.e., palindromes) and sequence specificities that could serve as binding sites for regulatory proteins. Both these features could affect assembly of the translational machinery on mRNA, the scan rate down an mRNA to the AUG start codon, and the efficiency of ORF translation and termination, including disassembly of the translational machinery. While there are a few examples of how such control mechanisms operate in plants, broad generalizations are not possible without additional research.

15.8   Problems

15.1   Assume a palindromic sequence has been associated with the relative translation efficiency of an mRNA. The first set of experiments introduced a mutation at one base in the stem structure. What would be your hypothesis for these experiments? Would you expect the mRNA translational rate to increase or decrease? Why? Assuming this first round of experiments is successful, and you have shown the palindrome does affect translation efficiency, can you think of another set of experiments to determine at what step in the translation process this structural feature has its effect?

15.2   One could imagine that a short ORF preceding the main ORF could enhance or inhibit the translation of the downstream ORF. Develop a cartoon model based on Figure 15.9 to explain this statement. Then provide a rationale for how such mechanisms might be beneficial to the host.

15.3   Congratulations on your new job with a new biotech company. Given your background in molecular genetics, your employer wants you to come up with a mechanism to control transgene expression *in planta* via a codon biased approach. Please describe your idea.

Further reading

Muench, D.G., Zhang, C. and Dahodwala, M. (2012) Control of cytoplasmic translation in plants. *Wiley Interdisciplinary Reviews: RNA*, **3**, 178–194.

Roy, B. and von Arnim, A.G. (2013) Translational regulation of cytoplasmic mRNAs. *Arabidopsis Book*, **11**, e0165.

References

Armache, J.P., Jarasch, A., Anger, A.M. *et al.* (2010). Cryo-EM structure and rRNA model of a translating eukaryotic 80S ribosome at 5.5-angstrom resolution. *Proceedings of the National Academy of Sciences USA*, **107**, 19748–19753.

Bailey-Serres, J. (1999). Selective translation of cytoplasmic mRNAs in plants. *Trends in Plant Science*, **4**, 142–148.

Browning, K.S. (1996) The plant translational apparatus. *Plant Molecular Biology*, **32**, 107–144.

Futterer, J. and Hohn, T. (1996) Translation in plants - rules and exceptions. *Plant Molecular Biology*, **32**, 159–189.

Gallie, D.R. (1996) Translational control of cellular and viral mRNAs. *Plant Molecular Biology*, **32**, 145–158.

Gallie, D.R. and Young, T.E. (1994) The regulation of gene-expression in transformed maize aleurone and endosperm protoplasts - analysis of promoter activity, intron enhancement, and messenger-RNA untranslated regions on expression. *Plant Physiology*, **106**, 929–939.

Geyer, B.C., Fletcher, S.P., Griffin, T.A. *et al.* (2007) Translational control of recombinant human acetylcholinesterase accumulation in plants. *BMC Biotechnology*, **7**: 27. doi:10.1186/1472-6750-7-27.

Michels, A.A., Kanon, B., Konings, A.W.T. *et al.* (2000) Cycloheximide- and puromycin-induced heat resistance: different effects on cytoplasmic and nuclear luciferases. *Cell Stress & Chaperones*, **5**, 181–187.

Reynoso, M.A., Blanco, F.A., Bailey-Serres, J. *et al.* (2013) Selective recruitment of mRNAs and miRNAs to polyribosomes in response to *Rhizobia* infection in *Medicago truncatula*. *Plant Journal*, **73**, 289–301.

Wang, L.J. and Wessler, S.R. (1998) Inefficient reinitiation is responsible for upstream open reading frame-mediated translational repression of the maize *R* gene. *Plant Cell*, **10**, 1733–1745.

Wang, L.J. and Wessler, S.R. (2001) Role of mRNA secondary structure in translational repression of the maize transcriptional activator *Lc*. *Plant Physiology*, **125**, 1380–1387.

Whittle, C.A., Malik, M.R. and Krochko, J.E. (2007) Gender-specific selection on codon usage in plant genomes. *BMC Genomics*, **8**: 169. doi: 10.1186/1471-2164-8-169.

Chapter 16
Protein folding and transport

16.1 The pathway to a protein's function is a complicated matter

As proteins are synthesized during the process of messenger RNA (mRNA) translation, two additional processes in assuring their proper functioning within the cell are their folding (Figure 16.1a) and intracellular addressing (Figure 16.1b). Protein folding includes the mechanisms by which different stretches of amino acids take on folds like α-helices or β-sheets, as well as how these secondary features then combine to provide unique three-diminsional structures (tertiary structures) that in turn allow the proteins to take on their specific catalytic or structural functions. Function in this context also means having proteins play their roles in the correct intracellular compartments. Plant cells, like all eukaryotic cells, have a complex organization due to the different compartments within the cells. These compartments are often delimited by membranes; for example, mitochondria and chloroplasts are separate spaces performing distinctive functions like energy generation and photosynthesis, respectively. The sequestration and separation of these functions likely minimize competition for resources between biochemical pathways, and hence may provide a selective advantage for plant cells. A fundamental challenge for any eukaryotic cell is how to direct proteins to specific compartments so those proteins can participate in select functions.

Localization has other particularities, like aligning proteins to surfaces and assuring they face a particular environment. A good example is the plasma membrane, the membrane that surrounds the cell, separating one cell type from another, and associating cells with specific capabilities to a tissue or organ, like the mesophyll cells responsible for photosynthesis in a leaf.

16.2 Protein folding and assembly

A key observation of protein folding was that certain types of amino acid sequence fold spontaneously, now commonly called self-assembly. This is especially true for the generation of secondary structures like **α-helices** and **β-sheet** configurations (Figure 16.2). This sort of folding occurs spontaneously because the sequences of amino acids can interact to stabilize the particular folds by hydrogen bonding. These hydrogen bonds drive the formation of folds because these structures are lower energy forms of the proteins and thus, are the thermodynamic impetus for their formation.

However, the formation of primary folds is not sufficient for a protein to take on its mechanistic functions, like catalyzing a chemical reaction or serving as a physical scaffold. It is the "folding of folds" and oligomerization (assembly of two or more proteins into a higher order structure) that gives proteins

Plant Genes, Genomes and Genetics, First Edition. Erich Grotewold, Joseph Chappell and Elizabeth A. Kellogg.
© 2015 John Wiley & Sons, Ltd. Published 2015 by John Wiley & Sons, Ltd.
Companion Website: www.wiley.com/go/grotewold/plantgenes.

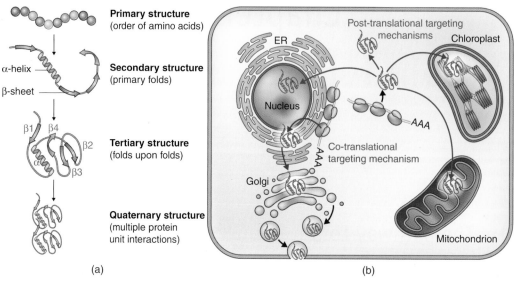

Primary structure
(order of amino acids)

Secondary structure
(primary folds)

α-helix
β-sheet

β1 β4
β2
α
β3

Tertiary structure
(folds upon folds)

Quaternary structure
(multiple protein
unit interactions)

(a)

Post-translational targeting
mechanisms
Chloroplast
ER
Nucleus
AAA
Co-translational
targeting mechanism
Golgi
Mitochondrion

(b)

Figure 16.1 As an mRNA is translated into a protein, the protein must fold in order to take on its specific functions (a), be it catalytic or structural. The first level of folding is often spontaneous based on the amino acid sequence within a region or domain. Tertiary structures can form spontaneously, but may also be facilitated by other proteins helping to bring the primary folds into proper interactions with one another. Most proteins require tertiary folding to become physiologically active. Quaternary structures of proteins are the highest level of organization that proteins can take on. Quaternary structures may be composed of two or more of the same protein (homopolymers), or two or more different proteins (heteropolymers). Then these properly folded proteins must get to the proper cellular locale where they play their particular function. The addressing or trafficking of proteins to their proper destinations can actually occur during mRNA translation (a co-translational event) or after the protein has been released into the cytoplasm (a post-translational event) (b)

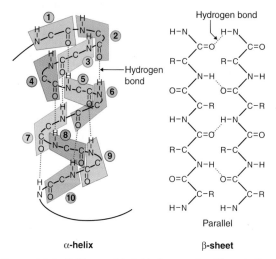

Hydrogen bond

Hydrogen
bond

α-helix

β-sheet

Parallel

Figure 16.2 Self-assembly folds in proteins like α-helices and β-sheets form spontaneously based on the amino acid sequence and ability of these structures to form hydrogen bonds (H-O). Hydrogen bonds are weak, electromagnetic bonds often in the case of proteins between the electronegative oxygen of the carbonyl group and the electropositive hydrogen of the amine

their distinctive attributes allowing them to perform equally distinct physiological functions. Improperly folded proteins are essentially unfolded or denatured proteins that can have serious consequences. We do not often observe this in plants, but improperly folded proteins in animals can lead to motor and cognitive dysfunctions. This assembly process, the same one that also recycles unfolded, denatured and indiscriminately aggregated proteins, eluded attention until it was noted that these assembly/folding mechanisms help organisms contend with adverse environmental conditions, something plants must do all the time for temperature, water and other biotic stresses. Equally important were observations that mutation in these processes also altered normal functions, so these mechanisms are not simply important for adverse situations. Perhaps more eye-opening has been the realization that these mechanisms are key to evolutionary processes (Tokuriki and Tawfik, 2009).

Figure 16.3 illustrates a general or generic model associated with folding processes (Boston *et al.*, 1996; Ellis, 2013). Proteins originally referred to as

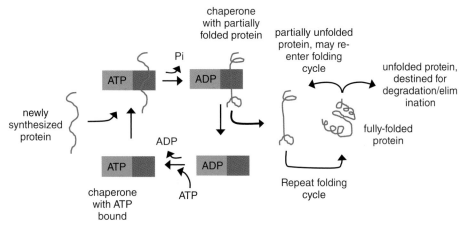

Figure 16.3 Secondary structure folding of proteins is facilitated by chaperone proteins. A newly synthesized protein can associate with a chaperone-ATP complex that then hydrolyzes ATP to ADP and inorganic phosphate (Pi), which in turn provides the energy for one iteration of protein folding. The complex may release the protein with additional secondary structure folds. The chaperone can then be recycled via an ADP for ATP exchange that repositions the chaperone for another round of folding. Adapted from Miernyk (1999)

chaperones can bind newly synthesized polypeptides, which along with ATP binding and other accessory proteins lead to ATP hydrolysis, followed by release of all the interacting components and yielding a more highly ordered protein. The ordered protein can go on to provide its physiological function, or it can be subject to iterative cycles of folding to yield the mature, active form. Under some conditions, the properly folded protein can unfold or aggregate, becoming physiologically inactive. These latter observations initially led to the discovery of chaperones. Investigators observed the induction and accumulation of several distinct protein species when a target organism was grown at elevated temperatures. These **heat-shock induced proteins** were given designations like HSP70 and HSP90, recognizing their size (i.e., 70 000 Da). Further studies have documented how ubiquitous these heat- or other environmentally induced proteins are and ultimately have correlated these with more general roles in protein folding/assembly.

Table 16.1 lists the various classes of **chaperones** and **foldases** found in plants to date. The greatest distinction between these classes is that the foldases like peptidyl prolyl isomerase have particular catalytic activities. Prolyl isomerase catalyzes the isomerization between the *cis* and *trans* configuration of proline residues in proteins (Schmid, 1993) (Figure 16.4). In contrast, catalytic activities have not been directly attributed to the chaperone and co-chaperone proteins

themselves, and these appear to work with other proteins to facilitate the proper folding of polypeptides.

The association of chaperones with heat shock and other biotic and abiotic stresses suggest that these proteins might participate in some sort of innate defense mechanisms. However, they are also essential for overall protein folding and assembly (Quinlan and Ellis, 2013). This is evident from observations that mutations within several of the *HSP* genes are lethal. This means that researchers have not been successful in isolating plants harboring knockout type mutations in several of the *HSP* genes. The lethality of these mutations is thus taken as evidence that these proteins play a normal role in protein folding and assembly, as well as having special function under certain stress conditions.

Chaperones and foldases are found in the cytoplasm and within organelles such as mitochondria, chloroplasts, endoplasmic reticulum (ER) and the nucleus. If one recalls that mitochondria and chloroplasts have genomes themselves and all the machinery necessary for gene expression and mRNA translation, it makes perfect sense that these organelles would import nuclear encoded HSPs to support protein translation. Perhaps somewhat more surprising is the finding that chaperones and foldases are also found in association with ER compartments and the nucleus, which speaks to the general importance of protein folding in all the cellular compartments.

Table 16.1   Example of foldase and chaperone classes in plants and their intracellular localization.

Protein class	Intracellular location
Chaperones	
Clp proteins	
HSP100 (ClpB)	Cytoplasm, Mitochondria
ClpA/C	Chloroplast
HSP90	
HSP80/90	Cytoplasm
GRP941	Endoplasmic Reticulum
HSP70	
HSP/HSC70	Cytoplasm, Nucleus, Chloroplast, Mitochondria
BiP/GRP78	Endoplasmic Reticulum
Chaperonins	
HSP60/Cpn60	Chloroplast, Mitochondria
Small HSPs	Cytoplasm, Mitochondria, Chloroplasts, Endoplasmic Reticulum
Co-chaperones	
HSP40	Cytoplasm, Mitochondria, Endoplasmic Reticulum
Cpn 10 (Cpn 60)	Mitochondria, Chloroplast
Foldases	
Protein disulphide isomerase	Endoplasmic Reticulum
Peptidyl prolyl isomerase	
Cyclophilin	Cytoplasm, Endoplasmic Reticulum, Mitochondria
FK506 binding protein	Cytoplasm, Endoplasmic Reticulum

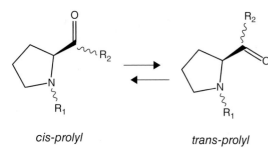

cis-prolyl   trans-prolyl

Figure 16.4 Illustration of *cis* and *trans* configurations of proline residues within a polypeptide chain. Isomerization between the forms occurs spontaneously, but can be controlled by specific prolyl isomerase enzymes. R1 and R2 refer to links with other amino- and carboxy-terminal amino acids within the protein

16.3   Protein targeting

Finding nuclear-encoded proteins, like the HSPs, targeted to organelles and membrane systems raises another important question about how targeting might occur and still maintain the proper fold and configuration of a protein for its activity (Heinig *et al.*,

2013). Figure 16.1b provides some perspective on this issue. First, the majority of nuclear genes generate mRNAs translated by "free" polysomes in the cytoplasm to yield "cytosolic" targeted proteins. If these proteins lack addressing signals, then they simply stay in the cytoplasm. However, if they contain addressing information, then they are targeted to a particular organelle or membrane. And this is exactly the premise that led to the discovery of leader or targeting signal sequences, amino terminal sequences which direct proteins to specific organelles or their insertion into select membranes.

16.4   Co-translational targeting

If a protein is targeted to the lumen side of the ER, or destined to be secreted, the mRNA for such proteins initially engages with the translational machinery assembled as free ribosomal complexes (Figure 16.5). Then as the nascent amino terminal peptide emerges,

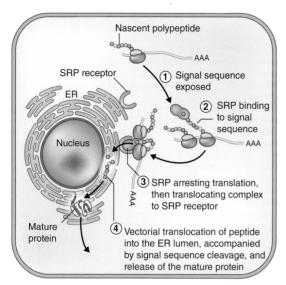

Figure 16.5 Steps in the co-translational targeting of proteins to the ER. Once the signal recognition particle (SRP) recognizes the emerging signal sequence polypeptide, SRP binds and facilitates migration of the complex to the SRP receptor bound to the ER membrane. Translation is then re-initiated, the targeted protein is vectorially (amino-terminal end of the emerging protein) directed across the membrane and into the lumen of the ER. As the polypeptide emerges on the lumen side of the ER, a protease cleaves the signal sequence releasing the mature (fully processed) form of the protein into the lumen space, where it must fold into its proper configuration

the **N-terminus signal sequence** is revealed and recognized by the **signal recognition particle** (SRP). Binding of the SRP halts translation, allowing the SRP-ribosomal complex to find its way and bind to a SRP receptor found on cytosolic face of the ER membrane. When translation re-initiates, the leader sequence directs the emerging protein through a protein complex spanning the ER membrane. As the newly synthesized leader sequence emerges on the lumen side of the ER, it is proteolytically cleaved from the remainder of the protein, which must then undergo folding to become fully processed to its mature form. If the protein is targeted for secretion, a dedicated vesicle system progresses from the ER through the Golgi system to be ultimately packed, then by a bulk flow mechanism the protein flows through the ER and Golgi systems and is ultimately packaged into secretory vesicles that can fuse with the plasma membrane, releasing the protein cargo to the

extracellular environment (Figure 16.1b). If the protein is to be retained in the lumen of the ER or Golgi apparatus, additional signal sequences are needed. For example, a **KDEL** (amino acids lysine, asparatic acid, glutamic acid and lysine) sequence found at the carboxy terminus of a protein serves as a retention signal for the protein within the ER compartment.

16.5   Post-translational targeting

Many nuclear-encoded proteins synthesized in the cytoplasm on free ribosomes have their ultimate function within organelles like mitochondria or chloroplasts. Their movement is also facilitated by amino-terminal signal sequences, but via a post-translational mechanism different from the co-translational mechanism (Figure 16.6). Post-translational targeting is initiated by binding of a chaperone to the distinct amino terminal signal sequence. The chaperone ferries the target protein to the surface of the target organelle, delivering the precursor protein to a channel traversing the

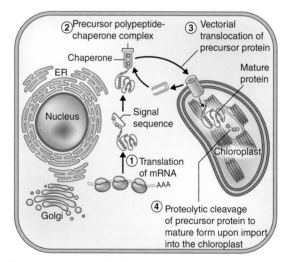

Figure 16.6 Steps in the post-translational targeting of proteins to the chloroplast. Once a precursor protein is synthesized, its signal sequence directs its movement to and into a particular organelle based on its signal sequence. Upon import, the signal sequence is removed by proteolytic processing and the mature form of the protein folds into its active configuration

Table 16.2 Summary of the different transport mechanisms targeting proteins to organelles (Adapted from Krebs *et al.*, 2010).

Organelle	Signal type	Transport mechanism	Accessory factors
Nucleus	Basic, internal peptide	Escort carrier	Nuclear pore complex
ER	N-terminus peptide	Delivered by cytoplasmic chaperone, fed through channel	Membrane translocases
Mitochondria	N-terminal amphipathic helix peptide	Delivered by cytoplasmic chaperone, fed through channel	Membrane translocases
Chloroplasts	N-terminal charged peptide	Delivered by cytoplasmic chaperone, fed through channel	Membrane translocases
Microbodies (Peroxisomes)	C-terminal peptide	Delivered by shuttled receptor	Receptors and translocases
Vacuole	Multiple, N-terminal and internal peptides	Golgi and non-Golgi dependent	Receptors and translocases

membrane. The precursor protein is fed through the channel, amino-terminus first, and as the signal sequence emerges on the inside of the organelle, the leader sequence is cleaved from the rest of the protein. Once the processed protein is released into the organelle compartment, it must then fold into its proper configuration, possibly facilitated by another chaperone, in order to become biologically active.

Table 16.2 provides a summary of the different mechanisms for transporting proteins into organelles. While the mechanisms for chloroplast and mitochondria are similar, the signal peptide sequence determines whether the protein goes to a chloroplast or a mitochondrion. Proteins targeted to the nucleus and microbodies differ significantly in the specific targeting signal sequence, its locale within the primary sequence (amino terminus versus carboxy terminus), and whether or not the signal sequence is retained in the fully mature proteins.

16.6 Post-translational modifications regulating function

While protein folding and targeting are essential for proper functioning of each protein, protein functions can be modulated by other **post-translational modifications** such as **phosphorylation, glycosylation, prenylations, acylation, acetylation, methylation, and other modifications** as well (Table 16.3 and Figure 16.7). Many of these modifications, in contrast to folding or intracellular targeting, are readily reversible, and hence can be used to regulate the catalytic activity of a protein or its physiological function. Phosphorylation is perhaps the best known of these mechanisms and has been documented as controlling enzyme activities and signal transduction cascades (del Pozo *et al.*, 2004; Bai *et al.*, 2007; Oh and Martin, 2011) in plants as well as many other organisms.

One of the best examples of phosphorylation playing an essential role in plants is that associated with disease resistance (del Pozo *et al.*, 2004; Oh and Martin, 2011) (Figure 16.8). In studies of gene-for-gene interactions between bacterial pathogens and host plant defense responses, bacteria challenging a plant cell may present an avirulent factor or Avr that is recognized by the plant cell resulting in an induced resistance response. The induced resistance response often ends with a programmed cell death phenotype. A particularly well studied example of this phenomenon is how the recognition response of tomato to a *Pseudomonas* is mediated by the activation of a specific protein kinase, Pto, an enzyme that in turn

Table 16.3 Examples of reversible post-translational modifications of plant proteins

Modification	Modifier	Donor	Residue
Phosphorylation	PO_3	ATP	Serine, threonine, tyrosine
Glycosylation	Sugar residue (i.e., glucose, N-acetylglucoamine)	UDP/GDP-sugar	Serine, threonine
Prenylation	5 carbon prenyl group or multiples of this	DMAPP, IPP, FPP, GGPP	cysteine
Acetylation	CH_3CO	Acetyl-CoA	lysine
Methylation	CH_3	S-adenosyl-methionine	lysine

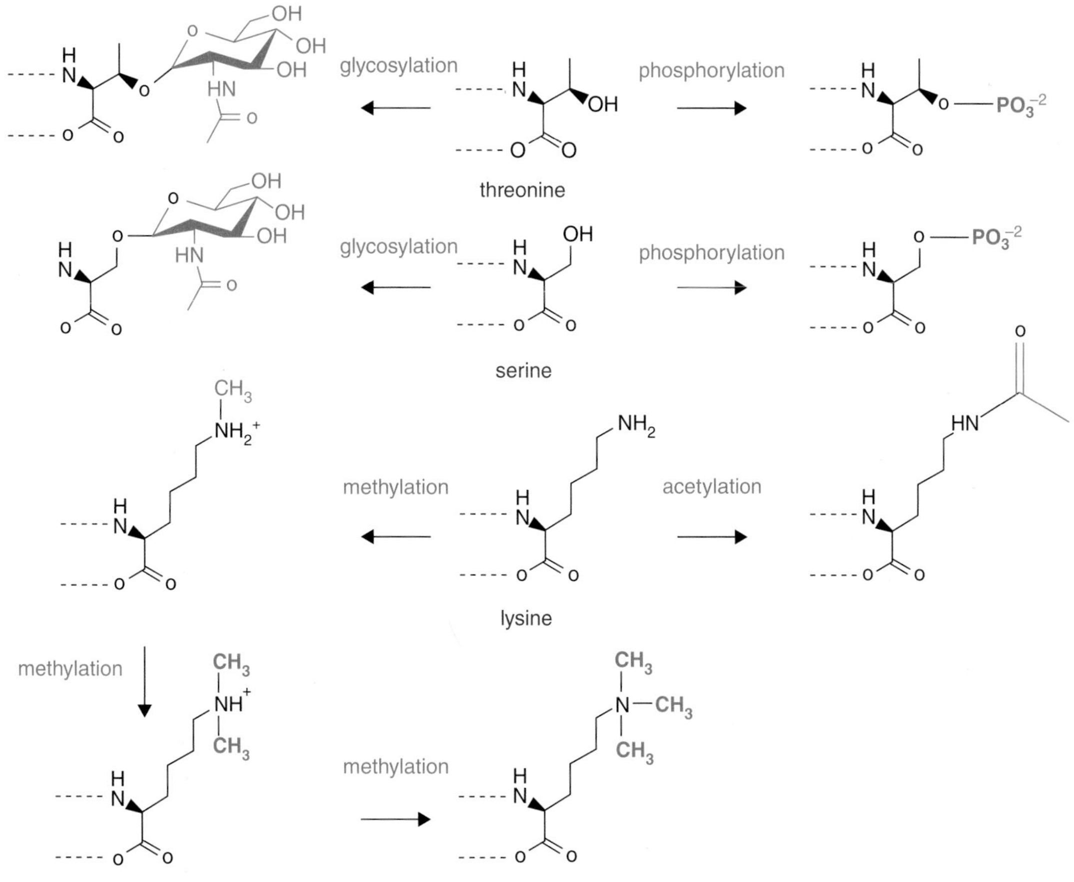

Figure 16.7 Illustrations of a few of the many possible reversible post-translational modifications of proteins occurring in plants. The particular amino acid residue within a larger polypeptide chain and the chemical modification associated with each are depicted

catalyzes the phosphorylation of another protein kinase. This kinase phosphorylates another kinase, which phosphorylates another. The phosphorylation of these particular kinases is referred to as a **MAP** (named after the first such kinase identified in animals as being <u>m</u>itogen <u>a</u>ctivated <u>p</u>roteins) **kinase cascade** and can be envisioned as a means for amplifying the initial recognition event (see also Chapter 10). Equally intriguing is the role other proteins play in these complicated processes. Proteins known as **14-3-3 proteins**

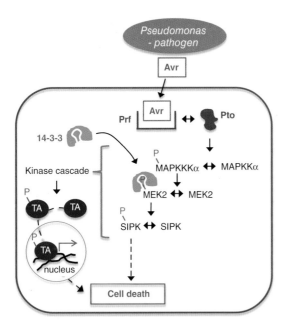

Figure 16.8 Illustration of a protein phosphorylation cascade proceeding from recognition of a potential pathogenic bacterium (via its Avr protein) by a plant cell receptor (Prf), that then activates the host cell protein kinase PtO. The "activated" PtO initiates an activation cascade of successive MAP kinases which phosphorylate one another, ultimately leading to a programmed cell death outcome that limits pathogen spread. The 14-3-3 proteins are resident within the plant cell and serve to lengthen the stability and longevity of the specific phosphorylated kinases, hence enhancing the host cell response to the pathogen. Adapted from Olga *et al.*, 2004, and Oh and Martin, 2011

can bind to specific phosphorylated proteins, stabilizing the phosphorylation state and thus enhancing the enzyme's kinase activity. A particular 14-3-3 protein in tomato was recently identified by its ability to bind to two of the MAP kinases activated by pathogen challenge, and mutational silencing of the gene encoding this particular 14-3-3 protein compromised the host plant's resistance response. Additional details for the subsequent events leading to the host cell's death are still being worked out, but there is evidence that the phosphorylation of *trans*-acting factors in the cytoplasm directs these proteins to the nucleus where they may alter the transcription of specific genes orchestrating the cell death program.

Phosphorylation and 14-3-3 proteins are important for other developmental programs besides plant–pathogen interactions and are commonly

used by plants for the regulation of metabolic enzymes (Bai *et al.*, 2007). However, such post-translational modification may act positively as well as negatively. Brassinolides, for example, are growth regulators associated with stem elongation, senescence and seed germination. These hormones are perceived by cell surface receptors that lead to the phosphorylation of cytoplasmic proteins that then migrate to the nucleus to activate specific gene expression patterns. Another set of specific 14-3-3 proteins has been discovered that can bind to these phosphorylated cytoplasmic proteins and impede their movement into the nucleus. Hence, these 14-3-3 proteins act as suppressors of brassinolide signaling. In contrast, the 14-3-3 proteins associated with plant–pathogen interactions facilitate the host response.

Histone acetylation and **methylation** are another set of post-translational modifications that have attracted significant attention for their role in the control of epigenetic events as was discussed in Chapter 10. And like other post-translation regulatory mechanisms, the specifics of methylation are more complicated than first imagined. Lysine residues can be mono-, di- and tri-methylated (Figure 16.7) and there are suggestions that the degree of histone methylation can serve to integrate metabolic states of plants with gene expression as illustrated in Figure 16.9.

16.7 Summary

While we have often noted in other chapters how mutations associated with a particular step in gene expression have revealed unique or unusual findings, such is not the case for the mechanism of protein folding and correct targeting. Mutations in these mechanisms appear to be lethal, or at least to be so severely handicapping that significant growth is not observed. As such, these mechanisms and especially those associated with folding of proteins via chaperones and foldases might contribute to an evolutionary trajectory influencing the evolution of the associated proteins (Tokuriki and Tawfik, 2009). In contrast, post-translational modifications of proteins are more associated with the modulation of protein function and thus serve as adaptations giving plant cells flexibility to contend with changing biotic and abiotic conditions.

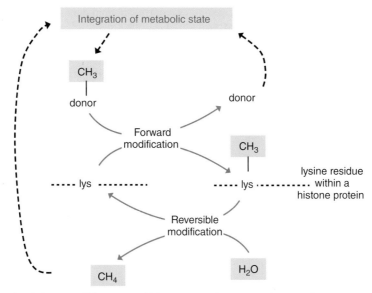

Figure 16.9 An example of the reversible nature of histone methylation. Monitoring of environmental cues and metabolic status of cells within an organism can result in mono-, di- and trimethylation of specific lysine residues within histone proteins, and this can have a direct effect on the expression of the gene(s) associated with these modifications. Histone methylation is dynamic and the methylation status of the histones associated with a gene is variable based on environmental and physiological cues. Adapted from Prabakaran *et al.* (2012). Reproduced with permission of John Wiley & Sons, Inc

16.8   Problems

16.1 Draw a cartoon diagram of a heterodimeric enzyme with one subunit consisting of three α-helices and the other subunit consisting of two antiparallel β-sheets.

16.2 Are α-helices or β-sheets predicted for the amino acid sequence given below? Use an internet accessible secondary structure prediction program such as http://www.compbio.dundee.ac.uk/www-jpred/index.html to predict the secondary structure fold(s) associated with the following protein sequence: MASAAVANYEEEIVRPVA-DFSPSLWGDQFLSFSIDNQVAEKYAKEIEAL-KEQTRNMLLATGMKLADTLNL.

16.3 Chaperones are said to facilitate protein evolution. How might a chaperone, a protein that facilitates folding of another protein, facilitate the evolution of a gene coding for a protein with one function to that coding for a protein with a new function (neofunctionalization)? Assume the gene undergoing evolution is present in the plant genome in multiple copies and only one copy of the gene is undergoing evolution.

16.4 Why would post-translational modifications like phosphorylation be an important control mechanism for protein function in environmentally stressed (i.e., water stress) plants?

Further reading

Chen, Z.J. (2013) Genomic and epigenetic insights into the molecular bases of heterosis. *Nature Reviews Genetics*, **14**, 471–482.

Kotak, S., Larkindale, J., Lee, U. et al. (2007). Complexity of the heat stress response in plants. *Current Opinion in Plant Biology*, **10**, 310–316.

Kwon, S.J., Choi, E.Y., Choi, Y.J. et al.. (2006) Proteomics studies of post-translational modifications in plants. *Journal of Experimental Botany*, **57**, 1547–1551.

Liu, C.Y., Lu, F.L., Cui, X. and Cao, X.F. (2010) Histone methylation in higher plants. *Annual Review of Plant Biology*, **61**, 395–420.

Malapeira, J., Khaitova, L.C. and Mas, P. (2012) Ordered changes in histone modifications at the core of the Arabidopsis circadian clock. *Proceedings of the National Academy of Sciences USA*, **109**, 21540–21545.

Miernyk, J.A. (1999). Protein folding in the plant cell. *Plant Physiology*, **121**, 695–703.

Oh, C.S., Pedley, K.F. and Martin, G.B. (2010) Tomato 14-3-3 protein 7 positively regulates immunity-associated programmed cell death by enhancing protein abundance and signaling ability of MAPKKK α. *Plant Cell*, **22**, 260–272.

Prabakaran, S., Lippens, G., Steen, H. and Gunawardena, J. (2012). Post-translational modification: Nature's escape from genetic imprisonment and the basis for dynamic information encoding. *Wiley Interdisciplinary Reviews-Systems Biology and Medicine*, **4**, 565–583.

Yuan, L., Liu, X., Luo, M. *et al.* (2013) Involvement of histone modifications in plant abiotic stress responses. *Journal of Integrative Plant Biology*, **55**, 892–901.

References

Bai, M.Y., Zhang, L.Y., Gampala, S.S. *et al.* (2007) Functions of OsBZR1 and 14-3-3 proteins in brassinosteroid signaling in rice. *Proceedings of the National Academy of Sciences USA*, **104**, 13839–13844.

Boston, R.S., Viitanen, P.V. and Vierling, E. (1996) Molecular chaperones and protein folding in plants. *Plant Molecular Biology*, **32**, 191–222.

del Pozo, O., Pedley, K.F. and Martin, G.B. (2004) MAPKKK alpha is a positive regulator of cell death associated with both plant immunity and disease. *EMBO Journal*, **23**, 3072–3082.

Ellis, R.J. (2013) Assembly chaperones: a perspective. *Philosophical Transactions of the Royal Society B: Biological Sciences*, **368**, 20110398.

Heinig, U., Gutensohn, M., Dudareva, N. and Aharoni, A. (2013) The challenges of cellular compartmentalization in plant metabolic engineering. *Current Opinion in Biotechnology*, **24**, 239–246.

Krebs, J.E., Goldstein, E.S. and Kilpatrick, S.T. (2010) *Lewin's Essential Genes*, Jones and Bartlett, Boston.

Oh, C.S. and Martin, G.B. (2011) Tomato 14-3-3 protein TFT7 interacts with a MAP Kinase Kinase to regulate immunity-associated programmed cell death mediated by diverse disease resistance proteins. *Journal of Biological Chemistry*, **286**, 14129–14136.

Quinlan, R.A. and Ellis, R.J. (2013) Chaperones: needed for both the good times and the bad times. *Philosophical Transactions of the Royal Society B: Biological Sciences*, **368**, doi: 10.1098/rstb.2013.0091.

Schmid, F.X. (1993). Prolyl isomerase - enzymatic catalysis of slow protein-folding reactions. *Annual Review of Biophysics and Biomolecular Structure*, **22**, 123–143.

Tokuriki, N. and Tawfik, D.S. (2009) Chaperonin overexpression promotes genetic variation and enzyme evolution. *Nature*, **459**, 668–U671.

Chapter 17

Protein degradation

17.1 Two sides of gene expression – synthesis and degradation

Chapter 16 describes several mechanisms for how post-translational modifications to proteins can alter their function or enzymological activities. However, post-translational regulation is a much broader topic and includes the mechanisms that control the actual degradation of a protein back to its component amino acids. If one thinks of protein functions as the outcome of gene expression, then both the synthesis and degradation of specific proteins are equally important for any phenotypic outcomes. Consider the development of a cell within an organ or tissue. At its inception, a recently divided cell will lack all the proteins necessary for its physiological function upon maturation. And as we have learned throughout this book, a big part of maturation and responding to developmental and environmental cues is the activation of selective gene transcription followed by the translation of the corresponding mRNA. So the complexion of proteins in cells is always changing. The paradigm of gene expression is paralleled by an equally challenging puzzle, how does a cell contend with proteins that have fulfilled their roles during a developmental or response phase? Do these cells have mechanisms that can selectively eliminate a protein or proteins associated with a function no longer needed?

These questions fall into an even larger arena of investigation. As noted in Chapter 16, proteins must fold properly in order to be functional. So, how do cells contend with proteins that inadvertently unfold or become non-functional? Scientists studying all kinds of organisms, including plants, have pondered such questions for quite some time. Unfortunately, the tools for studying protein turnover, degradation or catabolism are not as advanced as other areas of genetic research. However, recent findings have suggested that basic mechanisms are responsible for the regulated degradation of protein. Autophagy, apoptosis and senescence are terms referring to the more general processes leading to programmable degradation of a subset or all cellular proteins. The key here is the programmed nature of these processes. Selective protein degradation is often mentioned in the context of autophagy as well as in cells responding to particular signals or cues. The mechanisms associated with such selectivity are, however, better referred to as protein-tagging mechanisms.

17.2 Autophagy, senescence and programmed cell death

Senescence, autophagy and programmed cell death are common occurrences in plants. Cotyledons undergo programmed cell death during the course of seed

Plant Genes, Genomes and Genetics, First Edition. Erich Grotewold, Joseph Chappell and Elizabeth A. Kellogg.
© 2015 John Wiley & Sons, Ltd. Published 2015 by John Wiley & Sons, Ltd.
Companion Website: www.wiley.com/go/grotewold/plantgenes.

germination when stored reserves are mobilized to the growing embryo. Leaves on deciduous plants undergo senescence prior to leaf drop, a process that is thought to recycle macromolecules, such as proteins, to their constituent amino acids and export them to growth zones throughout the plant. Xylem development, the vasculature required for water movement in plants, entails a program wherein the intracellular contents of cells destined to becoming the xylem elements are broken down and recycled, a process referred to as catabolism.

Many, if not all, of these processes are correlated with protease activities. **Proteases** are enzymes that cleave peptide bonds either in an exo or endo fashion. Exo-proteases clip single amino acids one at time from the termini of proteins and can be distinguished as amino- or carboxy-proteases, depending on which end of the protein they work on. Endo-proteases are those that can cleave in the middle of the protein; these often prefer to cleave near a specific amino acid (Table 17.1). Serine proteases constitute the largest family of proteases in plants with well over 200 copies of these genes in each of the Arabidopsis and rice genomes, and these proteases cleave peptide backbones near serine amino acids. Distinct serine protease activities have been documented within organelles of tissues undergoing senescence. For instance, when the flag leaf of wheat (the uppermost leaf) begins senescence, serine protease activity is associated with the degradation of the photosynthetic machinery inside the chloroplast compartment. Aspartate proteases are the next largest family of these hydrolytic enzymes and utilize the acid side chain of aspartate residues to facilitate peptide bond cleavage. The aspartate proteases have also been associated with nitrogen recycling in nutrient limited plants.

Cysteine protease activities are frequently induced in senescing tissues and associated with the vacuole. Vacuoles are the lytic compartments of plant cells and contain a range of enzymes capable of recycling all the macromolecules found in a cell – protein, RNA, lipids, and so on. The role of vacuoles in autophagy is well appreciated from micrographs showing partially degraded organelles like mitochondria and chloroplasts engulfed in the vacuole.

Autophagy and senescence are also associated with the orderly breakdown of other macromolecular structures and molecules (Figure 17.1). Fragmentation of the genomic DNA found in the nucleus is yet another early sign of the onset of programmed cell death.

Table 17.1   Plant senescence and programmed cell death are associated with distinct classes of proteases. Some of these protease activities are induced by distinct treatments and are compartmentalized to unique organelles. Adapted from Roberts *et al.* (2012)

Family/member	Localization	Inductive treatment
Serine proteases carboxypeptidase	Vacuole	Senescence, girdling
ClpD	Chloroplast stroma	Senescence, dark, girdling
subtilisins	—	Senescence, dark, nitrogen
Aspartate proteases OsAsp1	—	Senescence, auxin, cytokinin
CND41	Chloroplast	Senescence, girdling, nitrogen
Cysteine proteases cathepsin	Vacuole, cytoplasm, secreted	Senescence, dark, girdling, programmed cell death
Papain-like	—	Senescence, girdling, nitrogen
metalloproteases FtsH	—	Senescence, nitrogen

17.3   Protein-tagging mechanisms

In contrast to the wholesale degradation of cellular contents by the autophagy/senescence programs, the selective degradation of proteins and enzymes appears to be mediated by a distinct class of proteases and an oligomeric structure, the proteasome, housed in the cytoplasm (Figure 17.2). The most distinguishing feature of this catabolic process is the "tagging" of proteins destined for this degradation pathway. In contrast to the post-translational modifications like phosphorylation or methylation, which are small chemical constituents, the tags used to mark a protein destined for turnover are small polypeptides of ~72–116 amino acids, discovered about 35 years ago (Table 17.2). The first and most common of these peptide tags is known as **ubiquitin**. While we now know that ubiquitin is found all across nature (hence it is ubiquitous, leading

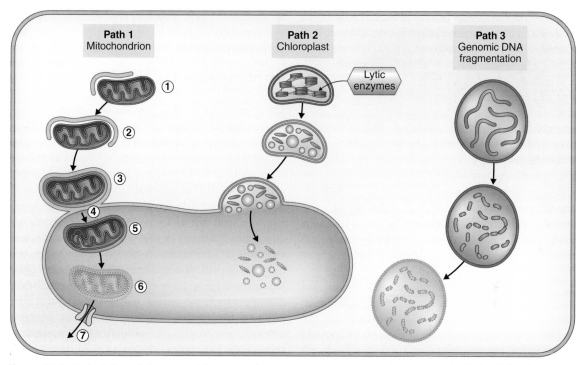

Figure 17.1 A depiction of the general features of autophagy, senescence and programmed cell death in plants. Path 1 illustrates the steps (1–6) of how an organelle like a mitochondrion is taken up by the vacuole (large blue structure) for destruction. Path 2 shows how chloroplast destruction commences by the insertion of lytic enzymes that begin the degradation of intra-organelle macromolecules like chlorophyll, prior to its fusion with the vacuole for complete recycling of its contents. Path 3 shows how the nucleus of the cell undergoes a fragmentation of the genomic DNA prior to its complete destruction

to its name), other protein tags have been identified recently as well. And interestingly and perhaps not too surprising, all have similar three-dimensional folds (Figure 17.3).

The ubiquitylation of target proteins occurs by the coordinated activity of three highly conserved proteins, **E1, E2** and **E3** (Figure 17.2). In a relatively complex cascade of events, E1 serves to activate ubiquitin and to pass on the ubiquitin tag to E2. This reaction hydrolyzes ATP to generate the energy needed to drive the reaction. The conjugation is between a cysteine residue of E2 and the carboxy-terminal glycine of ubiquitin. E3 is by far the most dominant and important player in the ubiquitin cascade because it is the enzyme responsible for transferring and conjugating ubiquitin from E2 to the target protein, which it may do as a direct transfer from E2 to the target or via acquiring the ubiquitin itself as an intermediate during the transfer process. E3 is a ligase enzyme, meaning it

transfers ubiquitin and ligates it to the target protein. Ubiquitin is thus covalently attached to the target protein, which is hence marked for degradation. Not surprising, the number of genes encoding for the E1, E2 and E3 proteins appear to reflect the specificity of their roles. For example, the Arabidopsis genome contains 2 genes coding for E1 proteins, 37 genes for E2, and more than 1400 genes for E3. This large number of E3 genes is a clear indication of the large number of proteins this enzyme marks for degradation.

The number of ubiquitin molecules conjugated to the target protein can be one to many, but in either case the "tagged" protein is then destined to the cytoplasmic **proteasome** (Figure 17.4). The proteasome is a cytoplasmic body consisting of the **regulatory particle** (RP) and the **central core particle** (CP), each composed of a mixture of proteins giving the proteasome specificity for the degradation of ubiquitinylated proteins. The RP unit serves as a regulatory gateway

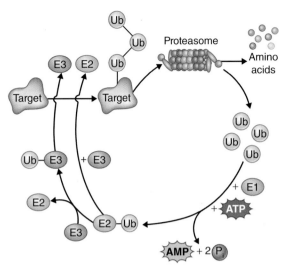

Figure 17.2 Depiction of the ubiquitin proteasome system for regulated degradation of target proteins. Ubiquitin (Ub or other polypeptide tags) must first be activated by conjugation to the E1 protein in an ATP dependent reaction. Ubiquitin is then transferred to E2, yielding the E2-ubiquitin conjugate. E2-ubiquitin then serves as the ubiquitin source for tagging a protein targeted for degradation in a reaction mediated by the E3 ligase enzyme. The ubiquitin-tagged target protein is then recognized by the proteasome, resulting in the degradation of the tagged protein to its constituent amino acids. Adapted from Vierstra (2009)

Table 17.2 Plants use several different polypeptide tags to mark proteins destined for rapid degradation/turnover

Polypeptide tag	Similarity to ubiquitin (%)	Number of isoforms	Amino acid length
Ub	100	1	76
RUB1	73	3	76
HUB1	35	2	72
SUMO	30	9	94
UFM	30	1	86

recognizing the tagged protein and threading the carboxy terminus of the target protein into the protease digestion chamber of the CP unit in an ATP dependent manner. The CP unit releases the amino acids from the digested targeted protein, leaving the oligomeric ubiquitin complex to be recycled by the RP unit for another round of protein tagging.

17.4 The ubiquitin proteasome system rivals gene transcription

The importance of these post-translation mechanisms controlling the levels of proteins and contributing to specific physiological outcomes comes from a wide variety of investigations. However, first recall that the sheer number of E3 genes in plant genomes is drastically greater than in mammalian genomes. Mammals have fewer than 100 E3 genes, whereas **Arabidopsis** has well over 1000. The plant E3 genes are simply duplicated versions of one another, but each gene encoding an E3 ligase contains additional protein domains providing for the specificity of that ligase. Given the large number of E3 genes and documentation of distinct roles for many in various facets of plant growth and development, it is not surprising that many investigators recognize the **ubiquitin proteasome systems** (UPSs) as a process rivaling gene transcription in terms of regulating processes, physiology and trait outcomes in plants.

Perhaps one of the most surprising roles of the UPS is its involvement in **hormone signaling**, in general. Involvement of the UPS with auxin, jasmonic acid, gibberellin, ABA and ethylene perception/signaling have all been reported. Figure 17.5 depicts the role of the UPS in the auxin signaling pathway as an example. Auxin responses are controlled by transcription factors referred to as auxin response factors, or ARFs. In turn, the ability of these ARFs to stimulate the transcription of the auxin response genes is controlled by suppressor proteins. When auxin is not present, proteins like AUX prevent the ARFs from stimulating transcription of the auxin response genes. When auxin is present, then suppression of the ARFs is relieved. How then does auxin relieve the suppression effect of AUX?

The surprise finding was that the ability of an E3 ligase to interact and tag AUX with ubiquitin was mediated by a third interacting protein, TIR1, and auxin itself. TIR1 interacts with one special class of E3 ligase, SCF-E3, encoded by four genes in **Arabidopsis**. Interestingly, the SCF-E3-TIR1 does not bind AUX in the absence of auxin, only in the presence of auxin. Furthermore, protein three-dimensional elucidation studies have now demonstrated that auxin serves as a kind of glue facilitating the binding of AUX to the TIR1 domain. This in turn leads to the transfer of

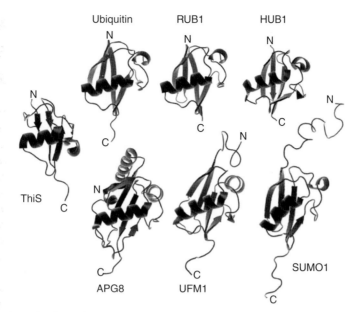

Figure 17.3 While a variety of polypeptide tags are used to mark proteins destined for degradation via the proteasome, all of the tags resemble the three-dimensional structure of ubiquitin, the most commonly used tag. The α-helices are shown as corkscrew structures, while β-sheets are shown as flattened arrows. Parts of the tags not in either of these structural motifs, α-helices or β-sheets, are shown as spaghetti strings. Adapted from Downes and Vierstra (2005). Reproduced with permission of Portland Press Ltd

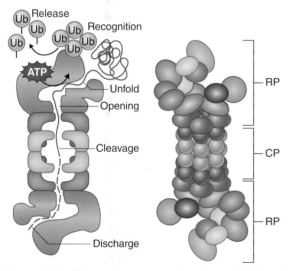

Figure 17.4 The proteasome consists of two functional units, the recognition particle (RP) and the central core particle (CP). Both units are composed of many different proteins. The RP unit serves to recognize ubiquitylated proteins, to feed the carboxy-terminus of the tagged protein into the degradation channel running through the CP unit. The RP unit also releases and recycles ubiquitin for subsequent rounds of protein tagging. The CP unit cleaves the peptide bonds of the targeted protein releasing the amino acids into cytoplasm so that they might be used for the synthesis of another protein

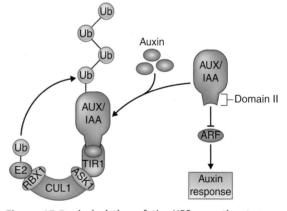

Figure 17.5 A depiction of the UPS operating to control the auxin response pathway in plants. This generalized model has been documented to be important for many of the other plant hormones, including GA and ABA. Transcription of auxin response genes is controlled by auxin response transcription factors (ARF), which are held in check by suppressor proteins like AUX. Only if auxin is present, AUX can be recognized and bound to TIR1, a putative auxin receptor protein that is also recognized by a specific class of E3 ligases. This results in the tagging of AUX with ubiquitin and it subsequent degradation by the proteasome, thus relieving transcriptional repression of the auxin response genes. In the absence of auxin, AUX is stable and acts to repress the auxin transcription factors from activating expression of the auxin response genes. Adapted from Vierstra (2009). Reproduced with permission of Macmillan Publishers Ltd

ubiquitin to the AUX protein via the E3 ligase, which leads to degradation of the AUX suppressor protein and results in expression of the auxin response genes. As auxin levels decline, degradation of the AUX protein subsides and suppression of auxin response gene expression is restored.

While this cascade of events is complicated, various permutations of the UPS system have been adopted for similar regulatory mechanisms in the hormone perception/response pathways for virtually all the major plant hormones. Equally impressive is how this system has been adopted for the control of many other traits in plants including plant–pathogen interactions (disease resistance mechanisms) and the developmental program of genetic self-incompatibility (Figure 17.6).

In **gametophytic self-incompatibility**, which is common in many solanaceous species such as petunia, if pollen from the same flower lands on the stigma, the pollen will germinate but pollen tube penetration through the pistillar tissue will be stopped. However, if pollen from a genetically compatible plant makes it way to the same flower, the pollen tube will grow through the pistil tissue to the ovule deep within the flower and pollinate the egg. This is known as gametophytic self-incompatibility because the outcome of pollination is dictated by the genotype of the pollen at the self-incompatibility locus. If the pollen has an allele in common with the stigma/pistil, then pollen growth is arrested and the interaction is referred to as self-incompatible.

Figure 17.6 illustrates a current understanding of how the self-incompatibility locus genotype affects the compatible versus incompatible reaction via a UPS mechanism. In the self-incompatible case, as pollen with a genetic allele in common with the female reproductive structure begins to migrate through the pistil tissue, ribonucleases (enzymes that degrade RNA) encoded by the maternal genome saturate the environment and are taken up into the growing pollen tube. Because the maternal RNases are not recognized by the pollen cell, the RNases will degrade all the pollen tube cell RNA, thus killing the pollen tube cell. If the germinated pollen does not have a self-incompatibility

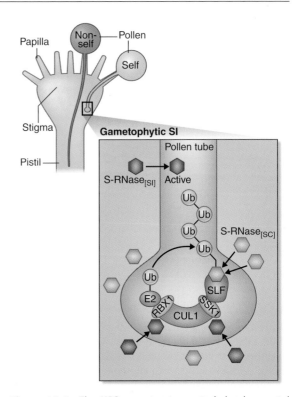

Figure 17.6 The UPS operates to control developmental programs as well as responses to hormones and biotic and abiotic cues. Gametophytic self-incompatibility (SI) is a process assuring out-crossing in many plant species and operates by the penetrating pollen tube being recognized as self or non-self in the pistil tissue of the style. Maternal-encoded RNases in the stylar tissue diffuse into the pollen tube of a germinated pollen. If the pollen arises from the same flower, then it harbors one of the alleles at the SI locus in common with the maternal tissue and thus cannot recognize the RNase as foreign. The RNase thus degrades the RNA within the germinated pollen tube and abolishes its chances of reaching the ovule tissue and fertilizing the egg. If the pollen reaching the stigma does not share an allele in common with the maternal tissue, the maternal S-RNase is recognized by a receptor-like protein, SLF, which is specifically associated with a distinct class of E3 ligases that mark the S-RNase with ubiquitin for degradation. In this case, pollen growth in not impeded and can lead to fertilization. Adapted from Vierstra (2009). Reproduced with permission of Macmillan Publishers Ltd

allele in common with the maternal tissue, then the RNases taken up into the pollen tube are recognized by a distinct E3 ligase–SLF complex, resulting in the tagging of the RNase with ubiquitin and targeting this enzyme for degradation. Hence, the pollen survives migration through the pistil tissue and can go on to reach the egg and consummate fertilization.

17.5   Summary

Two mechanisms for protein degradation were reviewed in this chapter. Autophagy, senescence and programmed cell death are similar mechanisms that lead to a complete loss of protein contents within a cell. They are not random processes, but are controlled and regulated, yielding very specific outcomes, like development of the xylem elements essential for water movement in plants. In contrast, the UPS is a selective process wherein the turnover rate of specific proteins is affected by a host of protein factors, thus providing selectivity and regulation of the overall process. Because the UPS is thought to operate throughout a plant's life cycle and in all cell types, this process might rival gene expression as a regulatory mechanism to control phenotypic outcomes.

17.6   Problems

17.1 Draw a graph depicting the time course for a plant response to a developmental signal or cue. It might be the induction time for flowering after transition to a shortened day length, or the development of a defense response (i.e., accumulation of a defense protein) after pathogen challenge. Add a second curve to the graph for the change you would expect if changes in the transcription rate of the corresponding gene controlled the response. You will be plotting the transcription rate of the gene encoding the protein responsible for flower induction or the pathogen defense protein. Add a third curve

to the graph assuming the transcription of the gene was constitutive, but the degradation rate of the protein was responsible for induction of the corresponding protein. Hence, your graph will contain three curves: one for the accumulation of a protein responsible for flowering or pathogen defense; a second curve that should precede the first, at least time wise, on the graph for the transcription rate of the gene; and a third curve which might be the inverted form of curve 2 and depicting the change in the degradation rate of the outcome protein.

17.2 Assume you have isolated a mutation in the auxin response pathway cascade resulting in no expression of the auxin response genes in response to auxin. You also suspect that the mutation is within the TIR1 protein. Make two predictions about what specific changes could account for this observation. Draw a cartoon depiction to help explain your answer.

Further reading

Bassham, D.C. (2009). Function and regulation of macroautophagy in plants. *Biochimica et Biophysica Acta-Molecular Cell Research*, **1793**, 1397–1403.

Downes, B. and Vierstra, R.D. (2005) Post-translational regulation in plants employing a diverse set of polypeptide tags. *Biochemical Society Transactions*, **33**, 393–399.

Reape, T.J. and McCabe, P.F. (2008) Apoptotic-like programmed cell death in plants. *New Phytologist*, **180**, 13–26.

Vierstra, R.D. (2009). The ubiquitin-26S proteasome system at the nexus of plant biology. *Nature Reviews Molecular Cell Biology*, **10**, 385–397.

Reference

Roberts, I.N., Caputo, C., Criado, M.V. and Funk, C. (2012) Senescence-associated proteases in plants. *Physiologia Plantarum*, **145**, 130–139.

Index

Plant Genes, Genomes and Genetics, First Edition. Erich Grotewold, Joseph Chappell and Elizabeth A. Kellogg.
© 2015 John Wiley & Sons, Ltd. Published 2015 by John Wiley & Sons, Ltd.
Companion Website: www.wiley.com/go/grotewold/plantgenes.